D1000853

# Immunochemical Methods in Cell and Molecular Biology

## Biological Techniques Series

J. E. TREHERNE
*Department of Zoology*
*University of Cambridge*
*England*

P. H. RUBERY
*Department of Biochemistry*
*University of Cambridge*
*England*

---

Ion-sensitive Intracellular Microelectrodes, *R. C. Thomas,* 1978
Time-lapse Cinemicroscopy, *P. N. Riddle,* 1979
Immunochemical Methods in the Biological Sciences: Enzymes and Proteins, *R. J. Mayer* and *J. H. Walker,* 1980
Microclimate Measurement for Ecologists, *D. W. Unwin,* 1980
Whole-body Autoradiography, *C. G. Curtis, S. A. M. Cross, R. J. McCulloch* and *G. M. Powell,* 1981
Microelectrode Methods for Intracellular Recording and Ionophoresis, *R. D. Purves,* 1981
Red Cell Membranes—A Methodological Approach, *J. C. Ellory* and *J. D. Young,* 1982
Techniques of Flavonoid Identification, *K. R. Markham,* 1982
Techniques of Calcium Research, *M. V. Thomas,* 1982
Isolation of Membranes and Organelles from Plant Cells, *J. L. Hall* and *A. L. Moore,* 1983
Intracellular Staining of Mammalian Neurones, *A. G. Brown* and *R. E. W. Fyffe,* 1984
Techniques in Photomorphogenesis, *H. Smith* and *M. G. Holmes,* 1984
Principles and Practice of Plant Hormone Analysis, *L. Rivier* and *A. Crozier,* 1987
Wildlife Radio Tagging, *R. Kenward,* 1987
Immunochemical Methods in Cell and Molecular Biology, *R. J. Mayer* and *J. H. Walker,* 1987

# Immunochemical Methods in Cell and Molecular Biology

R. J. MAYER

*Professor of Molecular Cell Biology*
*Department of Biochemistry*
*University of Nottingham, Nottingham, UK*

J. H. WALKER

*Lecturer in Biochemistry*
*Department of Biochemistry*
*University of Leeds, Leeds, UK*

1987

ACADEMIC PRESS

*Harcourt Brace Jovanovich, Publishers*

London San Diego New York Berkeley
Boston Sydney Tokyo Toronto

ACADEMIC PRESS LIMITED
24–28 Oval Road
London NW1 7DX

*U.S. Edition published by*
ACADEMIC PRESS INC.
San Diego, CA 92101



**British Library Cataloguing in Publication Data**

Mayer, R.J. (Roland John)
Immunochemical methods in cell and
molecular biology.—(Biological
techniques series).
1. Cytochemistry 2. Immunochemistry
3. Molecular biology 4. Immuno chemistry
I. Title II. Walker, J.H. (John Harold)
III. Series
574.87′6042 ~~QH611~~

ISBN 0–12–480455–7
LCCCN 87–71539

Typeset by Latimer Trend & Company Ltd, Plymouth
Printed in Great Britain by Galliard (Printers) Ltd, Great Yarmouth, Norfolk

# Preface

Immunochemical techniques have become increasingly important tools for molecular and cell biologists. Antibodies provide sensitive assays for specific antigens which may vary from small molecular weight haptens to macromolecules. The identification of clones from expression libraries links sensitive immunodetection methods to the powerful procedures of molecular biology and cloning. Furthermore, antibodies provide exquisite specificity for the isolation of antigens; careful selection of the antibodies can lead to one-step purification of enzymes retaining full activity.

We have aimed in this book to provide a readable introduction to the theory behind the methods and the choices available. We have also given extensive technical instruction to enable researchers to choose those procedures most suitable for their work.

We are especially grateful to Dr Ellen Billet of the School of Life Sciences, Trent Polytechnic for her contributions to the chapter on monoclonal antibody production and characterization. Part of this chapter was previously published by Elsevier Scientific Publishers (Ireland) Ltd as a review in "Techniques in the Life Sciences, Biochemistry—Volumes B1/11, Supplement Techniques in Protein and Enzyme Biochemistry, (Tipton K.S., editor)", and is reproduced here with permission of the publishers. We should also like to thank Dr Eric Blair of the Department of Biochemistry, University of Leeds, for making Fig. 4.27 possible and Ms C. M. Boustead for indexing the book.

R. J. MAYER
J. H. WALKER

# Contents

# 1
# Introduction

## 1.1. General Introduction

In recent years certain unifying forces have begun to pull together the diverse disciplines of the biological sciences. The most powerful of these forces has been the revolution in molecular biology which can supply exquisite tools for understanding the molecular basis of life. An important part of this revolution has been the development of monoclonal antibodies which provide us now with unlimited supplies of reagents for quantitating, localizing and isolating specific antigens. The development of human monoclonal antibodies perhaps sees us rapidly approaching the development of 'magic bullets' for curing cancer (Klausner, 1986).

The importance of antibodies and the immune response in health cannot be questioned and the clinical value of immunochemical reagents was appreciated long ago. Nevertheless, it is still true to say that biological scientists of various disciplines often find it difficult to adapt the methods of a related discipline to their own needs. This book has been written by two biochemists who now consider themselves to be cell biologists and hope that their change in perspective will help them to present the information in a manner which will make it easily understood by a wide readership.

## 1.2. Aims and Objectives

Immunochemistry is the study of the interactions between antigens (e.g. proteins) and antibodies. Immunization of an animal produces antibodies against proteins which the animal recognizes to be "foreign". The immunoglobulins which are produced in response to this challenge are specific to the antigen (or antigens) injected. The immunoglobulin population which is specific to an antigen is heterogeneous, in that subpopulations of immunoglobulin exist which interact with different regions (or antigenic determinants) on the surface of the antigen.

A problem in writing a book on this topic is that there is a tendency to generalize about immunochemical processes which are by definition not general but very specific phenomena. The production, processing and use of each antiserum will vary depending on the antigen of interest. Furthermore, biologists have very different interests in the properties of enzymes and proteins and therefore the requirements and objectives of each research group will be different. Naturally, this will lead to the use of different methods for the identification, isolation and quantitation of antigens from tissue extracts: the purity of antibodies needed for a piece of work may vary from crude antiserum (e.g. for immunotitration of enzyme activity) to carefully purified antibodies (for immunoadsorbent chromatography or antigen quantitation). Specific research requirements should be carefully considered before choosing from the methods and procedures which are outlined in this book. In particular, very careful thought should be given to the question of whether a polyclonal or a monoclonal antiserum should be prepared. For example, in many cases affinity-purified, polyclonal antibodies may be obtained more quickly and at a fraction of the cost of monoclonal antibodies.

Immunochemical (with humoral antibodies, i.e. in the serum) have been carried out on enzyme activity, enzyme evolution, the nature of antigenic determinants on enzymes, immunogenicity in relation to enzyme structure and function, the relationship between proenzymes and enzymes and between apoenzymes and enzymes, multiple forms of enzymes and allostery. The results indicate the complex relationship between factors responsible for immunogenicity of an antigen, e.g. the level of conformational organization, and the subsequent interaction of antibodies with the antigen (Arnon, 1973).

Sensitive immunochemical techniques are being used increasingly to measure the amount and rate of turnover (synthesis and degradation) of enzymes and proteins (Philippidis *et al.*, 1972; Speake *et al.*, 1976). Immunochemical methods have been developed to isolate polyribosomes containing immunoreactive nascent polypeptide chains, from which specific mRNA and cDNA libraries can be prepared (Palacios *et al.*, 1972). Antibodies can also provide extremely valuable reagents for the localization of specific antigens with the light microscope and with the electron microscope.

The purpose of this book is to review the methods involved in the production, processing and use of antisera. Each section is structured to emphasize the principles, practice, generality and problems of each technique. The book is not intended to be a completely comprehensive guide to all immunochemical techniques. The bibliography includes recommended reading for those interested in more extensive details of the methods mentioned here in brief. The book is intended to be a practical guide to anyone wishing to raise and use antisera and is based in part on the authors' own immunochemical studies on various antigens.

# 2
# Polyclonal Antiserum Production

Polyclonal antisera may be produced against a single protein antigen (monospecific antiserum) or against a mixture of protein antigens (polyspecific or multispecific antiserum). For many immunochemical studies it is advantageous to produce a monospecific antiserum (Walker *et al.*, 1976). However, polyspecific antisera may be produced deliberately to analyse complex protein mixtures (Bjerrum and Bøg-Hansen, 1976a). Often, antisera raised to an antigen from one species will cross-react with the same antigen in many other species. This is true, for example, for most cytoskeletal proteins, cytochrome oxidase (Hackenbrock and Hammon, 1975) and dopamine $\beta$-hydroxylase (Ross *et al.*, 1978). Antisera may also be prepared to commercially available antigens, e.g. glutamine synthetase (Koch and Nielsen, 1975), although it is advisable to test the purity of the antigen before beginning the immunization schedule. Such an assessment may indicate that further purification of an antigen is necessary before immunization can be started.

## 2.1. Preparation of Antigens

### 2.1.1. Antigen purity

As a working principle it should always be assumed that an antigen which is pure by biochemical criteria (e.g. single component in protein analytical systems) is not pure immunologically. This assumption can avoid disappointment after a lengthy immunization schedule. There are many examples of biochemically pure antigens giving rise to polyspecific antisera (Clausen, 1971) as a result of minor traces of highly antigenic contaminants.

The major dilemma is to decide how far to purify the antigen, measured in terms of time and effort, given that there is a variety of procedures for antiserum purification (Chapter 3). In general it is best to obtain the most purified preparation for use as antigen, knowing that the antiserum can subsequently be purified if necessary.

### 2.1.2. Nature of antigen preparation

Enzyme- and protein-purification procedures give rise to preparations in buffer solutions of different pH and ionic strength which can contain a variety of agents which have been included to protect the protein from inactivation (e.g. dithiothreitol, 2-mercaptoethanol, EDTA, glycerol, stabilizing ions and coenzymes). Work with membrane-protein antigens usually means the isolation and purification of a protein in the presence of detergents (often non-ionic) or at very high salt concentrations, e.g. 3 M KCl for plasma membrane antigens (Price and Baldwin, 1977). These agents may affect the microheterogeneity of antibody subpopulations (e.g. by masking antigenic determinants), but in general there is little evidence that they affect antibody production. Antisera to membrane antigens in detergents (e.g. Triton X-100) are not difficult to produce (e.g. Dennick and Mayer, 1977).

Although ionic detergents (e.g. sodium dodecyl sulphate) are known to disrupt antigen–antibody interactions (Crumpton and Parkhouse, 1972; Bjerrum and Bøg-Hansen, 1976b), antisera are routinely raised to protein subunits isolated by polyacrylamide gel electrophoresis in the presence of sodium dodecyl sulphate. For example, Strauss *et al.* (1975) homogenized antigen-containing gel in phosphate-buffered saline and injected it into rabbit foot pads. Similarly, immunization with single bands obtained by polyacrylamide gel electrophoresis in the presence of sodium dodecyl sulphate has been successfully used to raise monospecific antisera to actin (Blomberg *et al.*, 1977), microtubule-associated protein (Sherline and Schiavone, 1977) and myosin. It must be remembered that polypeptides of similar molecular weight may be present in a single piece (band) of polyacrylamide gel, e.g. tubulin and a polypeptide associated with neurofilaments (Gilbert, 1978). However, immunization with antigens prepared in this way has great potential for membrane antigens which are often very difficult to resolve, purify and identify in systems which do not contain sodium dodecyl sulphate. Protein subunits prepared in sodium dodecyl sulphate are frequently used as antigens, as more laboratories use two-dimensional polyacrylamide gel electrophoresis to purify membrane-protein subunits (Dale and Latner, 1969; O'Farrell, 1975). It is often easier to prepare a protein subunit reproducibly by these methods than to try and purify native proteins from some cellular organelle (e.g. synaptic vesicle membrane proteins).

It is tempting to suggest that all purified proteins should be subjected to a final purification step by polyacrylamide gel electrophoresis in the presence of sodium dodecyl sulphate; following staining, the subunit or subunits of the antigen of interest may be cut out, pooled and used for the preparation of samples for immunization. This approach could minimize the production of antibodies to contaminating antigens, except, of course, to those contaminants which have the same subunit size(s) as the antigen of interest. Antiserum to human liver monoamine oxidase has also been produced in this way (Russell *et al.*, 1978a).

Extensive details of gel electrophoretic procedures are given in the technical supplement together with the details of how to elute antigens from the polyacrylamide gel.

However, the problems of interpretation associated with the use of antisera to subunits of functional protein complexes in membranes must be recognized. For example, antisera raised to subunits isolated from multisubunit enzyme complexes may not cross-react with the holoenzyme (Werner, 1974). Nevertheless, for most antigens, antisera raised to single bands isolated from polyacrylamide gels provide reagents which can be used for the full range of immunochemical purposes, including immunoprecipitation and immunohistochemical localization.

### 2.1.3. Antigen mixtures

Polyspecific antisera have been prepared with great success to many protein mixtures, including subcellular fractions. Early work with such antisera was aimed at identifying and quantifying soluble (e.g. Weeke, 1973b) and membrane (e.g. Bjerrum and Bøg-Hansen, 1976a) antigens by immunoelectrophoretic procedures, which are still of value for studying changes in antigen populations (e.g. in membrane fusion or during chronic enzyme adaptation in tissues).

If the resolution of crossed-rocket immunoelectrophoresis is sufficiently good, immunoprecipitates can be cut from the agarose gel and used to produce monospecific antisera (Koch and Nielsen, 1975). However, it is very useful to be able to identify the biochemical nature of the antigen in a particular immunoprecipitate if the subsequent monospecific antiserum is to be of maximal use in biological experiments. Identification of an enzyme which retains activity in immunoprecipitates is relatively simple (Blomberg and Perlmann, 1971a). Similarly, proteins which bind calcium (Suttie *et al.*, 1977), epinephrine (Blomberg and Berzins, 1975) or which are phosphorylated (Gordon *et al.* 1977a,b) may be identified in immunoprecipitates after the incorporation of radiolabelled ligand or through changes in the electrophoretic mobility of the antigens after binding of the ligand.

Polyspecific antisera are also of great value when linked to immune blotting procedures, since they may enable minor antigens specific to subcellular fractions to be identified. This approach can be more successful than the production of monoclonal antibodies for the identification of the minor antigenic components of complex mixtures.

The relative immunogenicities of proteins in a mixture are of great importance when considering the outcome of an immunization schedule designed to obtain a polyspecific antiserum. The problem is that some proteins may be much better antigens than others, thus complicating the subsequent use of the antiserum (e.g. in immunoprecipitation analyses in gels).

Indeed, in some cases it may not be possible to obtain precipitating antibodies to some of the major protein species in an antigen mixture, while minor protein species may give rise to high titres of antibodies. Surprisingly, it may be that a single protein in a protein mixture is the only effective antigen in the mixture. Injection of whole chromaffin granules seems to give rise to a monospecific antiserum to dopamine $\beta$-hydroxylase (Helle *et al.*, 1979).

In the preparation of polyspecific antisera to membrane antigens, contaminating cytosolic proteins may act as excellent antigens, giving much better antibody production than the membrane antigens. Furthermore, even when the intention is to produce a monospecific antiserum (e.g. to purified integral mitochondrial membrane enzymes) contaminants in the preparations can give much larger titres of antibodies than the antigens of interest (e.g. Walker *et al.*, 1976).

## 2.2. Preparation of Antigens for Immunization

### 2.2.1. Adjuvants

Protein antigens are usually mixed with material that will increase the concentration of circulating antibodies, i.e. adjuvants. There are a variety of adjuvants which may themselves be antigenic (e.g. tubercule bacilli) or not (mineral oil) but which can improve the humoral immune response by mechanisms which include increasing the number of cells involved in antibody formation, ensuring a more efficient processing of antigens and prolongation of the duration of the antigen in the immunized animal. The most commonly used adjuvant is complete Freund's adjuvant, which consists of killed mycobacteria (e.g. tubercle bacilli), an oil and an emulsifier. In incomplete Freund's adjuvant the mycobacteria are omitted (Clausen, 1971). Many immunization schedules for proteins have been used but usually Freund's complete adjuvant is used for the first series of injections (i.e. the injections on day 1), and subsequently incomplete adjuvant (e.g. Mason *et al.*, 1973) is used.

Equal volumes of the antigen preparation and the adjuvant are vigorously mixed to give a stable emulsion which can be used as the preparation for injection. A stable emulsion may be produced by repeated aspiration of the mixture into and out of a hypodermic syringe with a large-gauge needle, by homogenization with a loose-fitting pestle, or by sonication. Recently, a synthetic adjuvant, muramyl dipeptide, has become available (Institut Pasteur). This compound apparently has several advantages over Freund's adjuvant, including an increased ability to induce the humoral immune response. Precipitation of antigens on to aluminium oxide is also frequently used to immobilize antigens (Hudson and Hay, 1976). Samples of *Bordetella pertussis* may be added to enhance the immune response.

### 2.2.2. Amount of antigen

Humoral antibody production is a function of the nature (e.g. with or without adjuvant) and amount of the antigen and the immune response of the individual animal. A poor immune response (immune tolerance) can occur to small or large amounts of an antigen (Mitchison, 1968). Enzymologists and protein chemists are often faced with the problem of obtaining enough of their purified protein for biochemical analysis. Fortunately, this does not present a problem immunologically, where the small amounts of protein antigens available may be quite adequate. For example, excellent precipitating titres have been obtained in sheep with a total of 400 μg of acetyl-CoA carboxylase (Walker *et al.*, 1976). Minute amounts of protein antigens have been used to raise antisera. The fact that antisera have been raised to immunoprecipitates cut out from agarose gels testifies to the sensitivity of the immune response. Similarly, very small amounts of antigenic protein (e.g. 1 μg) bound to solid supports will give rise to antisera (Stevenson, 1974). In general, however, a total of 0.5–4.0 mg of protein per rabbit and 1–10 mg of protein per sheep over the entire immunization schedule is recommended. For polyspecific antisera the upper limits of antigen amount suggested above are probably preferable, although it must be stressed that there is no definite answer to the question of how much antigen should be used in an immunization schedule.

### 2.2.3. Weakly immunogenic antigens

Some proteins are poor antigens, particularly those which show little evolutionary divergence between mammalian species (e.g. cytochrome c, actin, myosin). Two different approaches have been used to improve the immune response of these proteins. Antisera may be raised to poor mammalian antigens in non-mammals (e.g. myosin in fowl). Denaturation of antigens with sodium dodecyl sulphate may also increase the antibody response to weakly immunogenic antigens. Alternatively, proteins may be covalently modified in some way to improve their immunogenicity. This may be achieved by cross-linking with glutaraldehyde to give homopolymers

(Avrameas and Ternynck, 1969) or heteropolymers, e.g. casein with ovalbumin (Houdebine and Gaye, 1976), or S-100 protein with methylated bovine serum albumin (Levine, 1967). One of the most popular proteins for preparing heteropolymers is the highly antigenic protein haemocyanin. Essentially the same approach is used to raise antibodies to haptens, including phosphotyrosine, which has been coupled to IgG and used successfully for immunizing rabbits (Ek and Heldin, 1984).

A further approach which has been used to increase immunogenic properties of weak antigens is the coupling of proteins to Sephadex or Sepharose (Al-Sarraj *et al.*, 1978; Orlov and Gurvich, 1971). The protocol for obtaining excellent precipitating titres to rabbit casein (12 mg) in 8–12 weeks is shown in Fig. 2.1.

Sepharose 2B (8 ml)

↓ CNBr activation (moderate conditions; Porath *et al.*, 1973)

"activated" Sepharose 2B

↓ Bovine serum–albumin (50 mg)
Overnight at 4 °C

Sepharose 2B–albumin (5 mg albumin/ml Sepharose)

↓ Recombined casein polypeptides (41 mg)
+glutaraldehyde (final concentration, 5% v/v)
Overnight at room temperature

Sepharose 2B–albumin–casein

↓ Extensive washing with 20 mM sodium phosphate buffer, pH 7.0, containing 0/15 M NaCl

Sepharose 2B–albumin–casein (3.4 mg recombined casein/ml of Sepharose–albumin)

**Fig. 2.1** Preparation of sepharose–albumin–casein.

Interestingly, many workers have tried to reduce or abolish the immunogenicity of enzymes to use them in enzyme-replacement therapy. Here studies have been carried out with protein heteropolymers where the enzyme (e.g. pig liver uricase) was coupled to homologous rabbit albumin for injection into rabbits. The heteropolymer in these conditions was found to be non-immunogenic and non-antigenic (Remy and Poznansky, 1978). Alternatively, immunogenicity of bovine serum albumin and bovine liver catalase has been abolished in rabbits by coupling of these proteins to polyethylene glycols (Abuchowski *et al.*, 1977a,b). These authors envisage that the linear, flexible, uncharged hydrophilic polymers which are attached to the proteins may abolish immunogenicity by providing a shell around the enzyme that covers antigenic determinants; by presenting a flexible, unbranched hydrophilic surface for inspection by the immune process, the interiorized enzyme may escape being recognized as a foreign substance.

In principle, however, it should be possible to obtain antibodies to any protein by presenting it to the animal in an immunogenic form. Since antibodies can be obtained to many small molecules, e.g. acetylcholine (Spector *et al.*, 1978) and a wide variety of biologically active compounds (Yalow, 1978) it seems likely that antibodies could be obtained to all proteins, e.g. by careful choice of species for immunization and by presenting the antigen in the correct form. Some poor protein antigens may behave as haptens, with antibodies only being produced against a single antigenic determinant. In such cases the antibodies may only be detectable by immunoaffinity or immunofluorescent techniques.

Finally, it is possible that for some antigens a very long immunization schedule is required before antibodies may be detected, e.g. choline acetyltransferase (Rossier, 1975).

## 2.3. Immunization Procedures

### 2.3.1. Selection of species

Rabbits, guinea-pigs, goats, sheep and fowl are commonly used for antisera production. Naturally, the best antisera are raised in a species which is unrelated to the species from which the antigen was prepared. It should be remembered that the immune response will vary from animal to animal in a given species. This consideration may be important if the antigenicity of a protein is unknown or known to be poor. In such a case it may be better to inject several rabbits instead of one sheep in order to be able to use the best antiserum produced. This approach presumes that sufficient antigen is available for multiple immunization schedules.

Several other considerations affect the choice of species for immunization, including expense of animals, housing of animals, source of antigen, volumes of antiserum required (e.g. very large volumes of 1 litre can be obtained in a single bleed from a goat or sheep), and ease of bleeding.

### 2.3.2. Sites of immunization

Antigen preparations may be presented to animals by several routes.

Antigen–adjuvant mixtures may be injected subcutaneously and intramuscularly. Antigen may be injected intravenously or into foot pads (e.g. of rabbits) and in these cases complete Freund's adjuvant may be injected separately at subcutaneous sites (Clausen, 1971). There are two distinct types of immunization procedure; single-site injections and multi-site injections. Multi-site injections can be given both subcutaneously and intramuscularly, and offer the advantage of presenting the antigen by a variety of routes with the expectation of provoking the maximum immune response. It may be advantageous to immunize rabbits intradermally in the thicker part of the skin above the scapulae (Harboe and Ingild, 1973).

### 2.3.3. Immunization schedules

Animals should be bled before immunization to obtain non-immune (control) serum and to ensure that the animal does not already contain high titres of unwanted antibodies. Many rabbits, for example, contain antibodies to intermediate filaments in their sera. Many types of immunization schedule have been attempted but in principle antigen is presented at regular intervals on several occasions with trial bleeding after the second and subsequent injections. Details of these procedures are given in Chapter 7.

## 2.4. Trial and Preparative Bleeding Procedures

The methods for carrying out these bleeding procedures are described in Chapter 7. Procedures for the bleeding of rabbits and sheep (also suitable for goats) are described in detail.

# 3
# Polyclonal Antiserum Processing

In this chapter methods will be described for assessing the specificity of polyclonal antisera and for purifying antisera to prepare the most convenient fractions for immunochemical studies. Although most immunochemical reactions will occur with unfractionated antiserum, there are many advantages in using purified immunoglobulins, in particular the reduction of non-specific interactions. The optimal reagent which can be obtained from a polyclonal antiserum is usually affinity-purified immunoglobulin fraction obtained by affinity chromatography on an immunoadsorbent column of immobilized antigen.

## 3.1. Purification of Immunoglobulins

An early scheme for the classification of proteins defined albumins as proteins soluble in water and globulins as proteins sparingly soluble in water but soluble in dilute aqueous salt solutions. An alternative definition classed globulins as those proteins salted out of aqueous solution by 50% saturated ammonium sulphate. Sera contain serum albumin as a major component and so considerable purification of immunoglobulins can be achieved by separating albumins from globulins. The basic isoelectric points of the immunoglobulins provide a second important criterion for the isolation of these proteins.

Techniques for serum fractionation include ethanol fractionation, salt fractionation, gel filtration and ion-exchange chromatography. These techniques are described in detail elsewhere (Williams and Chase, 1967; Harboe and Ingild, 1973; Hudson and Hay, 1976) but a general procedure for mammalian sera is described in Chapter 7. This chapter also contains methods for the preparation and use of antigen columns for the affinity purification of specific antibodies. Immunoglobulin G may also be purified by chromatography on Protein A Sepharose (Kronvall *et al.*, 1970) using commercially available columns (e.g. Biorads Affigel Protein A and Affigel Protein A MAPS). It is important to note, however, that immunoglobulins of different subclasses and species of animal bind to Protein A to varying extents (see Table 7.7).

## 3.2. Assessment of Antiserum Specificity and Titre

Although relatively slow and insensitive, immunodiffusion in agar or agarose gels still provides a useful first step for the assessment of an antiserum. Immunoelectrophoresis can also provide useful information about the specificity of an antiserum, although it is also not particularly sensitive and requires very large amounts of antiserum. It does, however, have two advantages: (a) multi-subunit proteins will form single immunoprecipitin lines which can be shown to contain several polypeptides (see below); and (b) the areas of the immunoprecipitin "rockets" are proportional to antigen amount, making the method quantitative. Nevertheless, electrophoresis procedures have now been superseded by immune-blotting techniques in which antisera are tested against nitrocellulose blots of proteins analysed by SDS-PAGE.

Antisera should be tested against the crudest preparation available (such as a detergent-solubilized homogenate) in addition to the purified fraction. It is also advisable to test the antiserum against fractions obtained during the purification of the antigen so that the immunoprecipitin line, or the band seen on the nitrocellulose blot, can be seen to copurify with some other marker of the antigenic fraction (e.g. enzyme activity).

For specific antisera the titre is defined as the amount of antigen in mg needed to precipitate all of the specific antibody from 1 ml of antiserum, so that no antigen nor specific antibody remains in the supernatant (Sewell, 1967). The titre may be obtained by various procedures, including single radial immunodiffusion (Brecker, 1969), immunoprecipitation of labelled antigen or enzyme of known specific activity, and enzyme-linked immunosorbent assays (see below). Titres may range from 50 to 100 $\mu$g/ml of antiserum. An alternative definition of titre is also often used in the literature to refer to the highest dilution at which an antiserum still gives a positive response. Thus, for example, in an ELISA test an antiserum has a titre of 10 000 if a 1 in 10 000 dilution of the antiserum still produces a positive response against the antigen.

### 3.2.1. Immunodiffusion

Double-diffusion analyses (Ouchterlony, 1968) in agar or agarose gels can be used to identify the number of antigen–antibody systems with a tissue extract and the antiserum (Fig. 3.1a). The analyses are particularly useful for obtaining reactions of identity, comparing a purified antigen (in one well) with components of an antigen mixture (in a second well) by their reaction with an antiserum (in a third well).

The concentration of the antigens which are in opposition to an antiserum in an immunodiffusion analysis may be varied by serial dilution, e.g. dilutions of a tissue extract around a central antiserum-containing well (Ouchterlony, 1968). Alternatively, the amounts of immunoreactants (antigens and antibodies) may be varied by opposition of a range of different volumes of tissue extract and antiserum (Piazzi, 1969 and Fig. 3.1b). The advantage of these types of immunodiffusion analyses is that, by varying the amounts of the immunoreactants, latent contaminating antigen–antibody systems may be revealed, e.g. for 6-phosphogluconate dehydrogenase (Fig. 3.1b). In general, the resolution of immunodiffusion analyses is such that they are only suitable for assessment of antisera which contain antibodies to a few macromolecular antigens.

It should be noted that multiple immunoprecipitate lines can occasionally be artefacts, e.g. due to refilling wells (Kabat, 1971). Tissue extracts which contain membrane–protein antigens may give multiple lines in immunodiffusion analyses. These multiple lines have been interpreted to be the result of differential phospholipid binding to the antigen (Hackenbrock and Hammon, 1975 and Fig. 3.2b) or to the presence of subunit and holoenzyme in the tissue extract (Poyton and Schatz, 1975). Detergents are usually present in tissue extracts containing membrane-protein antigens and can obviously influence the behaviour of the antigens in immunodiffusion analyses. The choice of buffer used for immunodiffusion analysis may be varied to provide conditions which most favour the native state of the antigen. Rabbit antisera give sharp immunoprecipitation lines when the gel buffer contains 0.15 M salt, while antisera from fowl give good precipitation lines when the gel buffer contains 0.6 M salt. The addition of protease inhibitors to the gel and sample buffers (e.g. phenylmethyl sulphonylfluoride, EDTA or Trasylol) are to be recommended for all forms of immunochemical analysis.

The problems of studies on detergent-solubilized membrane proteins in immunodiffusion analyses are shown clearly by work on monoamine oxidase. This enzyme is an integral protein in the outer mitochondrial membrane. Enzymological studies on monoamine oxidase have generated much controversy and interest by indicating that multiple forms of the enzyme may exist, as demonstrated electrophoretically (Sandler and Youdim, 1972) and by pharmacological means (Johnston, 1968). The molecular basis for the electrophoretic heterogeneity has been reported to be based on differential phospholipid binding (Houslay and Tipton, 1973) possibly to a single gene product (Tipton *et al.*, 1976; Dennick and Mayer, 1977). However, recent data suggest that the two forms of the enzyme (A and B) may have different

Double immunodiffusion (a)

1. Unadsorbed antiserum

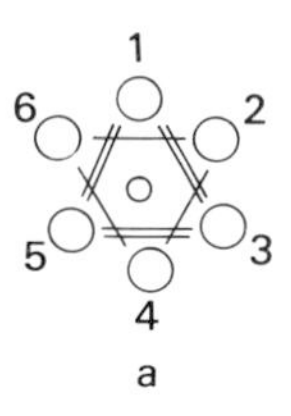


2. Adsorbed antiserum

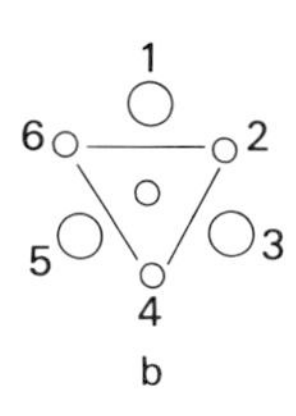


Piazzi immunodiffusion (b)

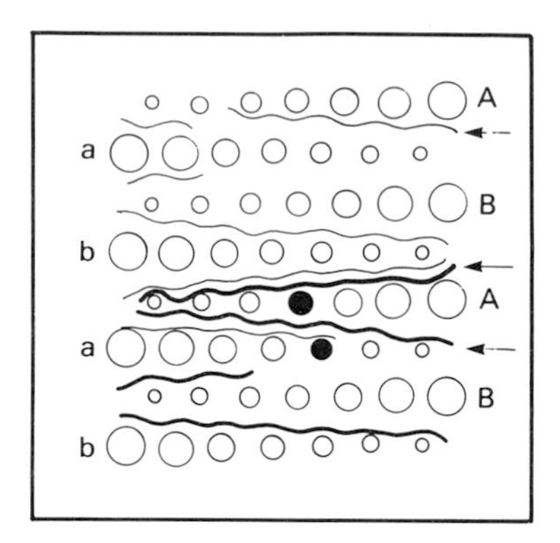


Rocket immunoelectrophoresis (c)

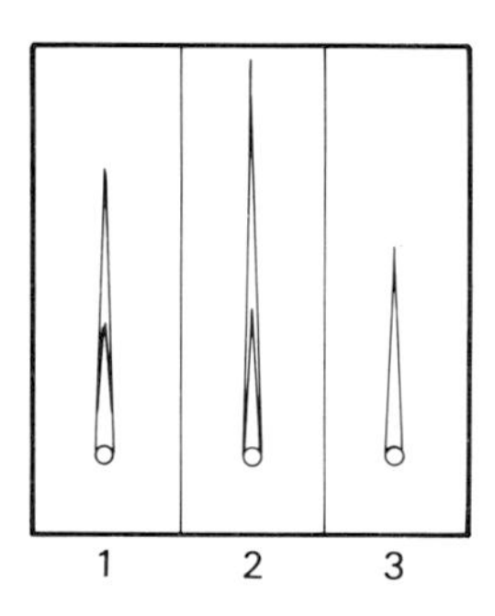


Crossed-rocket immunoelectrophoresis (d)

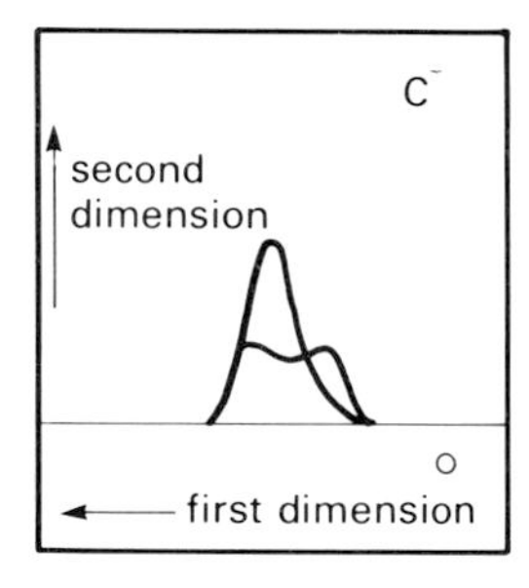


C = Cytochrome Oxidase

**Fig. 3.1** Immunodiffusion and immunoelectrophoresis analyses. Figs. 3.1 a–c show the analyses of antisera before and after adsorption. (a) Double immunodiffusion: wells 1, 3 and 5 contained purified cytochrome oxidase, and wells 2, 4 and 6 contained samples of a particle-free supernatant prepared from a homogenate of rat liver. Immunodiffusion was carried out for 2 days at room temperature in a moist atmosphere. (b) Piazzi immunodiffusion: wells contained particle-free supernatant prepared from mammary gland of mid-pregnant, a, and 15 day lactating, b, rabbits. Opposing wells contained unabsorbed antiserum, A, and completely absorbed antiserum, B. The arrowed immunoprecipitate lines did not stain for 6-phosphogluconate dehydrogenase activity. Immunodiffusion was carried out for 24 h at room temperature in a moist atmosphere. (c) Rocket immunoelectrophoresis: each well contained 10 $\mu$l of a particle-free supernatant prepared from mammary glands of 15 day lactating rabbits. Immunoelectrophoresis of acetyl-CoA carboxylase was carried out for 16 h at 10 V/cm at 15 °C with 0.1% (v/v) antiserum in 1% (w/v) agarose gels. Unadsorbed antiserum was present in tracks 1 and 2.

FAD-containing polypeptides distinguished by limited proteolytic digestion and reactivity with monoclonal antibodies (see Chapter 5).

The molecular basis of the pharmacological heterogeneity may reside in the fact that the enzyme may be distributed on both sides of the outer mitochondrial membrane (Russell *et al.*, 1978b). Many authors (e.g. Minamuira and Yasunobu, 1978) have shown that after detergent solubilization the enzyme fractionates, in many analytical procedures, into multiple species, particularly molecular weight species. These species are the result of enzyme aggregation in the detergent solution.

It should be clear from this description that monoamine oxidase has all the enzymological and physicochemical properties in detergent solution to be a model of potential immunochemical problems with membrane enzymes. It is no surprise, therefore, that it does present immunochemical problems (Russell *et al.*, 1978a) which are interesting and from which useful generalities may be noted.

In early immunochemical work with this enzyme (Dennick and Mayer, unpublished observations) it was observed that in immunodiffusion analyses detergent must be present in the gel; otherwise multiple immunoprecipitation lines near to the antigen well were commonly seen. This phenomenon results from the aggregation of the enzyme on entering the detergent-free agarose or agar gel. More detailed immunodiffusion analyses have been carried out with the enzyme in order to try to establish the immunochemical relationship of the enzyme in different human tissues (Russell *et al.*, 1978a).

Even with the purified human monoamine oxidase, multiple immunoprecipitation lines were generally observed on immunodiffusion; two immunoprecipitation lines (one of which stained for enzyme activity) were seen with freshly prepared enzyme, and after storage at −20 °C three or four lines were commonly observed (Fig. 3.2a and b).

Interestingly, although two or more immunoprecipitation lines were given with preparations of the purified enzyme, only one line was given with the enzyme in Triton X-100 extracts of liver mitochondria (i.e. the subcellular fraction from which the enzyme was prepared). This indicates that the physicochemical state of the enzyme is different in the two detergent preparations; this state may be dictated by several factors, including the protein/detergent ratio and the phospholipid content of the preparation. It is important to note that multiple immunoprecipitation lines are obtained with

---

Adsorbed antiserum was present in track 3. (d) Crossed rocket immunoelectrophoresis: a mixture of purified cytochrome oxidase and particle-free supernatant proteins was separated by electrophoresis at 20 V/cm for 2 h at 15 °C in the first dimension. Immunoelectrophoresis in the second dimension was carried out for 16 h at 10 V/cm at 15 °C into a gel containing unadsorbed antiserum (0.25% v/v) to cytochrome oxidase.

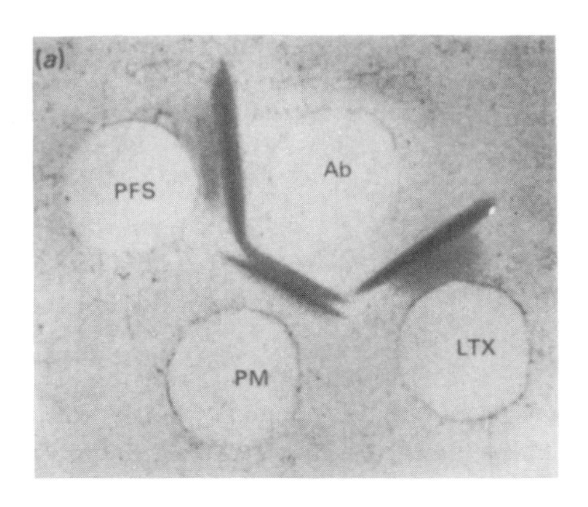


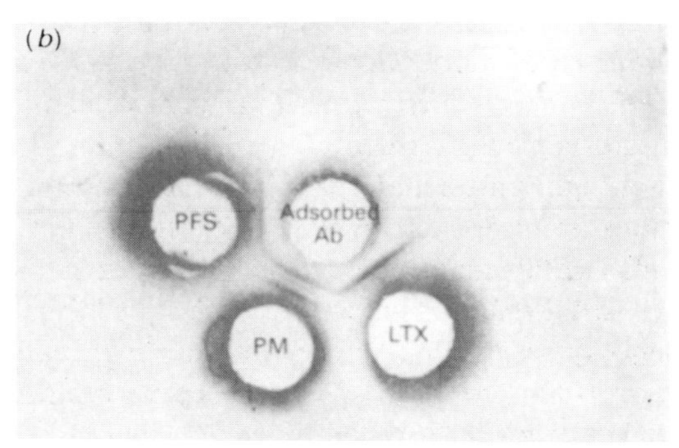


**Fig. 3.2** Immunodiffusion of human liver particle-free supernatant (PFS), purified liver monoamine oxidase (PM) and Triton X-100 extracts of liver mitochondria (LTX) against unadsorbed (a) and adsorbed (b) antiserum to human monoamine oxidase. Immunodiffusion was carried out for 2 days at room temperature in 1% (w/v) agarose in 20 mM potassium phosphate, pH 7.0, containing 0.15 M NaCl and 1.5% (w/v) Triton X-100. Immunoprecipitation lines were stained for protein with Coomassie brilliant blue.

unadsorbed antiserum (Fig. 3.2a) and with antibodies to monoamine oxidase which have been purified by mitochondrial adsorption (Fig. 3.2b). Adsorption was performed by binding antibodies to monoamine oxidase on to mitochondria in iso-osmotic conditions, followed by elution with 0.2 M glycine-HCl, pH 2.2, and subsequent adjustment to pH 7.0 with 1 M potassium phosphate (Fig. 3.12).

The facts that multiple lines were obtained with the purified enzyme and not with the enzyme in a detergent extract of mitochondria and that multiple lines (Fig. 3.2b) were obtained with purified antibodies as well as crude antiserum (Fig. 3.2a) clearly distinguish the observed phenomena from those that could be expected if the multiple lines were due to contaminating antigens. These types of phenomena on immunodiffusion or immunoelectrophoresis can clearly complicate the interpretation of reactions. Although, as might be expected, multiple lines were obtained on immunodiffusion analysis of detergent-solubilized extracts of human liver, placenta, brain cortex and platelets, clear complete reactions of identity were seen for the enzyme from these different tissue sources (Russell *et al.*, 1978a).

A further problem shown in Fig. 3.2 is that a subcellular fraction ("particle-free supernatant" prepared by centrifugation of liver post-mitochondrial fractions at $6 \times 10^6 g$ for 1 min), which should not contain mitochondria, does contain immunodetectable monoamine oxidase. This can also be shown by radiochemical assay of enzyme activity (Russell *et al.*, 1978a). This phenomenon may result from detachment of the enzyme from the outer mitochondrial membrane, but this is most unlikely for an integral outer-membrane protein. It is much more likely that the "particle-free supernatant" contains microsomal monoamine oxidase, or enzyme attached to outer mitochondrial membrane, or membrane fragments which have sloughed off during homogenization of the tissue.

This observation highlights a major difficulty experienced when examining subcellular fractions immunochemically, namely impurity of the subcellular fractions which may result from artefacts of homogenization or centrifugation. Careful analyses of reactions of identity are called for to distinguish the unexpected presence of an enzyme (or precursor or degradation product) in a subcellular fraction from the possibility of a contaminating antigen–antibody system (Walker *et al.*, 1976).

### 3.2.2. Immunoelectrophoresis

In classical immunoelectrophoresis, antigens are separated in agar or agarose gels by electrophoresis in one dimension and are then allowed to interact by diffusion with the antiserum in a second dimension (Grabar and Williams, 1953). This provides a much more sensitive means of assessing the specificity of an antiserum than immunodiffusion. Another technique with excellent resolving power for multispecific antisera is cross-over electrophoresis (Moody, 1976), where tissue extract and antiserum are opposed in a gel with the antigen at the cathode and the gel at pH 8.6, so that most antigens but not the antibodies are negatively charged. Electrophoresis forces the antigens towards the antiserum well. Electroendo-osmosis causes the antibodies to move towards the antigens and fine immunoprecipitation lines are rapidly produced (within 40 min).

Electrophoresis of antigens at pH 8.6 into a gel which contains antibodies has been routinely used. The antigen may be forced into the antibody-containing gel directly (Fig. 3.1c; Laurell, 1966) or after electrophoretic separation of the protein mixture at pH 8.6 in the first dimension (Fig. 3.1d; Weeke, 1973a). Several variations of the latter procedure, including tandem-crossed immunoelectrophoresis (Kroll, 1973a) and crossed-line immunoelectrophoresis (Kroll, 1973b), have been developed to add the facility of identification of immunoprecipitation lines to the advantages of resolution and quantitation which electrophoresis into antibody-containing gels has over classical immunoelectrophoresis.

A whole variety of methods for fractionating antigen mixtures has been used in the first dimension of two-dimensional immunoelectrophoresis. Proteins may be separated in the first dimension by means of isoelectric focusing (Soderholm *et al.*, 1975) or electrophoresis in polyacrylamide gels in the presence (Converse and Papermaster, 1975; Webb *et al.*, 1977; Chua and Blomberg, 1979) or in the absence (Loft, 1975; Lundahl and Liljas, 1975; Ekwall *et al.*, 1976) of sodium dodecyl sulphate. As well as providing better antigen resolution these techniques permit the determination of valuable protein parameters such as p*I* and antigen molecular weight, although in recent years these procedures for the determination of p*I* and $M_r$ have been completely overtaken by the immunoblotting procedures outlined below.

Any antigen-fractionation procedure may be immunochemically monitored by means of fused-rocket immunoelectrophoresis (Svendsen, 1973), although dot-blotting or ELISA procedures are more convenient than immunoelectrophoresis for this purpose.

Immunoelectrophoresis into antibody-containing gels is also subject to certain artefacts. Multi-subunit proteins may dissociate in the low ionic strength buffers used for electrophoresis, giving rise to complex immunoprecipitate lines (Paskin and Mayer, 1976). A further complication, studied in some detail by Bjerrum and Bøg-Hansen (1976b), is the proteolytic modification of antigens and the effect this has on immunoprecipitation lines. Phenomena such as multiple precipitation lines, "split" precipitation lines, "flying" precipitation lines and "skewed" precipitation lines may occur if proteolytic degradation takes place. These effects may result from contaminating proteolytic activity in the purified immunoglobulin (e.g. plasmin) or in the tissue extract. Membrane-protein antigens are particularly susceptible to proteolysis when extracted from membrane in either ionic or non-ionic detergents, and care should be taken to include inhibitors of proteolysis in extraction buffers. For proteins which require the presence of denaturing agents at all times (e.g. keratins) it is possible to perform immunoelectrophoresis in the presence of sodium dodecyl sulphate or urea (Lee *et al.*, 1978).

If more than one rocket is obtained, a major problem is to identify the immunoprecipitation line of interest.

Enzymes which retain activity in immunoprecipitation lines make identification of the precipitation lines relatively easy, assuming that a sufficiently sensitive histochemical stain is available (e.g. dehydrogenases (Fig. 3.1b) or esterases). In this way it is easy to identify the antigen–antibody system of interest (Walker *et al.*, 1976). Alternatively, radioligand binding may be used to identify an immunoprecipitation line of interest, e.g. with radiolabelled neurotransmitter receptor agonists or antagonists (Blomberg and Berzins, 1975; Teichberg *et al.*, 1977).

However, if an enzyme is inactive in an immunoprecipitation line, or the antigen of interest is a catalytically inactive protein, then some other reaction of identity must be employed. Many methods have been designed which rely on the use of purified antigen to identify the immunoprecipitation line which is given by the same antigen, among the multiple lines which may be given by a tissue extract. This approach relies on the immunochemical purity of the purified antigen and also requires the availability of purified antigen in sufficient quantities for the analyses. Purified antigens can be used for reactions of identity with the antigen of interest in tissue extract by immunodiffusion (Ouchterlony, 1968) or immunoelectrophoretic analyses (Axelsen *et al.*, 1973). Monospecific antiserum to some component of interest in an antigen mixture, or alternatively a polyspecific antiserum to all components of the mixture except the component of interest, can be interposed in an intermediate gel between the antigens resolved by electrophoresis in the first dimension and the multispecific antiserum to be used for second dimensional immunochemical analysis. In this way a component of interest can be positively identified (Weeke, 1973a).

If the subunit molecular weight of an antigen of interest is known, then the gel immunoprecipitate for this antigen may be identified by analysing individual immunoprecipitates by polyacrylamide gel electrophoresis in the presence of sodium dodecyl sulphate (Norrild *et al.*, 1977; Walker, unpublished observations). Antigen mixtures labelled radioactively *in vivo* or *in vitro* may be analysed by immunoelectrophoresis, and the individual immunoprecipitates analysed by polyacrylamide gel electrophoresis in the presence of sodium dodecyl sulphate (Fig. 3.3) followed by autoradiography (Norrild *et al.*, 1977), or fluorography (Witzemann and Walker, 1981; Fig. 3.4). The practical details of these methods are given in Chapter 7.

Finally, since the area underneath a rocket immunoprecipitate is proportional to the amount of antigen, crossed-rocket immunoelectrophoresis may be used to monitor antigen purification. The area beneath a rocket divided by the amount of protein in the sample well (at a fixed antiserum concentration) may be used as an equivalent to a classical "specific activity" in antigen purification schedules (Fig. 3.5).

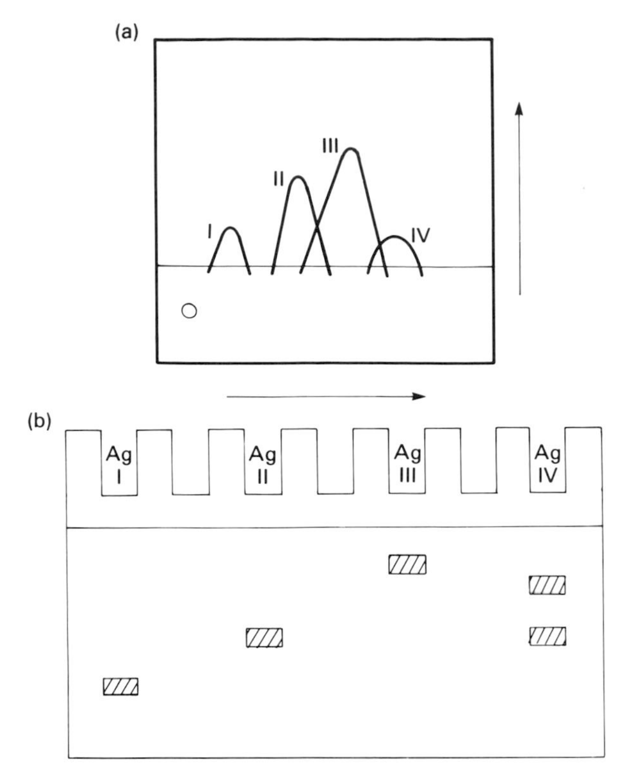


**Fig. 3.3** Determination of antigen subunit molecular weight. (a) A radiolabelled antigen mixture is analysed by crossed-rocket immunoelectrophoresis. (b) The individual immunoprecipitates are analysed by polyacrylamide gel electrophoresis in the presence of sodium dodecyl sulphate. After staining for protein, and drying the gel, the subunits of each antigen may be identified by autoradiography or fluorography. Ag = antigen.

### 3.2.3. Gel overlay and immunoblotting procedures

(Gershoni and Palade, 1982, 1983; Towbin and Gordon, 1984; Gershoni, 1985)

Polyclonal antisera usually cross-react strongly with SDS and urea-denatured antigens even after these antigens have been analysed by 1D (Fig. 3.6) and 2D gel (Fig. 3.7) electrophoretic methods. Initially, gel overlay procedures were devised in which polyacrylamide gels were treated with 25% isopropanol 10% acetic acid to fix proteins in the gel. The antigenic components were then identified by sequential incubation in antiserum and then either fluorescently labelled second antibody (Valderrama *et al.*, 1976),

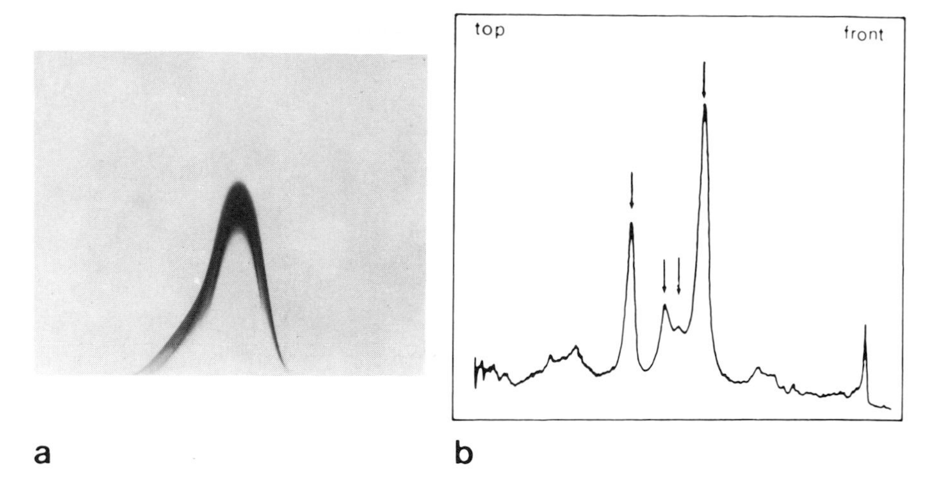


**Fig. 3.4** Determination of antigen subunit composition. (a) Crossed-rocket immunoelectrophoresis. A Triton X-100 extract of freshly prepared membrane fragments containing acetylcholine receptor (AChR) was separated by electrophoresis in agarose in the first dimension followed by electrophoresis into a section of gel containing antibodies specific for AChR. (b) Densitometric scan of the immunoprecipitate containing radiolabelled antigen after analysis by SDS-PAGE. The polypeptides were visualized by means of fluorography. Arrows indicate polypeptides of apparent $M_r$ 40 000, 50 000, 60 000 and 65 000.

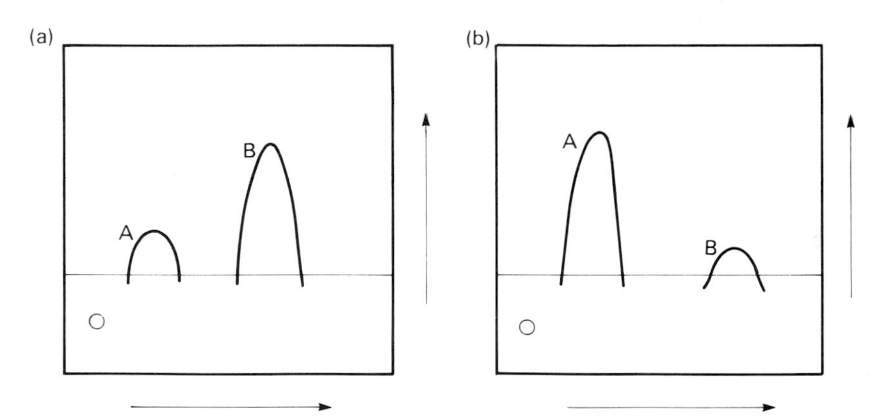


**Fig. 3.5** Analysis of antigen mixtures during antigen purification. The same amount of protein is applied to the sample well from a preparation before (a) and after (b) a step in the purification of antigen A. Antiserum concentration in the gel is the same in both immunoelectrophoresis plates (a and b).

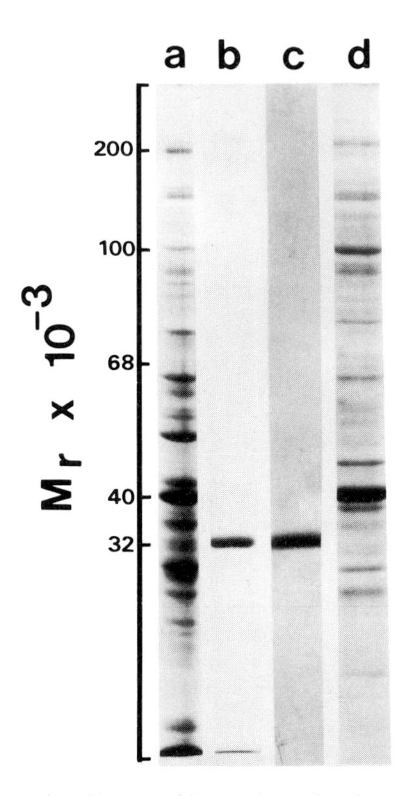


**Fig. 3.6** Use of one-dimensional immune blots to determine the specificity of an antiserum to calelectrin isolated from the electric organ of *Torpedo marmorata*. Lane a, Coomassie-stained total soluble proteins; lane b, purified calelectrin stained with Coomassie; lane c, an immunoblot of the total electric organ proteins, which are shown stained with Coomassie Blue in lane d. Reproduced from Sudhof *et al.* (1985) with permission of the publisher.

or enzyme-coupled antiserum (Olden and Yamada, 1977) or $^{125}$I-labelled Protein A (Burridge, 1978). These methods suffer, however, from the disadvantage that the antibody reaction could only take place on the gel surface. Also, the ability of antibodies to recognize an antigen may be reduced as a result of denaturation by sodium dodecyl sulphate and the fixing solution.

Transferring proteins separated by SDS–PAGE on to nitrocellulose paper or nylon sheets (blotting) considerably improves the detection of the antigens, since some antigen renaturation probably occurs and since all of the protein in the gel is transferred on to the surface of the blot. Furthermore, the blots can be handled with far greater ease than the polyacrylamide gels.

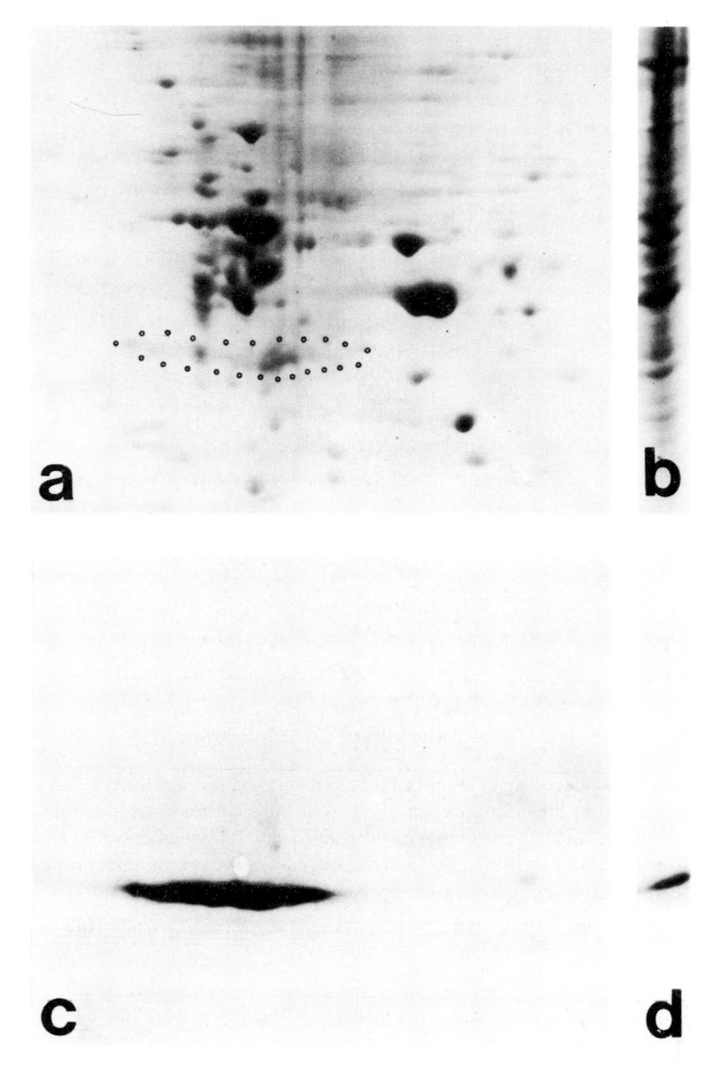


**Fig. 3.7** Use of a two-dimensional immune blot to identify calelectrin after two-dimensional electrophoresis. (a) and (b) are 2D and 1D gels stained with Coomassie. (c) and (d) are immune blots stained for calelectrin.

Proteins are usually transferred on to nitrocellulose or nylon by diffusion (Southern, 1975) or by electrophoresis (Towbin *et al.*, 1979). The proteins bind due to hydrophobic interactions. Electrophoretic transfer can be accomplished rapidly using one of several transfer buffers and simply constructed apparatus. Battery chargers intended for charging 12 V car batteries work admirably in transferring the proteins on to nitrocellulose.

The proteins transferred on to nitrocellulose may be stained in several ways to check that the transfer has been successful, and it is advisable to stain both the gel and the blot to assess the extent of transfer. Using the standard transfer buffer (Towbin *et al.*, 1979), which contains methanol, some proteins (e.g. myosin) transfer poorly. Also some proteins (e.g. calmodulin) may bind only poorly to nitrocellulose, in which case it is advisable to use blotting media which bind proteins covalently or to bake the nitrocellulose filter at 80 °C for 2 h to bind the proteins more firmly. One of the most useful stains for protein transferred on to nitrocellulose is Ponceau S which, although relatively insensitive, can be completely removed from the stained protein bands by washing with buffers of neutral pH (e.g. PBS, see Chapter 7). Another very useful stain which can be used prior to immune staining is Fast Green (Eichner *et al.*, 1984), which also is lost from stained protein bands. The blot remains bluish-green to some extent but this does not interfere with immune staining and provides an aesthetically pleasing background to the brown staining obtained for peroxidase activity with diaminobenzidine. Fast Green is as sensitive as Coomassie for detecting proteins, and even greater sensitivity may be obtained with India ink (Hancock and Tsang, 1983) which cannot, however, be removed subsequently from the blot. Recently, a colloidal iron stain has also been introduced for staining proteins transferred on to nylon sheets (Ferrodye, Janssen Pharmaceuticals).

The blot must now be treated with specific antibody to identify the antigen(s) of interest. An important preliminary to this is to saturate the protein-binding sites on the transfer paper to prevent non-specific binding of antibody. A variety of proteins have been used for this purpose, including serum albumin, ovalbumin, and heat-inactivated sera. Surprisingly, however, one of the cheapest and most successful solutions available is low-fat dried milk reconstituted in phosphate-buffered saline (Johnson *et al.*, 1984). Incubations as short as 5 min in this solution are reported to be sufficient to block all the protein-binding sites on the nitrocellulose paper, and in our experience dried milk is vastly superior to the much more expensive and more commonly used bovine serum albumin. Alternatively, non-specific protein-binding sites may be blocked by using non-ionic detergent in the blocking and washing buffers. Thus, 0.1% Tween 20 or Triton X-100 have been used and Tween 20 has been shown to be as effective as serum albumin in blocking non-specific binding to nitrocellulose (Batteiger *et al.*, 1982). It has, however, been reported that non-ionic detergent may remove some of the protein bound to the nitrocellulose.

Antigens are detected by sequential incubation in antibody and an antibody-detection system (Fig. 3.8). The detection systems used are com-

Electrophoresis (1D or 2D)

↓

Transfer antigens on to nitrocellulose (or nylon)

↓

Block sticky sites (Tween 20 and/or BSA)

↓

Incubate with first antibody (or labelled lectin, or protein)

↓

Second antibody incubation (enzyme or radiolabelled)

↓

Detect second antibody

**Fig. 3.8** Immune-blotting protocol.

monly peroxidase-labelled or $^{125}$I-labelled second antibody or Protein A. Alternatively, more sensitive methods are available such as the peroxidase–antiperoxidase method, and methods which utilize biotin-labelled second antibody and avidin or streptavidin-conjugated third stages (Fig. 3.9). Another very sensitive method involves the use of gold-labelled second antibodies and subsequent intensification with silver enhancement. Interestingly, silver-intensification kits are also available from Amersham International to intensify the diaminobenzidine reaction product obtained with peroxidase-labelled reagents. A summary of the various different detection systems available is given in Table 3.1 and extensive technical details and references are given in Chapter 7.

It is of considerable interest that protein blotting and gel overlay procedures may be used to detect various protein–protein and protein–ligand interactions (Gershoni, 1985) in addition to antibody and lectin-binding proteins.

**Fig. 3.9** Streptavidin staining methods Antigen (—▲—) is detected by incubation with biotin-labelled antibody (Y) followed by (a) streptavidin-labelled horseradish peroxidase (O-HRP); (b) streptavidin (O) and then biotin-labelled horseradish peroxidase (•—HRP); or (c) a soluble complex containing many molecules of streptavidin and biotin-labelled horseradish peroxidase.

**Table 3.1** Antibody detection methods on blots

| *Label* | *Reference* |
|---|---|
| Peroxidase | Tsang *et al.* (1983) |
| Alkaline phosphatase | Turner (1983) |
| Colloidal gold | Moeremans *et al.* (1985) |
| Biotin/Streptavidin/label | Ogata *et al.* (1983) |
| $^{125}I$ | Burnette (1981) |

Some proteins retain enzymic or biological activity even after gel electrophoresis in the presence of sodium dodecyl sulphate (Petrucci *et al.*, 1983). Similarly, even severely denatured antibodies can be renatured as, for example, is the case with trichloroacetic acid-precipitated IgG, which can be solubilized in 6 M guanidine hydrochloride. After slowly removing the guanidine-HCl by dialysis (6 M to 4 M to 1 M to phosphate-buffered saline) the full antibody activity of the IgG may be restored (Biorad bulletin 1099). It is, however, often important to avoid the use of reducing agents if renaturation is to occur as, for example, is the case with certain growth factors. Gel overlay and protein-blotting procedures have successfully identified proteins which bind calcium (Maruyama *et al*, 1984), calmodulin (Fig. 3.10; Glenney and Weber, 1980; Walker *et al.*, 1984a; Billingsley *et al.*, 1985), actin (Snabes *et al.*, 1981; Walker *et al.*, 1984b), vinculin (Otto, 1983), S100, Troponin C, DNA (Burgess *et al.*, 1984), vimentin and tubulin (Belin and Boulanger, 1985), spectrin and ankyrin (Baines and Bennett, 1985) and fibronectin (Bell and Engvall, 1982). Similarly, GTP-binding proteins and the $\alpha$-bungarotoxin-binding subunit of the acetylcholine receptor have been identified (Gershoni, 1985). Specific components of the extracellular matrix have even been identified by allowing living cells to bind nitrocellulose blots of proteins analysed on SDS gels (Hayman *et al.*, 1982) and interactions between cytoskeletal proteins and adenovirus particles have also been monitored (Belin and Boulanger, 1985). The method has usually been performed using radiolabelled probes, although enzyme-conjugated probes and biotin-labelled probes have also been used, in particular for calmodulin (Billingsley *et al.*, 1985) and DNA (Leary *et al.*, 1983).

These protein-blotting methods may be used in conjunction with antibody blots to further characterize a particular antigen.

It is also of value to mention here that several authors have been able to use nitrocellulose blots as affinity matrices for the isolation of specific antibodies (Olmstead, 1981; Talian *et al.*, 1983; Ungewickell, 1985; Perrin and Aunis, 1985). A protocol for this procedure is given in Chapter 7, and clearly has the advantage in principle that specific antibodies may be

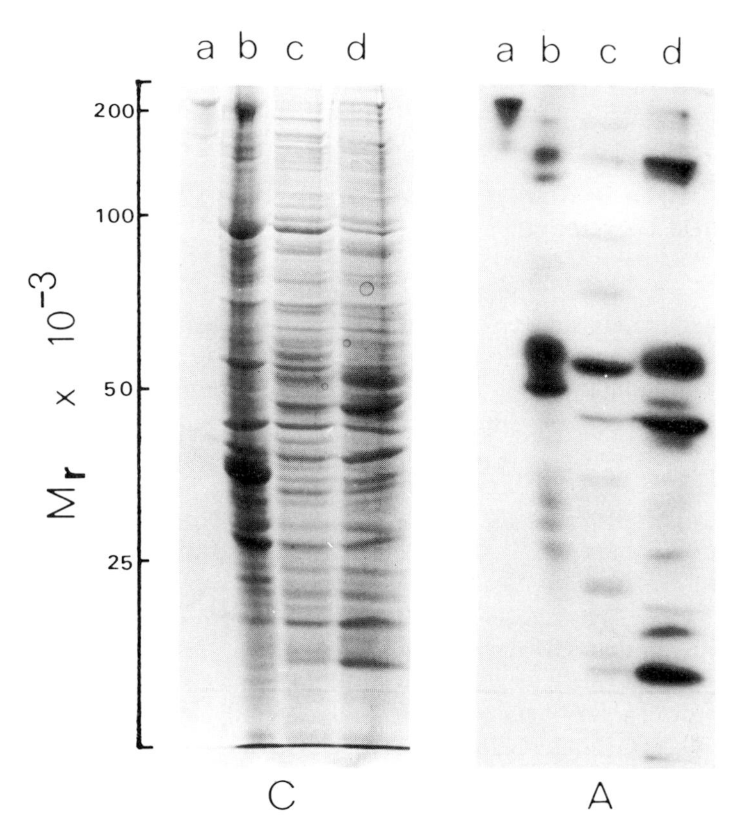


**Fig. 3.10** Gel overlay to demonstrate calmodulin-binding proteins. Samples of (a) purified fodrin, (b) electric organ, (c) electric lobe, and (d) brain minus lobe were analysed by SDS-PAGE on 8.75% gels. The gel was incubated with radiolabelled calmodulin in the presence of 1 mM $CaCl_2$ and then washed and stained with Coomassie to stain proteins (C). The gel was then dried down and subjected to autoradiography (A) to localize calmodulin-binding proteins. Reproduced from Walker *et al.* (1984a) with permission of the publisher.

obtained when neither pure antigen nor monospecific antisera are available. The method is sometimes referred to as the "poor man's monoclonal".

Protein blotting and gel overlay procedures have been reviewed recently (Gershoni, 1985) and supply extremely valuable methods for identifying various protein–ligand and protein–protein interactions. Gel overlay procedures involve incubating fixed gels with labelled probes and are well known for the detection of lectin-binding and antibody-binding proteins. Protein blotting is essentially the same except that proteins are transferred from gels on to nitrocellulose paper which is then incubated with the probe (Gershoni and Palade, 1982, 1983).

## 3.3. Purification of Polyclonal Antibodies to the Antigens of Interest

Polyclonal antisera may be rendered more specific by treating the antisera with contaminating antigens to remove unwanted antibodies (Kwapinski, 1972). Alternatively, specific antibodies may be purified by affinity chromatography on columns (or other immobilized preparations) of the antigen (Kabat, 1967).

### 3.3.1. Adsorption of an antiserum with contaminating antigens

Adsorption of a polyclonal antiserum can be performed with soluble or immobilized contaminating antigens. Antisera to enzymes can be conveniently adsorbed with contaminating antigens in subcellular fractions which contain little or none of the antigen of interest, with contaminating antigens in subcellular fractions prepared from a tissue in a developmental stage at which there is little or no antigen of interest, and with fractions which are discarded during the purification of the antigen of interest (Walker *et al.*, 1976).

The proportion of contaminating antigens required for complete adsorption of an antiserum can be determined by quantitative immunoprecipitation of the contaminants, quantitative rocket immunoelectrophoresis, the sharpness and intensity of immunoprecipitation lines on immunodiffusion analysis (Walker *et al.*, 1976) or by Sewell titration (Sewell, 1967).

Another convenient procedure is the use of an ELISA to monitor the removal of antibodies to unwanted antigens (Walker *et al.*, 1985). As shown in Fig. 3.11, samples of antisera are adsorbed with increasing amounts of an unwanted antigen which is also used to coat the wells of the ELISA tray. As increasing antigen is used to adsorb the antiserum, less antibody remains to react with the antigen on the ELISA wells. Thus the correct proportion for removal of all antibody to unwanted antigens can quickly be established. This method may also be used to eliminate specific antibodies from sera to generate control serum, especially for immunohistochemical experiments.

These analyses should be performed when either soluble or immobilized adsorbents are to be used. Adsorption in solution should be performed at equivalence, whereas immobilized contaminants may be used in excess.

Preparations of soluble contaminating antigens can be conveniently used as adsorbents (Walker *et al.*, 1976). Immunoglobulins can be purified from the adsorbed antiserum by techniques described in Chapter 7. This should ensure removal of most of the adventitious proteins in the preparation after adsorption.

Insolubilized contaminating antigens may be used for antiserum adsorption. Contaminating antigens may be insolubilized by polymerization (e.g. with glutaraldehyde). Since gel formation is dependent on protein concentration, it is sometimes advisable to add a protein such as bovine serum albumin (50–100 mg/ml) to the dilute preparation of contaminating antigens to ensure rapid gel formation.

When polymerized gels (glutaraldehyde heteropolymers) are used for column chromatography, problems may arise because of the poor flow properties of the material. Adsorption in batches may therefore be more convenient. Alternatively, contaminating protein antigens may be immobilized on to solid supports (e.g. to Sepharose with cyanogen bromide) for use in adsorption procedures.

Sepharose-immunoadsorbents, prepared by coupling proteins from a subcellular fraction enriched with a contaminating antigen, are very conveniently used. These immunoadsorbents have good flow properties or alternatively can be easily used in batches.

Some problems may occur with immunoadsorbents prepared by polymerization of proteins, or coupling proteins to solid supports. Polymerization or immobilization of antigens may reduce their antigenicity and this should be estimated and allowed for when determining the quantity of an adsorbent to be used. Another problem is related to the use of protein preparations enriched with contaminating antigens which also invariably contain some of the antigen of interest. This means that the concentration of antibodies to the antigen of interest will be depleted by the immunoadsorption procedures. However, loss of some antibodies to the antigen of interest is a small price to pay for removal of all the contaminating antibodies.

The production of tissue-specific antisera is a relatively easy process. A polyspecific antiserum (e.g. multi-tissue antiserum) may be conveniently mixed with glutaraldehyde heteropolymers or acetone powders derived from tissues other than the one of interest. Antisera specific for different developmental stages in the life-span of an organism may be prepared in an analogous manner.

### 3.3.2 Affinity isolation of specific antibodies

Purified antigens may be immobilized in one of several ways to produce affinity columns for the isolation of specific antibodies. In particular,

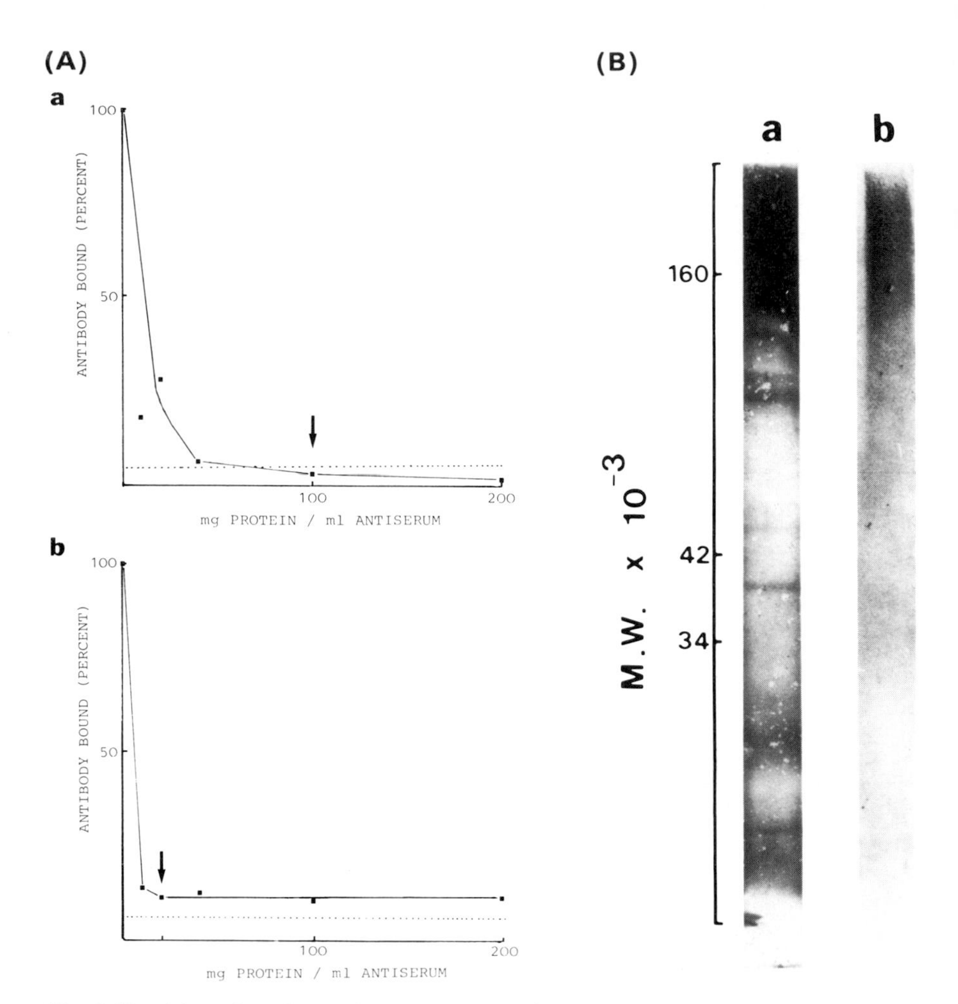


**Fig. 3.11** Adsorption of an antiserum and testing of its specificity. A: Plastic wells of ELISA trays were coated with (a) liver membrane fragments or (b) electric organ particle-free supernatant. Samples of antisera were adsorbed with these fractions and the residual antibody assayed in the ELISA wells. Arrows indicate the proportion of the fraction used subsequently for purification of the antiserum. (– – –), level of non-specific binding obtained with unadsorbed control serum. B: Immune blots of fractions containing synaptic vesicles using (a) unadsorbed antiserum and (b) adsorbed antiserum. C: Immunofluorescence of antisera to vesicles on 5 μm paraffin sections of *Torpedo* electric organ. (a) Unadsorbed antiserum; (b) adsorbed antiserum; Ventral (v) and dorsal (d) membranes are indicated. Scale bar = 10 μm.

(c)

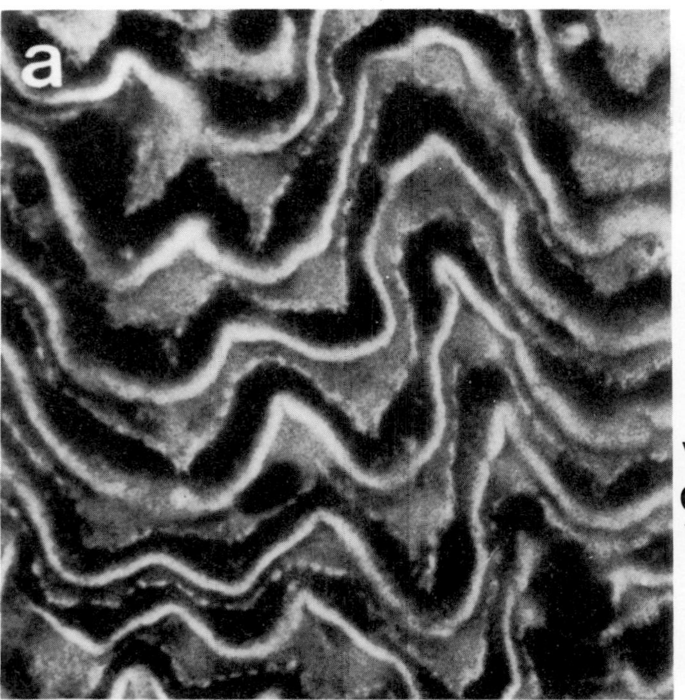


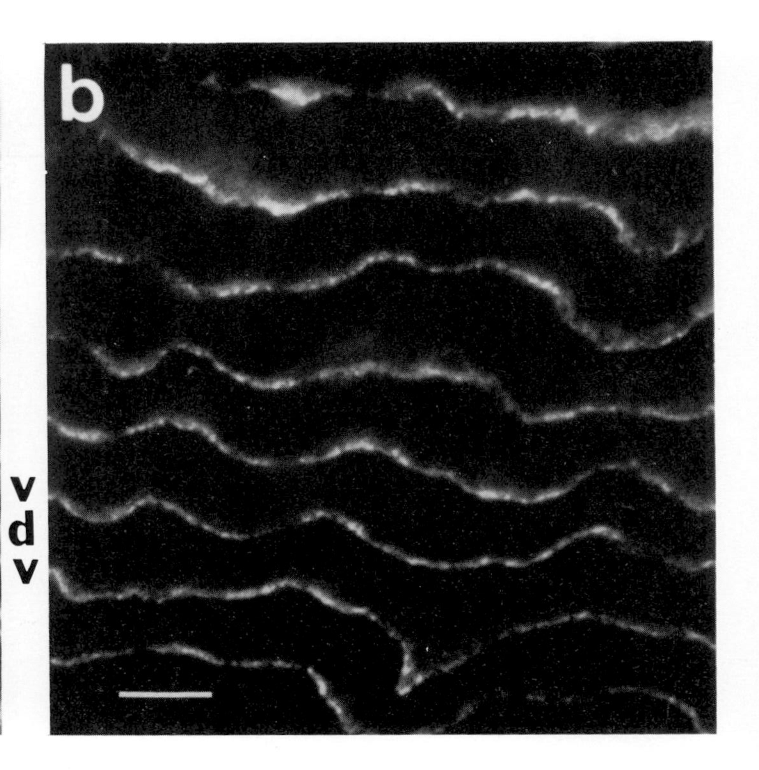

cyanogen bromide-activated Sepharose and glutaraldehyde-activated Ultrogel provide useful immobilization matrices, and methods for activating and binding proteins to these matrices are given in Chapter 7. These and other activated matrices (e.g. Affigels 10, 501, 502 from Biorad; Reactigel from Pierce) are available commercially.

Once prepared, the immunoadsorbent is used to bind specific antibodies which can subsequently be eluted. For example, fibrinogen and ovalbumin have been coupled to Sepharose (Bouma and Fuller, 1975) and then used to purify their respective antibodies. Alternatively, antibodies can be bound to polymerized antigen, e.g. casein (Houdebine and Gaye, 1976).

From these examples it is apparent that these techniques are often used for proteins which can be obtained in relatively large amounts so that enough antigen is available to adsorb a significant volume of antiserum. Antiserum to fatty acid synthetase has been purified with Sepharose-fatty acid synthetase (5 mg fatty acid synthetase; Alberts *et al.*, 1975). Re-utilization of immunoadsorbents increases the usefulness of small quantities of immobilized antigen.

The main problem with the use of immobilized antigen of interest as immunoadsorbent is the quality of the antigen. If a purified antigen is immunologically impure (Chapter 2), and the same antigen is used to prepare the immunoadsorbent, then contaminating antibodies could be bound to the immobilized antigen. The situation is complex and depends on the degree of purity of the antigen and the relative immunogenicities of the antigen of interest and the contaminants. Probably the best way of overcoming this problem is to purify the antigen by preparative electrophoresis on polyacrylamide gels in SDS. The antigen can be eluted from the polyacrylamide gel as described in Chapter 7, and then immobilized.

Finally, binding of antibodies (or antigens, Chapter 4) to immunoadsorbents is complicated by non-specific binding of proteins, e.g. non-specific immunoglobulins. This necessitates the use of washing procedures to try and elute these proteins before elution of the antibodies of interest, e.g. with 0.15 M NaCl or 0.1 M $NaHCO_3$ (Alberts *et al.*, 1975) or with detergent–salt mixtures (Smith *et al.*, 1978) or with 0.2 M glycine-HCl, pH 6.0. With immunoadsorbents made from CNBr-activated Sepharose it is recommended that only IgG fractions prepared by ammonium sulphate fractionation should be applied to the column to minimize non-specific binding problems.

Convenient partial adsorption procedures (i.e. not with purified antigen of interest) for membrane antigens can be carried out if the membrane fraction can be easily isolated. Antibodies to monoamine oxidase have been partially purified by adsorption to mitochondrial preparations and elution by low pH. The method is shown in detail in Fig. 3.12. With this procedure, approxi-

IgG (containing antibodies to the enzyme) is mixed with a mitochondrial preparation (in 100 mM sodium phosphate buffer, pH 7.2, in the proportion of 2 mg of IgG protein to 1 mg of mitochondrial protein. Suspension incubated at 4 °C for 2–3 h.

↓

Centrifuge for $10^5 g_{av}$ min. Pellet is washed twice with 20 mM sodium phosphate buffer containing 0.15 M NaCl.

↓

Washed pellet incubated with 0.2 M glycine-HCl, pH 2.8, volume equal to the original volume of the IgG preparation) for 1 h at 0 °C with constant stirring.

↓

Centrifuge for $225 \times 10^3 g_{av}$ min. Supernatant is immediately neutralized with 1 M potassium phosphate, pH 7.2.

↓

Solution dialysed overnight at room temperature against 20 mM sodium phosphate, pH 7.0, containing 0.15 M NaCl.

**Fig. 3.12** Partial purification of antibodies to monoamine oxidase by mitochondrial adsorption.

mately 4% of the protein in the IgG fraction bound to the mitochondrial preparation. The eluted antibodies to monoamine oxidase retained approximately 60% of the immunoinhibitory capacity of the unadsorbed IgG fraction.

Elution of antibodies of interest can be achieved with a variety of agents. The aim of these treatments is to break the electrostatic and hydrophobic interactions which bind antigens to antibodies. Low pH (e.g. 0.2 M glycine-HCl, pH 2.6), high pH (e.g. 1 M ammonium hydroxide, pH 11.5), high salt concentration (e.g. 4.5 M $MgCl_2$), chaotrophic salts (e.g. 3 M NaSCN), and 1 M propionic and acetic acids have been used to release antibodies (or antigens) from immunoadsorbents.

Comparisons of several eluants for the removal of casein antibodies from a Sepharose-casein immunoadsorbent have been carried out (Al-Sarraj, White and Mayer, unpublished observations). The effects of 2.5 M sodium iodide, pH 9.0, and 3 M sodium thiocyanate, pH 6.8, were compared with those of 0.2 M glycine-HCl, pH 2.8. The sodium iodide and sodium thiocyanate were removed from the antibody preparations by Sephadex G-25 chromatography, and the glycine-HCl was rapidly neutralized with 1 M Tris solution. In all cases the preparation of eluted antibodies was cloudy, and centrifugation of each suspension resulted in sedimented protein. The ratio of supernatant protein concentrations in the centrifuged eluants was approximately 4:2:1 for glycine-HCl, thiocyanate and iodide respectively. It is probable that the combination of differential elution and differential denaturation resulted in the different concentrations of eluted antibodies. The effectiveness of the antibody preparations in immunoprecipitating casein is shown in Fig. 3.13. There is a marked reduction in the capacity of each preparation of purified antibodies to immunoprecipitate casein.

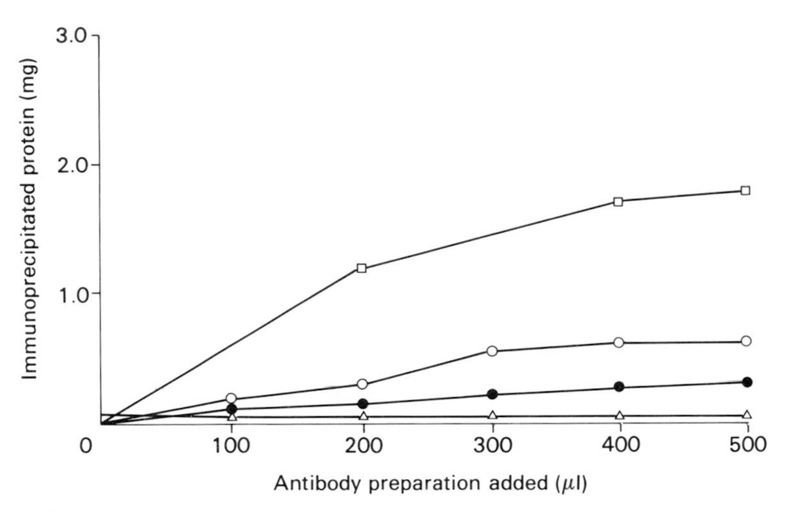


**Fig. 3.13** Immunoprecipitation of casein by adsorbed antibody preparations. Antibody preparations were prepared by eluting samples of Sepharose-casein immunoadsorbent with 0.2 M glycine-HCl, pH 2.8 (○); 2.5 M sodium iodide, pH 9.0 (●); and3 M sodium thiocyanate, pH 6.8 (△). Antibody preparations were processed as described in the text. Antibodies in unadsorbed antiserum (IgG) are shown for comparison (□). Casein (6 μg) was added to each sample, the volume was adjusted to 1 ml with phosphate-buffered saline, and immunoprecipitation was allowed to proceed for 4 days at 4 °C.

Antibodies eluted with glycine-HCl maximally precipitate about 30% of the casein immunoprecipitated by unadsorbed antiserum. This contrasts with the 60% of immunoinhibitory capacity shown by the monoamine oxidase antibody preparation which was eluted from a mitochondrial preparation by glycine-HCl. The ratio of maximally immunoprecipitated casein was 4:22:3 with antibody preparations eluted with glycine-HCl, thiocyanate and iodide respectively, which is in very good agreement with the ratio of protein concentrations in these antibody preparations (see above).

It is probable that denaturation of eluted antibodies results in antibody precipitation. As shown in Fig. 3.13, the capacity of the antibody preparations to precipitate antigen is a function of their antibody concentrations. Again, the individuality of each antigen–antibody system and immunoadsorption method is illustrated (cf. antibodies to casein and monoamine oxidase) in terms of the effects of eluants on immunochemical behaviour of the eluted antibodies.

Attempts have been made to separate antigen–antibody complexes by milder procedures, including ion-exchange chromatography (McCauley and Racker, 1973) and electrophoresis (Dean *et al.*, 1977). The affinities of the eluted antibodies must be considered, namely whether all the antibodies are eluted in the described conditions or whether the eluted antibodies are those with weaker affinities for the macromolecular antigen. The latter situation could be advantageous if the eluted antibodies were to be used to prepare an antigen-binding immunoadsorbent. It should be possible to elute antigen from such an immunoadsorbent under carefully defined conditions.

Antibody subpopulations responding to different antigenic determinants can be purified by techniques including immunoadsorption (Arnon, 1973; McCans *et al.*, 1975) and in the same way antibodies to catalytic and non-catalytic sites on enzymes or different oligomeric forms of proteins (Pages *et al.*, 1976) may be fractionated by immunoadsorption procedures.

In spite of the technical problems, purified antibodies to an antigen of interest are of great value in immunochemical studies (Chapter 5), especially since they considerably reduce non-specific phenomena in immunoprecipitation and immunoaffinity procedures.

Purified antibodies are much easier to use and give much clearer results in immunoprecipitation reactions in gels. Problems of background staining are avoided and this means that the quantitative measurements can be made more easily. Furthermore, if immunoprecipitation or immunoadsorption (immunoaffinity) techniques are to be used to immunoisolate antigens from tissue extracts, purified antibodies avoid many of the problems of non-specific precipitation and non-specific adsorption respectively, which can plague these techniques when unpurified IgG is used.

Criticisms of adsorption of antisera with immunoadsorbents prepared with the antigen of interest (i.e. based on antigen quality, see above) may be overcome if adsorption with preparations enriched with contaminating antigens precedes adsorption with the antigen of interest. In this way antibodies to contaminating antigens could be removed first, and would therefore not bind to those contaminating antigens which may contaminate the antigen of interest which has been used to prepare the immunoadsorbent.

### 3.3.3. Reassessment of antiserum specificity

The success of adsorption procedures with either contaminating antigens or the antigen of interest must be assessed qualitatively and quantitatively. The techniques for assessment are those described previously, and the procedures are used to test for the monospecificity of the antiserum (Fig. 3.1).

Cross-rocket immunoelectrophoresis is a very sensitive technique for observing the effects of adsorption on the specificity of an antiserum qualitatively. Even in a case where an antiserum is very specific before adsorption, e.g. casein (Al-Sarraj *et al.*, 1978), cross-rocket immunoelectrophoresis can be used to provide evidence for minimal qualitative alteration in antibody specificity, during antibody purification from the IgG fraction (Fig. 3.14) by immunoaffinity chromatography.

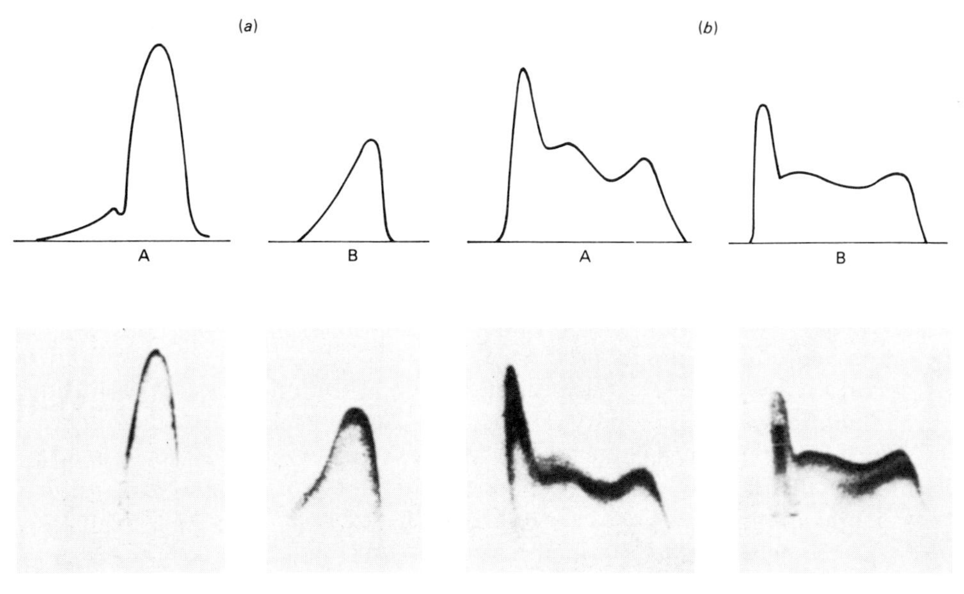


**Fig. 3.14** Cross-rocket immunoelectrophoresis of $^{32}P$-labelled recombined casein and $^{32}P$-labelled tissue extract from mammary gland of lactating rabbit. Immunoelectrophoresis was carried out in 1% (w/v) agarose gels. The $^{32}P$-labelled preparations were prepared as described by Al-Sarraj *et al.* (1978). Coomassie Blue staining of the gels of $^{32}P$-labelled recombined casein (a) and $^{32}P$-labelled tissue extract (b) tested against unadsorbed antiserum (A) and adsorbed antiserum (B). Radioautograms of each crossed-rocket are shown beneath their respective Coomassie-stained rockets.

Simple Ouchterlony double diffusion is often adequate for demonstrating the complete removal of contaminating antibodies by immunoadsorption. Immunodiffusion is best when a small number of antigen–antibody systems are known to be detectable by a multispecific antiserum. Clear demonstration of the removal of contaminating antibodies by adsorption of a multispecific antiserum to cytochrome oxidase with a fraction enriched in contaminating antigens is shown in Fig. 3.15 (Walker *et al.*, 1976).

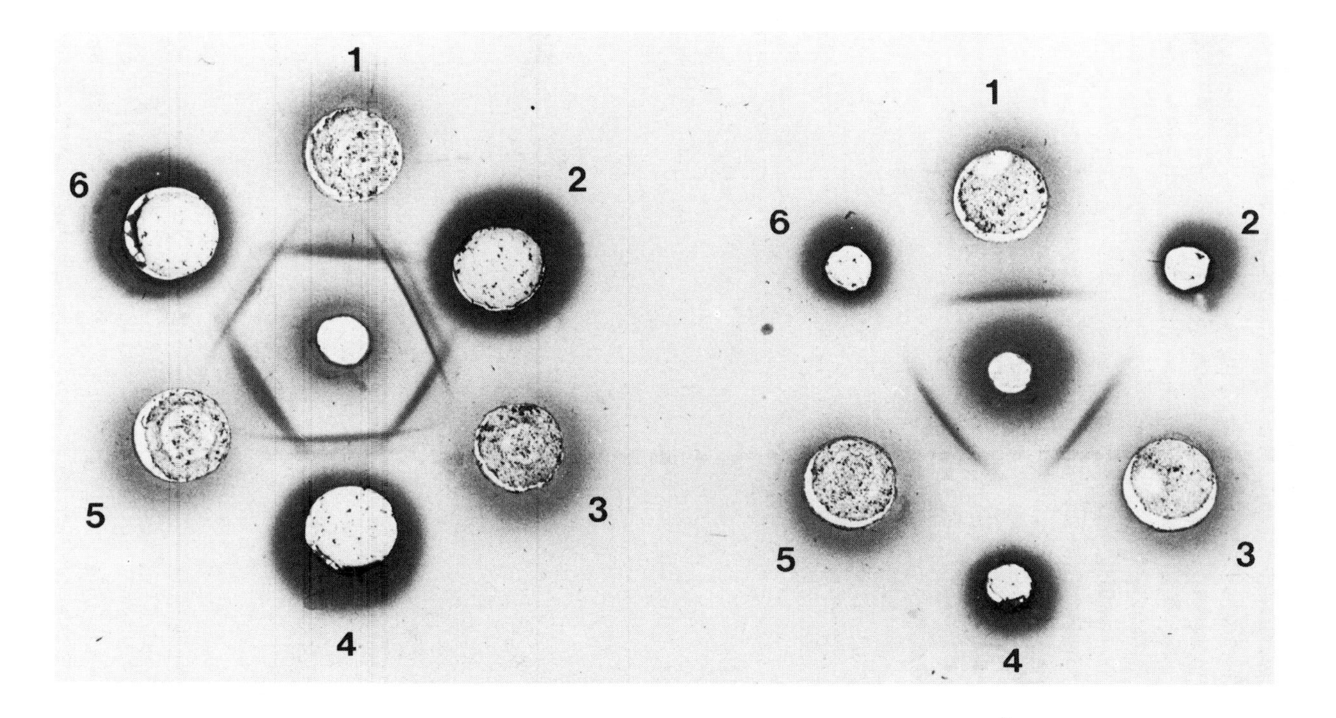


**Fig. 3.15** Ouchterlony double-diffusion analysis with unadsorbed and adsorbed antiserum. Diffusion was performed in 1% (w/v) agarose gels containing 20 mM sodium phosphate buffer, pH 7.0, 150 mM sodium chloride and 1% (w/v) Triton X-100. Wells 1, 3 and 5 contained purified enzyme, and wells 2, 4 and 6 contained samples of a particle-free supernatant prepared from homogenate of rat liver. The central wells contained (a) unadsorbed antiserum; (b) completely adsorbed antiserum to cytochrome oxidase. Immunodiffusion was carried out for 2 days at room temperature in a moist atmosphere.

Since the purification techniques may result in some loss of antibody titre, the relative titre of the adsorbed antiserum should be estimated, e.g. by rocket immunoelectrophoresis (Fig. 3.1c and d).

Alternatively, quantitative immunoprecipitation may be used to assess loss of antibodies during an adsorption procedure. An example of this method is shown for casein antibodies in Fig. 3.13.

Immunoblotting experiments are also extremely valuable for comparing unadsorbed and adsorbed antisera, and also ELISA procedures can provide sensitive methods for determining the antiserum specificity and titre. Furthermore, in some cases it may be valuable to check the specificity of the antiserum by immunohistochemical procedures to ensure that the adsorbed antiserum reacts with the correct cell compartment, and produces the expected staining pattern on tissue sections.

# 4
# Uses of Polyclonal Antisera

In this chapter the uses of antisera which are of value to cell and molecular biologists, enzymologists and protein chemists will be discussed. Antisera may be used to determine the amounts of protein antigens and to isolate antigens. Antisera may also be used to identify protein antigens in different subcellular compartments, in different cells and even in different species. Antisera are routinely used to isolate and purify protein antigens from tissue extracts.

Antisera are convenient probes that can be used to localize antigens by immunohistochemical techniques. The vectoral orientation of enzymes in the membrane can also be conveniently established with antisera. Immuno-inhibition of catalytic activity can be measured to establish if an active site of an enzyme is oriented externally or internally on a cell surface membrane (i.e. plasma membrane) or organelle membrane (e.g. mitochondria and microsomes). Finally, antisera often recognize antigens from various species, which considerably extends their value.

## 4.1. Determination of Antigen Amount

Methods for the estimation of the amount of a protein antigen include quantitative precipitation of an antigen from solution, quantitative immuno-precipitation in gels, measurements dependent on immunotitration of enzyme activity, radioimmune assays and enzyme-immune assays.

Quantitative precipitation was developed by Heidelberger and Kendall (1929) and is based on the observation that the addition of increasing amounts of a soluble antigen to a series of tubes containing a constant volume of antiserum results initially in an increase in the amount of precipitate formed until a maximum is reached. At equivalence of antigen and antibody an estimate can be obtained of the antibody content of the serum expressed in terms of added antigen.

A standard curve which relates the amount of precipitate to the amount of antigen in the presence of excess antibodies can be produced. The amount of antigen in a preparation can therefore be measured. However, methods for measuring the amount of precipitated protein are not very sensitive and other procedures have been designed to estimate the amount of an antigen. These procedures will be described and compared in terms of their relative value to biological scientists.

### 4.1.1. Quantitative immunoprecipitation in gels

Agar or agarose gels are frequently used as support media in which antigens and antibodies may react, and the most valuable techniques are those in which antibodies are incorporated into the support medium (e.g. into 1% w/v agarose in an appropriate buffer system). Antigen amount can be estimated by single radial immunodiffusion (Mancini *et al.*, 1965), where antigen diffuses into an antibody-containing gel. Initially, antigen concentration around the antigen well is high so that the antigen diffuses away from the well and forms soluble immune complexes until the concentration of antigen falls to a value at which immunoprecipitation occurs. This gives rise to an immunoprecipitation ring around the well. The area of this ring gives a measure of the amount of antigen in the preparation being tested. Naturally, this technique can be used with a monospecific antiserum but if residual enzyme activity can be used to mark the immunoprecipitation ring then a multispecific antiserum can be used. In this case the single immunodiffusion analysis is carried out with a tissue extract or biological fluid, and the immunoprecipitation line of interest is identified by means of a histochemical stain for enzyme activity. Multispecific antisera have been frequently used in this way, e.g. by geneticists to estimate the amount of a gene product in extracts from a number of genetically different organisms (Lanzerotti and Gullino, 1972).

An alternative method of estimating antigen amount is rocket immunoelectrophoresis. The principle of immunoprecipitation is similar to that described for single radial immunodiffusion, except that the antigen is forced into the antibody-containing gel by electrophoresis. In this case an immunoprecipitation "rocket" is produced, the height (or area) of which is proportional to the amount of antigen in the sample of the tissue extract. Histochemical techniques can also be used with this method where appropriate.

Immunoprecipitation lines of interest can also be visualized by radioligand binding and autoradiography, e.g. binding of radioactive bungarotoxin to immunoprecipitated postsynaptic cholinergic receptor (Teichberg *et al.*, 1977).

Rocket immunoelectrophoresis can be a very sensitive and convenient technique for estimating the amount of an antigen. Sensitivity will naturally depend on the specific antigen–antibody system under study, but 1–5 $\mu$g/ml of antigens in tissue extracts or biological fluids can be readily estimated. The amount of casein in mammary explants at different times after hormonal stimulation (Al-Sarraj *et al.*, 1979) can be conveniently measured by this technique. Immunoelectrophoresis was carried out as described by Axelsen *et al.* (1973). Agarose gel (1% w/v) containing adsorbed antiserum (76 $\mu$g

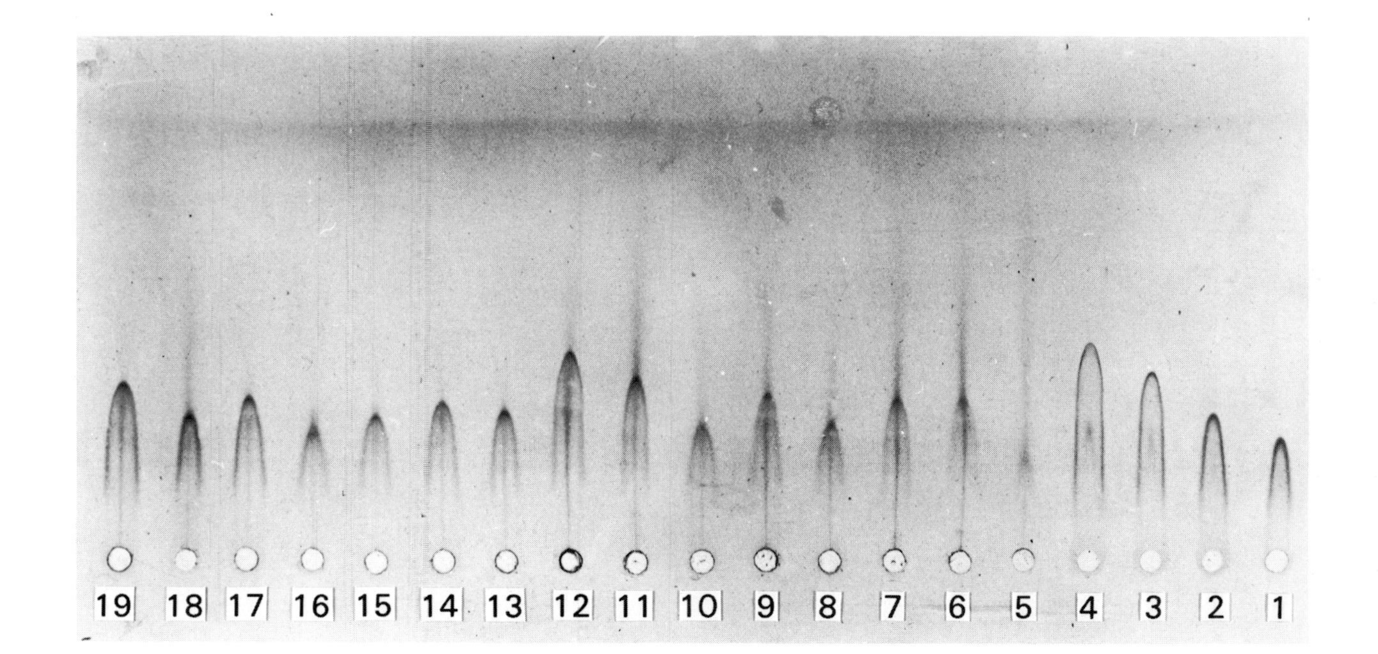


**Fig. 4.1** Quantitative determination of casein by means of rocket immunoelectrophoresis. Wells 1–4 contained 18, 36, 54 and 72 ng of rabbit recombined casein polypeptides as standards. The other wells contained particle-free supernatant (2–10 $\mu$l) from freshly prepared explants (well 5); from explants cultured for 24, 48, 72 and 96 h with hormones (insulin, prolactin and cortisol) wells 6, 12, 17 and 19; from explants cultured after the removal of hormones for 24 h (wells 7, 8, 9 and 10) or 48 h (wells 13, 14, 15 and 16); and from explants cultured throughout in the absence of hormones (wells 11 and 18). Electrophoresis was carried out as described by Weeke (1973b) at 0.5 V/cm for 16 h at approximately 15 °C.

protein/ml of gel; Al-Sarraj *et al.*, 1978) was used. Adsorbed antisera and purified immunoglobulin fractions are recommended, since they often eliminate background staining when gels are stained with Coomassie Blue after electrophoresis. This increases the accuracy of measurement of peak area (or height) which is proportional to the amount of antigen in each sample. Purified antigen (recombined casein polypeptides) was used as standard (Fig. 4.1). Often, "flying rockets", i.e. not dropping to the sample well, are obtained; the reasons for this are not understood. Nevertheless, measurement of rocket height from the centre of the sample well gives a good measure of antigen amount. A linear relationship between rocket height and amount of casein is obtained (Fig. 4.2). The reason for the line not passing through the origin is again not understood, but this phenomenon is often found in rocket immunoelectrophoresis (Weeke, 1973b).

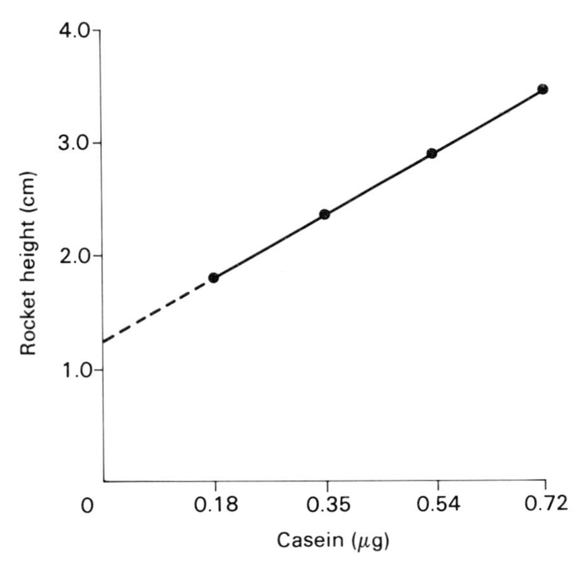


**Fig. 4.2** Standard curve relating rocket height to amount of antigen. The validity of this method of antigen quantitation can be conveniently verified by an alternative technique, e.g. radial immunodiffusion.

## 4.1.2. Measurements coupled to enzyme activity

The interaction between enzymes and their respective antibodies generally leads to a reduction in enzyme activity (Arnon, 1973). The enzyme may be completely inhibited, partially inhibited or in some cases stimulated. A correlation has been found between substrate size and the extent of immunoinhibition, i.e. greater inhibition is achieved with larger substrates (Cinader and Lafferty, 1964). Often, stimulation occurs only when enzyme activity is assayed with poor substrates or when mutant enzymes with poor catalytic activity are studied.

Immunotitration of enzyme activity is commonly used as a measure of the amount of an enzyme in a tissue extract or biological fluid. The titrations are based on the assumption that the volume of antiserum required to completely immunoinhibit enzyme activity is proportional to the amount of the enzyme in the tissue extract or biological fluid. The specific activity of an enzyme (e.g. activity per unit protein or per unit DNA) may change after some physiological stimulus as a result of activation or inactivation, or because the concentration (i.e. number of protein molecules) has either increased or decreased. This distinction is fundamental to the acute (changing activity of pre-existing enzyme) and chronic (changing the number of enzyme molecules) regulation of metabolic pathways. Demonstration of changing activity or amount of an enzyme is of paramount importance in the interpretation of all experiments in which enzyme activity is measured in biological systems in non-steady-state conditions.

Two alternative immunochemical approaches have been used. In one approach, increasing volumes of antiserum are added to fixed volumes of tissue extract (obtained from a living organism in a defined physiological state) in separate incubation tubes, and the volume of antiserum (or the extrapolated volume) required for complete inhibition of enzyme activity is used as a measure of enzyme–antibody equivalence (Fig. 4.3). Alternatively, increasing enzyme activities from a tissue extract are added to a fixed volume of antiserum and the extrapolated point at which enzyme activity can first be measured is taken as a measure of the equivalence of antigen and antibody (Figs 4.4 and 4.5). In the first approach, equal enzyme activities from tissues in different physiological states should require the same volume (or extrapolated volume) of antiserum for complete inhibition when the enzyme amount is changing. In the second approach, equal enzyme activities from tissue extracts should be inhibited to exactly the same extent when the enzyme amount is changing.

Both approaches have been successfully used to distinguish enzyme amount changes from activity changes and are therefore described.

Immunochemical measurements of enzyme amount have been carried out by determination of the volume of antiserum required to completely inhibit the enzyme activity. A common method is to mix increasing volumes of antiserum with fixed volumes of tissue extract. Each mixture is usually incubated at 30–37 °C for some time, e.g. 30 min (not if the enzyme is heat-labile) and then incubated at 4 °C for a prolonged period (e.g. overnight).

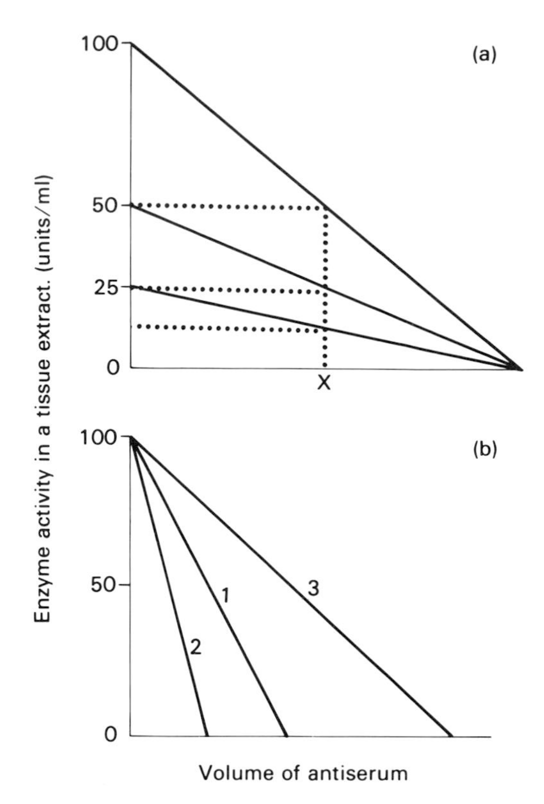


**Fig. 4.3** Immunotitration of enzyme activity. (a) Titration of the same amount of an enzyme of different specific activities. Volume of antiserum (X) required for 50% inhibition of enzyme activity. (b) Titration of activated or inactivated enzyme in a tissue extract relative to enzyme activity before change. Activity before change (1); two-fold activation of enzyme (2); two-fold inhibition of enzyme (3).

This technique has been varied so that incubations have been carried out for much shorter periods (e.g. Peavy and Hansen, 1975). Each incubation mixture is then centrifuged to remove immunoprecipitated enzyme, and enzyme activity in the supernatant is assayed. Alternatively, enzyme activity has been measured in preparations which have not been centrifuged, where the rapid anti-catalytic effect of antibodies is seen (Hizi and Yagil, 1974). The volume of antiserum required for complete inhibition of enzyme activity must often be an extrapolated volume, or alternatively the volume required for 50% inhibition of enzyme activity can be measured. This is necessary even when measurements are made on supernatants prepared by centrifugation of suspensions to remove immunoprecipitates, which may contain

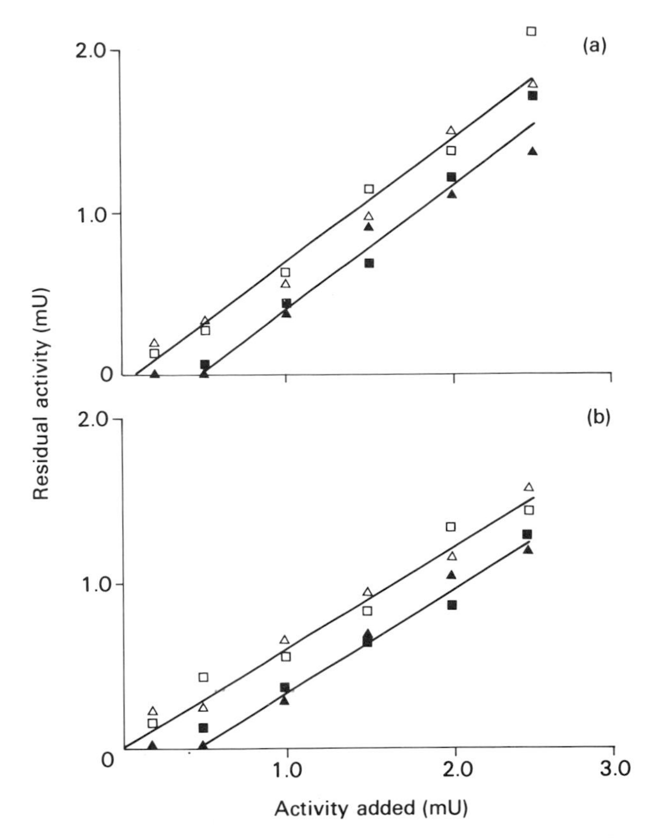


**Fig. 4.4** Immunotitration of pyruvate dehydrogenase from rat adipose tissue. Incubation of fixed volumes of antiserum with increased enzyme activity were carried out for 15 min at 4 °C. Incubation mixtures were centrifuged for 6 min at 14 000$g_{av}$ before assay. Total pyruvate dehydrogenase was assayed by the method of Stansbie *et al.* (1976). Each symbol represents the mean of six measurements at 2 days *prepartum* and three measurements at 2 days *postpartum*. Lines were determined by linear regression analyses. Immunotitration of pyruvate dehydrogenase from (a) parametrial, (b) subcutaneous adipose tissue. Animals killed at 2 (■) days *prepartum* and 2 (▲) days *postpartum*. Titration with control sera are shown by open symbols (□, △).

residual enzyme activity (Betts and Mayer, 1977). This may be explained by the relative activities of free enzyme and enzyme–antibody complexes in conditions of equivalence or antibody excess.

The measurement of enzyme amount is based on the existence of a linear relationship between the amount of an enzyme and the volume of antiserum required to completely immunoinhibit enzyme activity. This assumes that changes in specific activity of an enzyme for any reason are not associated with loss of a significant number of antigenic determinants (Fig. 4.3a).

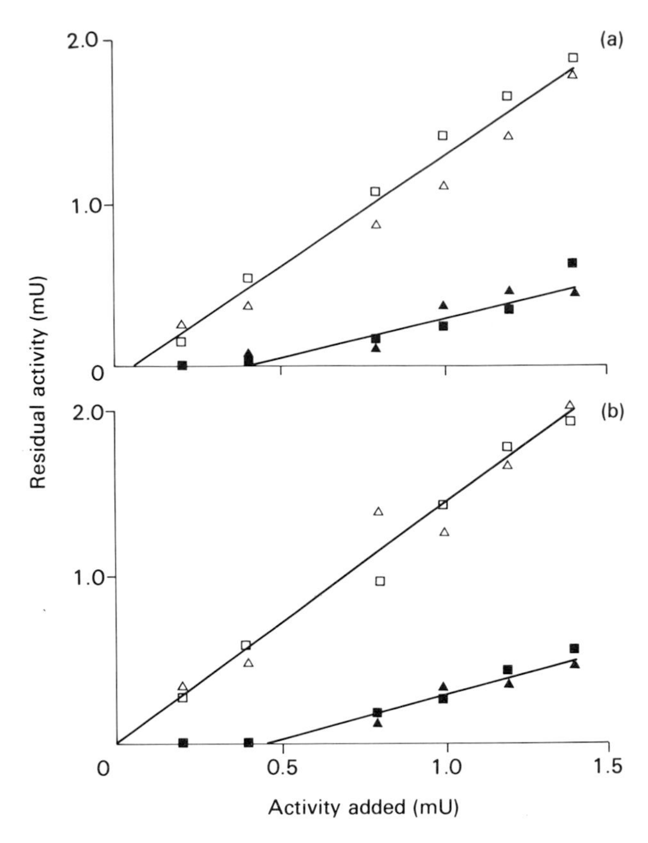


**Fig. 4.5** Immunotitration of fatty acid synthetase from rat adipose tissue. Incubations of fixed volume of antiserum with increasing enzyme activity were carried out for 2 h at 4 °C. Incubation mixtures were centrifuged for 6 min at 14 000$g_{av}$ before assay. Fatty acid synthetase was assayed by the method of Speake *et al.* (1975). Experimental conditions, expression of results and symbols are the same as those described in Fig. 4.4.

Immunotitration of an enzyme can be used to examine the nature of the change in enzyme activity in a cell after some physiological stimulus. It is possible to tell whether an enzyme has been activated or inactivated or if its amount has changed. If the change in enzyme activity is due to activation or inactivation of a pre-existing amount of enzyme then immunotitration of enzyme activity would be as shown in Fig. 4.3b. If the amount of the enzyme changes as a consequence of the stimulus then immunotitration of the same number of units of enzyme activity in tissue extracts before and after the stimulus should require the same volume of antiserum for complete inactivation. In such cases the volume of antiserum required for complete inhibition of enzyme activity can be used as a measure of the amount of the enzyme in the tissues (e.g. Betts and Mayer, 1977).

There are some complications in the use of these procedures. Two considerations deserve mention here. Firstly, the volume of antiserum required to completely inhibit enzyme activity is that required to quantitatively precipitate the enzyme. It is advisable to measure enzyme activity in supernatants after centrifugation to remove immunoprecipitates if problems of interpretation of immunotitrations are to be avoided, as in studies on glucose 6-phosphate dehydrogenase in mouse (Hizi and Yagil, 1974) and rat liver (Peavy and Hansen, 1975). However, in some cases immunoinhibition of enzyme activity is not linked to precipitation of the enzyme (e.g. Fig. 4.6; Manning and Mayer, unpublished observations). Here enzyme activity is completely inhibited by the addition of a very small volume of antiserum relative to that required to precipitate the enzyme. This phenomenon may be due to depolymerization of the highly polymerized form of the enzyme which is required for the activity of acetyl-CoA carboxylase (Walker *et al.*, 1976). This type of phenomenon may occur quite frequently.

Secondly, as previously noted it is usual for incomplete inhibition of enzyme activity to occur on adding increasing volumes of antiserum to a fixed enzyme activity. This means that a residual enzyme activity can be measured irrespective of how much antiserum is added. To avoid this problem, calculation of the volume of antiserum for 50% inhibition of activity can be carried out (see above). However, for better accuracy the addition of increasing enzyme activities to a fixed volume of antiserum is often preferred because it minimizes the extrapolative errors used in calculating antigen–antibody equivalence.

Immunotitration of pyruvate dehydrogenase and fatty acid synthetase in rat adipose tissue in the perinatal period is shown in Figs 4.4 and 4.5. Clearly, in all cases an identical equivalence point is obtained which shows that the change in the activity of pyruvate dehydrogenase (3–5-fold) and fatty acid synthetase (20–60-fold) in the perinatal period is caused by a fall in the amounts of the enzymes in the tissue. This is a very important observation which could not be obtained without immunochemical methods.

Several points of immunochemical interest arise from these studies. Short incubation periods were used for both antigen–antibody interactions. This is necessary because both enzymes are inactivated considerably on remaining at 4 °C for any length of time: Obviously, a compromise is necessary between loss of enzyme activity in tissue extracts and the time required for immunoinhibition of enzyme activity. Preliminary experiments were therefore carried out to find the minimal time required for maximal immunoinhibition in antiserum excess. This experiment leads to the choice of times indicated in Figs 4.4 and 4.5. This type of preliminary experiment is necessary for all immunoenzymological studies with unstable enzymes. Two further interesting points arise from these studies. Firstly, the immunotitrations were carried out with crude sera, prepared by taking the supernatant from clotted blood. Immunotitration is probably the only immunochemical technique where such a preparation could be recommended. Secondly, antisera raised in sheep against enzymes from pig heart and rabbit mammary gland were used to precipitate the corresponding rat enzymes.

Sometimes antisera cross-react very well with antigens from other species, but they sometimes fail to react at all (e.g. anti-rabbit casein does not react with human or rat casein). Finally, it should be noted that the non-specific effects of control serum and antiserum on the activity of fatty acid synthetase should be the same, i.e. the lines in Figs 4.5a and b should be parallel. The reason for their divergence is probably that the products of the fatty acid synthetase reaction, long-chain fatty acids, inhibit the enzyme reaction. However, serum albumin binds fatty acids and therefore prevents this inhibition. The protein concentration of the control serum was twice that of the antiserum, and therefore more inhibition of enzyme activity should occur in the latter case. Even activation of the enzyme may occur in the presence of control serum, because of binding of fatty acids to serum albumin and possibly other proteins. Activation or inhibition of enzymes by serum components may be a common occurrence in immunotitration studies. This type of problem may be overcome by the use of purified specific antibodies.

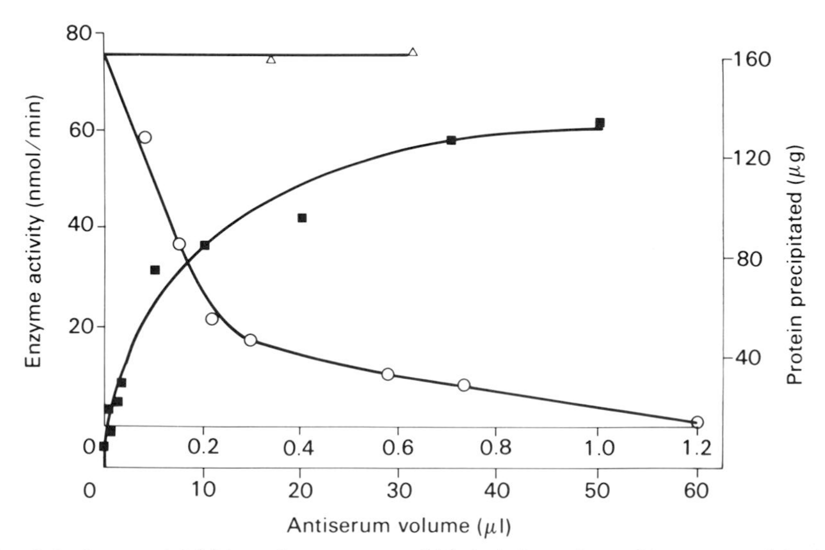


**Fig. 4.6** Immunoinhibition of an enzyme which is independent of immunoprecipitation. Immunotitration of acetyl-CoA carboxylase with monospecific antiserum. Enzyme activity after treatment with control serum (△), and antiserum (○). Immunoprecipitated protein (■).

### 4.1.3. Radioimmune and enzyme-immune assays

Immune assays which involve the separation of bound from free label have been described using reagents labelled with radioactivity (Berson and Yalow, 1959; Miles and Hales, 1968a,b), with enzymes (Wisdom, 1976; DeSavigny and Voller, 1980) and with fluorescent dyes (Soini and Hemmila, 1979; O'Donnell and Suffin, 1979). Fluorescent antibody assay kits for immunoglobulins are available from Biorad. These assays are based on the use of micrometre-sized polystyrene beads coated with antibody as an immobilized phase which binds antigen. This is then detected with fluorescently labelled antibody (Fig. 4.7). High sensitivities are obtained with these methods, which are routinely used clinically. Interestingly, methods have also been described which do not require the separation of bound from free label. These methods are discussed elsewhere (Jolley *et al.*, 1984) and may be used to process many samples rapidly with high sensitivity.

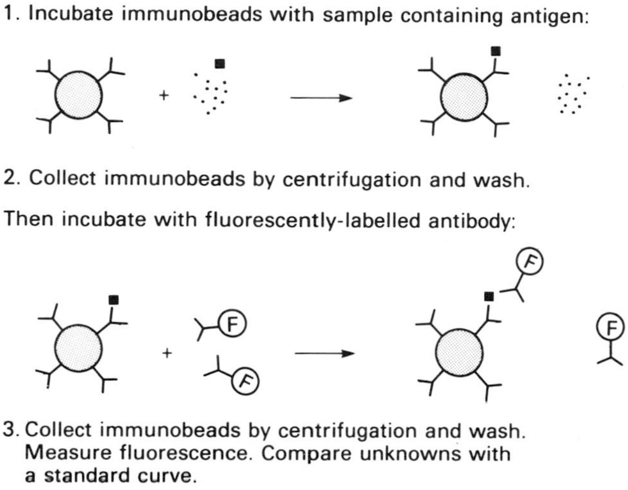

1. Incubate immunobeads with sample containing antigen:

2. Collect immunobeads by centrifugation and wash.
Then incubate with fluorescently-labelled antibody:

F

3. Collect immunobeads by centrifugation and wash.
Measure fluorescence. Compare unknowns with a standard curve.

**Fig. 4.7** Fluorescent immune assay.

#### 4.1.3.1. Radioimmune assays

Radioimmune assays (Fig. 4.8), which employ the principle of competitive binding of antibodies to radioiodinated antigen and to antigen in biological fluids, have been used for many years (Hunter, 1967). Such systems are currently used for the assay of many peptide hormones (Collins and Hennam, 1976) and for protein (Bolton and Hunter, 1973a,b) and enzyme (Roberts and Painter, 1977) antigens.

Ag and Ab = antigen and antibody

Long period

1. nAg unknown (e.g. Ag = $Ag^{125}I$) $+ nAg^{125}I +$ labelled antigen nAb limiting amount (e.g. $\frac{1}{2}[Ag + Ag^{125}I]$) $\longrightarrow \frac{n}{2}Ag + \frac{n}{2}Ag^{125}I + Ab\frac{n}{2}Ag + Ab\frac{n}{2}Ag^{125}I$

(Simplified to assume high-affinity antibody and one combining site per antibody molecule)

2. $\frac{n}{2}Ag + \frac{n}{2}Ag^{125}I + Ab\frac{n}{2}Ag + Ab\frac{n}{2}Ag^{125}I \longrightarrow \underbrace{\frac{n}{2}Ag + \frac{n}{2}Ag^{125}I}_{\text{measure radioactivity (free)}} + \underbrace{Ab\frac{n}{2}Ag + Ab\frac{n}{2}Ag^{125}I}_{\text{measure radioactivity (bound)}}$

separate bound and free species

3. Plot bound/free versus known antigen concentration to obtain standard curve so that unknown antigen concentration can be estimated.

**Fig. 4.8** Radioimmune assays.

Radioimmune assays for proteins or enzymes are extremely sensitive and specific but need to offer a good working range over which the amount of antigen can be measured. Iodinated antigens are extensively used in radioimmune assays. However, iodination of proteins may cause modification of their molecular properties (Krohn *et al.*, 1977) which could affect any of the parameters needed for a good radioimmune assay of a protein. Indeed, the highest specific radioactivities of a radioiodinated protein may only be achieved with considerable antigen modification (Hunter, 1967). The use of lactoperoxidase-catalysed iodination may cause less antigen modification, in that it labels predominantly tyrosine residues in certain proteins (Krohn *et al.*, 1977).

A major consideration in the design and use of a radioimmune assay for proteins is the availability of a method for separating free protein antigen from antigen–antibody complexes, since this is fundamental for competitive binding assays. Separations by electrophoresis, ion-exchange chromatography, gel filtration and solvent and salt precipitation have been used. Several authors have used double-antibody techniques, where an antiserum to the IgG fraction which contains the antibodies of interest (e.g. rabbit antiserum to sheep IgG) may be used to precipitate the antibody–antigen complexes of interest (Hunter, 1967). Antigen–antibody complexes are also routinely isolated by binding to *Staphylococcus aureus*, containing Protein A

(Jonsson and Kronvall, 1974; see Section 7.5.3), binding to Protein A-Sepharose, or by precipitation with polyethylene glycol (Creighton *et al.*, 1973). Naturally, a procedure must be specifically designed for the antigen and antiserum of interest, since there are so many differences in the molecular and physicochemical properties of protein antigens.

Two interesting and extremely useful developments are the immunoradiometric (Miles and Hales, 1968a, b; Readhead *et al.*, 1973; Woodhead *et al.*, 1974) and two-site assays (Catt and Tregear, 1967; Addison and Hales, 1971; Ling and Overby, 1972). These differ from classical radioimmune assays in that antibodies instead of antigen are iodinated and immunoadsorbents are used to great advantage.

The immunoradiometric assay is shown in Fig. 4.9. The technique was developed for polypeptide hormones. The advantages of the method over radioimmune assays are that all the unknown antigen reacts with antibodies at least once and the immunocomplex product is assayed against a low background. Furthermore, since the antigen is not modified by iodination, little interference in the reaction of the antigen with the specific antibodies should occur. The sensitivity of the technique is substantially increased, since specific antibodies are attached to an immunoaffinity matrix before iodination. In this way at least one antigen-binding site on each antibody molecule

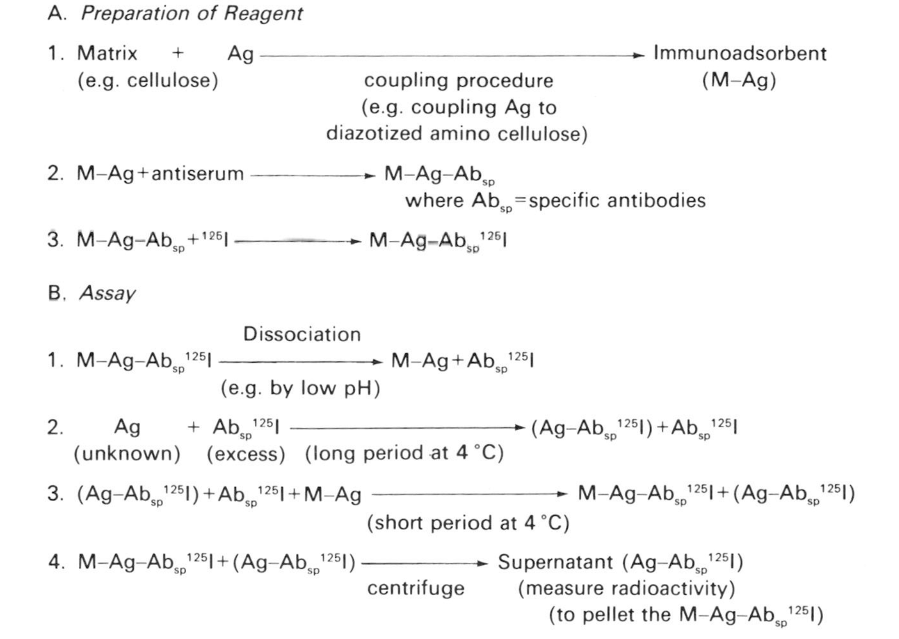


**Fig. 4.9** Immunoradiometric assay procedure.

is protected during the iodination procedure, i.e. the site which is interacting with antigen. The other sites may or may not be damaged by the procedure. The iodinated reagent is usually eluted from the immunoadsorbent with low pH. Low-avidity antibodies are first removed by washing with dilute HCl at pH 3, and the required high-avidity antibodies are then eluted with dilute HCl at pH 2. The assay depends on the selection of iodinated antibodies with a very high affinity for antigen so that very low concentrations of antigen in biological fluids can be detected. The most satisfactory storage procedure for iodinated antibodies (up to 2 months) has been found to be recombination of the antibodies with the immunoadsorbent and storage at −20 °C. Antibodies are then eluted as described previously when needed for assay.

Non-specific protein effects are minimized if standards are prepared in the biological fluid containing the antigen (e.g. serum). The principle of the assay is similar to a radioimmune assay in that it involves an initial prolonged incubation (e.g. up to 5 days) and subsequent separation of immunocomplex from unreacted reagent (in the case of immunoradiometric assays this is iodinated antibody). Again, this is nicely achieved with an immunoadsorbent to the antibodies of interest. Shortly after mixing a large excess of immunoadsorbent with the assay mixture (e.g. 30 min) the antibody–immunoadsorbent complex is removed by centrifugation, and the supernatant radioactivity (which is proportional to the amount of antigen in the biological fluid) is measured.

The disadvantage of the technique centres on the recurrent problem of a requirement for a relatively large amount of antigen for the preparation of immunoadsorbents. For many proteins there would be no problem in this respect but many tedious enzyme purification schemes result in very small quantities (e.g. 0.5–1.0 mg) of purified enzyme, most of which may be consumed in an immunization schedule. However, methods of this type, where antibodies are iodinated, will clearly avoid problems associated with iodinated protein (e.g. decreasing immunogenicity, difficulties in iodinating protein antigens, problems caused by the presence of detergents in membrane-protein preparations).

An alternative assay of antigen amount which needs less antiserum and may offer potential increases in assay sensitivity, precision and specificity is the two-site assay procedure (Fig. 4.10). The method involves the coupling of unlabelled antibody to an insoluble matrix such as Sepharose (Bolton and Hunter, 1973a,b), cellulose or the surfaces of plastic tubes. The immobilized antibodies can then be used as a means of extracting antigen from biological fluids. The major proviso of the method is that the antigen should have more than one antigenic determinant so that its uptake on to the immunoadsorbent can be measured by the subsequent reaction of a second labelled antibody.

A. *Preparation of Reagent*

1. Matrix+Ab ⟶ Immunoadsorbent reagent
   (e.g. cellulose paper or polypropylene or polyethylene centrifuge tubes) ($M–Ab_{sp}$)

B. *Assay*

1. $M–Ab_{sp}$+Ag $\xrightarrow{\text{incubate}}$ $M–Ab_{sp}$–Ag
   Ab excess unknown (e.g. 24 h at 4 °C)
2. $M–Ab_{sp}$–Ag+$Ab_{sp}{}^{125}I$ ⟶ $M–Ab_{sp}$–Ag–$Ab_{sp}{}^{125}I$
   (e.g. 24 h at 4 °C)
3. Remove immunoadsorbent from tube or preferably wash tubes carefully and measure radioactivity.

**Fig. 4.10** Two-site assay procedure.

This method has been refined with monoclonal antibodies by the use of two monoclonal antibodies. One monoclonal antibody binds to the surface and subsequently to the antigen at one epitope, and the other monoclonal antibody binds to a second epitope.

The major advantages of the two-site technique are increased sensitivity resulting from the immunological extraction of the antigen from the biological fluid, low non-specific radioactivity and the fact that when antibodies are immobilized on tubes no separation stage of the assay is necessary. Disadvantages centre on the fact that repeated washings are required, necessitating the use of antibodies with the very low antigen–antibody dissociation constants. In practice this is unlikely to be a problem.

The two-site assays can be conveniently performed in plastic centrifuge tubes or ELISA plates (see below). These methods probably offer the greatest potential for cell biologists, since antigen modification is not required and great sensitivity may be ensured by the multiple antigenic determinants which are often present on the surfaces of macromolecular antigens.

### 4.1.3.2 Enzyme-linked immunosorbent assays (ELISA)

(Engvall and Perlmann, 1971; Engvall *et al.*, 1971; Van Weemen and Schuurs, 1971a, b)

One of the most convenient and versatile procedures for immunoassay is the enzyme-linked immunosorbent assay, which has already been discussed in

Section 3.2 as a means of detecting antibody response, and for determining the titres of antisera. The method is also mentioned in Chapter 5, and extensive practical details are given in Chapter 7.

Most ELISA procedures are based on the use of plastic 96-well microtitre trays to which antigen or antibody molecules adsorb as a result of hydrophobic interactions. The binding capacity of the microtitre tray may be increased by coating the plastic wells with nitrocellulose dissolved in acetone. The antigen is usually diluted in a sodium carbonate/bicarbonate buffer at pH 9.6, and is used at a concentration of 0.5–2 $\mu$g/ml. Various different assays can be performed. Firstly, if antigen is immobilized on the microtitre plate then samples of antisera may be incubated in the wells, and antibody binding may be detected by subsequently incubating with enzyme-labelled second antibody (Fig. 4.11a). This method is routinely used for screening hybridoma culture supernatants (see Chapter 5 for details). The method may be extended to provide a quantitative assay for the antigen based on antibody competition for the immobilized antigen and free antigen in the sample being assayed (Fig. 4.1b). This method is dealt with in detail in Chapter 7. A second approach, the sandwich method, is based upon the use of antibodies immobilized on the microtitre plate to bind specific antigen by one antigenic determinant. The amount of antigen bound to immobilized antibody is then

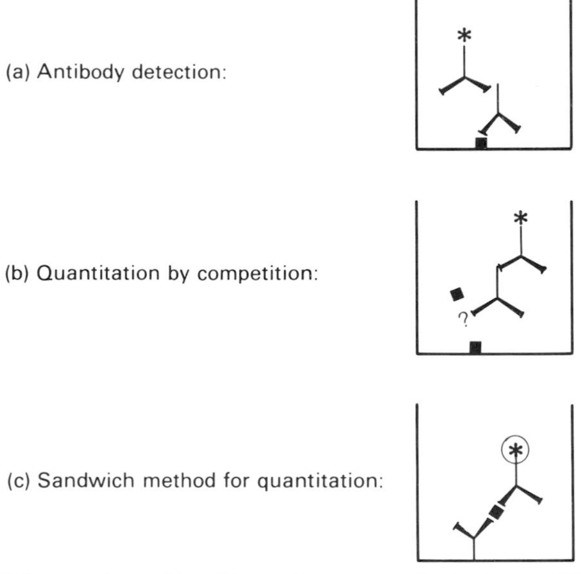


**Fig. 4.11** ELISA procedures. Key: ■ = antigen; ⅄ = antibody (e.g. rabbit); ⅄ = enzyme-conjugated second antibody (e.g. peroxidase-conjugated goat anti-rabbit; ⅄ = enzyme-conjugated antibody (rabbit—same as ⅄).

detected using the same antibody preparation labelled by conjugation with enzyme (e.g. horseradish peroxidase) to bind to a second antigenic determinant. This sandwich method is identical in principle to the fluorescent immune assay (Fig. 4.7) and the two-site radioimmune assay (Fig. 4.10) described above.

Enzyme-immune assays have also been developed which use antigen immobilized on cyanogen bromide-activated Sepharose (Hamaguchi *et al.*, 1976a, b; Papermaster *et al.*, 1976). The assay involves two stages. In the first stage competitive binding of antibodies to immobilized antigen and to free antigen in the biological fluid occurs. As the amount of free antigen in the biological fluid increases, less antibodies bind to the immobilized antigen. In the second stage of the assay, antibody binding to the immobilized antigen is estimated with a second enzyme-conjugated antiserum. The proportion of enzyme activity which is immobilized is therefore an indirect and amplified measure of the amount of antigen in the biological fluid. The sensitivity of assays for the enzyme can be optimized with fluorogenic substrates (e.g. methylumbelliferone phosphate).

Enzyme-immune assays have also been developed with reagents attached to glass rods (Hamaguchi *et al.*, 1976b). The principle of the method is outlined in Fig. 4.12. The binding of the $F_{ab}$-$\beta$-galactosidase complex is proportional to the amount of antigen. Therefore the quantity of the 4-methylumbelliferone produced by the galactosidase is proportional to the amount of antigen. The use of glass rod immunoadsorbents decreases problems of non-specific binding of the $F_{ab}$-$\beta$-galactosidase complex compared to particulate matrices (e.g. Sepharose) and improves reproducibility considerably by decreasing the handling problems encountered with particulate reagents. The specificity and sensitivity of the immunoadsorbents can be significantly improved by the use of purified antibodies (i.e. purified by immunoaffinity chromatography) for the preparation of the glass rod absorbents. Reasons for the development of enzyme-immune assays include cost relative to radioimmune assays, danger associated with radioimmune assays, and the short half-life of radioactive isotopes of iodine.

#### 4.1.3.3. Radioimmune assays of small molecules: cyclic nucleotides

Many radioimmune assays for small molecules have been developed and used to great advantage in clinical situations (e.g. for steroids, prostaglandins and many drugs). Biological scientists increasingly want to assay very small quantities of metabolites. These metabolites may be very unstable or subject to rapid alteration of concentration following some physiological stimulus. Immunochemical methods theoretically offer a very specific system for the

This method uses glass rods as solid phase.

A. *Preparation of Immunoadsorbent*

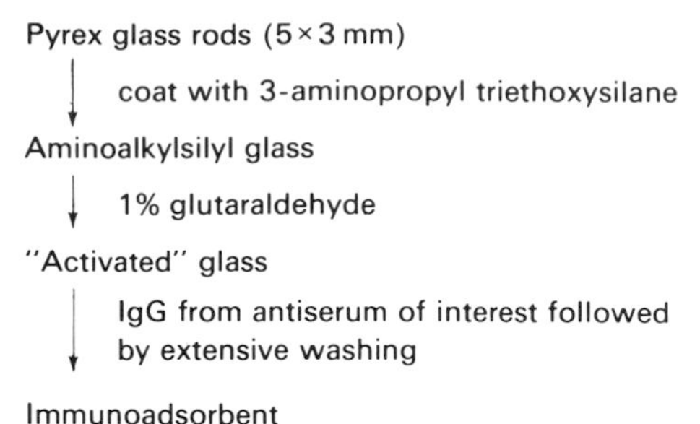


B. *Preparation of Antibody–Fab–β-D-galactosidase complex*

The complex is prepared by coupling Fab fragment and β-D-galactosidase with $N_1N^1$-O-phenylenedimaleimide.

C. *Assay*

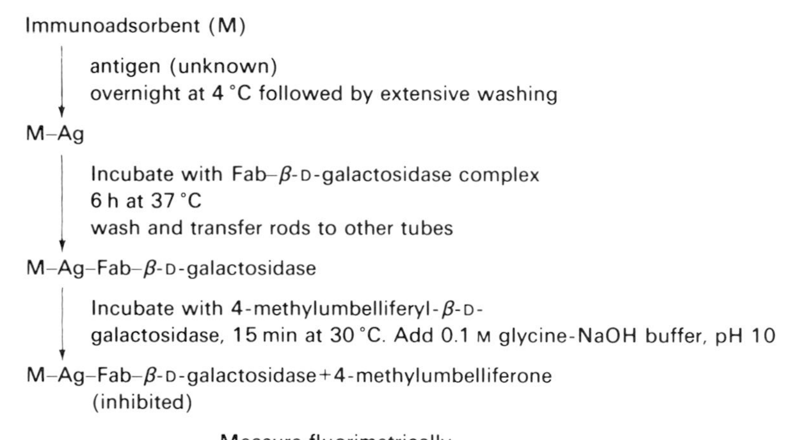


Measure fluorimetrically

**Fig. 4.12** An enzyme-immune assay.

rapid isolation and quantitation of these substances. Much interest has been shown in the last few years in determining the concentrations of cyclic nucleotides in cells in different physiological states. Assaying the concentration of these substances is difficult, involving radioactive prelabelling of nucleotide pools or use of preparations of specific binding proteins. Antibodies to cyclic nucleotides provide ideal "specific binding proteins". Production of antisera to these molecules involves rendering them antigenic by coupling to an immunogenic protein such as serum albumin or haemocyanin (see also Chapter 5, Fig. 5.10). Radioimmune assays are subsequently carried out by competitive-binding techniques.

Very specific polyclonal antisera to cyclic GMP have been produced and used in this manner (e.g. Steiner, 1974). Cyclic GMP is succinylated at the 2′-*O*-position (Cailla *et al.*, 1976) and subsequently conjugated with human serum albumin (Steiner, 1974). Several rabbits are immunized with a total of 0.25 mg of the succinyl–cGMP–albumin complex emulsified in Freunds complete adjuvant by injection at two subcutaneous sites (shoulders) and two intramuscular sites (hind legs). Each animal further receives a set of four injections at fortnightly intervals for 2 months and then at monthly intervals. The immune response of the animals varies somewhat but after 10 weeks of immunization high titres of specific antibodies have been obtained. For example, after this time specific antibodies were given with high affinity for cGMP and low cross-reactivities with cAMP, ATP, GTP, GDP, 5′-GMP, cCMP, cIMP and cUMP. Such an antiserum has been used in a radioimmune assay of cGMP (Fig. 4.13); the assay was developed in order to measure the concentration of cGMP in mouse neuroblastoma cells after treatment with various agonists. The measurement of cGMP concentration was in very good agreement with measurements made using a [$^3$H]guanine prelabelling method for measurement of cGMP (Strange, 1978). This specific example illustrates the potential of immunochemical methods for the measurement of the concentrations of cellular metabolites.

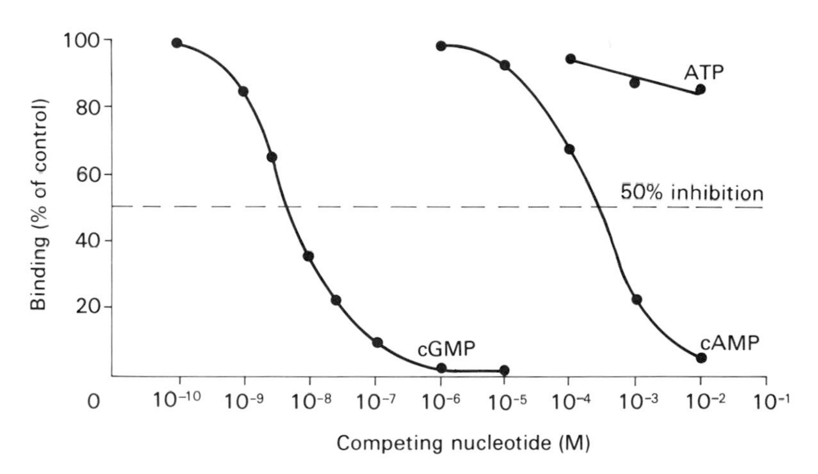


**Fig. 4.13** Binding of [$^3$H]cGMP to antibodies in the presence of unlabelled cGMP, cAMP or ATP. [$^3$H]cGMP (2 nM) was incubated with antiserum (final dilution 1/40) and known concentrations of competing nucleotides in 50 mM Tris-HCl buffer, pH 7.5, containing disodium EDTA (4 mM) in a final volume of 200 $\mu$l for 2 h in an ice/water bath. Bound [$^3$H]cyclic GMP was determined by precipitation with ammonium sulphate followed by centrifugation.

### 4.1.4. General conclusions

The biological scientist must decide on one or more of the methods to measure the amount of a protein antigen. The choice will ultimately be tailored to the specific antigen of interest, and consideration should be given to the molecular and physicochemical properties of the protein and the nature of the tissue extract or biological fluid to be used.

The amounts of antigens in tissue extracts vary considerably, so the same antigen can be present in biological fluids or tissue extracts at markedly different concentrations. The method of quantitation used will vary according to the concentration of antigen present. For example, the amount of $\alpha$-foetoprotein may be estimated in foetal plasma or amniotic fluid ($10^{-6}$–$10^{-1}$ g/ml) by immunodiffusion or immunoelectrophoretic methods, but estimated in normal plasma ($10^{-9}$–$10^{-8}$ g/ml) only by radioimmune assay (Leek and Chard, 1974). Enzymes ($10^{-6}$–$10^{-4}$ g/ml) can be measured by immunodiffusion or immunoelectrophoretic methods, e.g. acetyl-CoA carboxylase (Manning *et al.*, 1976; Walker *et al.*, 1976). The sensitivity of immunoassays which are coupled to enzyme activity depends on the molecular activity of the enzyme and sensitivity of the assay method (e.g. radiochemical assays are often most sensitive and allow $10^{-7}$–$10^{-6}$ g/ml of enzyme to be measured). Immunoassays linked to enzyme activity, immunodiffusion and immunoelectrophoresis are limited by the volume of biological fluid or tissue extract which can be used for the estimations, e.g. 5–20 $\mu$l for immunoelectrophoresis and immunodiffusion, depending on well size, 10–500 $\mu$l for assays linked to enzyme activity, depending on the nature of the assay (e.g. spectrophotometric or radiochemical) and reactant concentration (e.g. affecting the specific radioactivity of the substrate). These practical limitations probably mean that these methods are less sensitive (e.g. 1–2 orders of magnitude) than radioimmune or enzyme-immune assays, where $10^{-10}$–$10^{-8}$ g/ml can usually be estimated (Leek and Chard, 1974; Hamaguchi *et al.*, 1976a).

## 4.2. Reactions of Identity

Antisera have been used in phylogenetic studies where the existence of conformation homology has been corroborated by immunological cross-reaction (Arnon, 1971). The nature of antigenic determinants on macromolecules has been examined by conformational alteration of proteins (Arnon, 1971) or by inspection of the immunological relationships of peptide fragments obtained from proteins (Arnon, 1971; Beeley, 1976). The uses of reactions of identity are illustrated in several sections of this book, particularly in Chapter 3.

## 4.3. Immunoisolation Procedures

Monospecific antisera provide rapid and highly specific methods for isolating antigens either by immunoprecipitation or by affinity chromatography on columns of immobilized antibodies. Affinity chromatography on immobilized monoclonal antibodies can often purify antigens in one step and it is possible to select for a monoclonal antibody which will release the antigen under relatively mild conditions (e.g. 2 M NaCl or pH 5.0) so that the antigen retains its biological activity. Since large amounts of monoclonal antibody can be obtained, this can provide a very useful approach for antigen purification. This approach is far less useful for polyclonal antisera where it is usually impossible to elute the antigen from the column in an undenatured form.

Immunoisolation procedures are very useful for cell and molecular biologists, enzymologists and protein chemists. For example, enzymologists frequently wish to study the acute or chronic regulation of enzyme activity in tissue. Rapid (acute) changes in enzyme activity may be mediated by some covalent modification of a protein (e.g. phosphorylation). Slower (chronic) changes in enzyme activity are mediated by changes in the amount of an enzyme in a tissue. Changes in the amount of an enzyme can be achieved by altering its rate of synthesis or degradation in response to a stimulus. Measurement of covalent enzyme modification or rates of enzyme synthesis or degradation requires isolation of the enzyme from tissue extract. For all of these measurements it is best to isolate an enzyme rapidly in order to avoid possible modification of the enzyme in a laborious purification procedure. Antibodies provide a rapid way to isolate an enzyme from a tissue extract. The specificity of antibodies is fundamental to their use and is possibly only equalled by some affinity-chromatography systems (Don and Masters, 1975).

All of these developments rely on the quality of techniques for the immunoisolation of antigens. Two alternative procedures, immunoprecipitation and immunoadsorption (immunoaffinity chromatography) are extensively used. A method of immunoisolation involving *S. aureus* as adsorbent is also described in Section 7.5.3.

### 4.3.1. Immunoprecipitation methods

In some of these methods it is usual to add carrier enzyme or protein to a radiolabelled tissue extract or biological fluid. Subsequently, enough monospecific antiserum is added to give an excess over antigen (1–2-fold) and the mixture is incubated in conditions to immunoprecipitate the enzyme. Incubation conditions vary but often consist of a short period (e.g. 30 min) at 30–

37 °C followed by prolonged (overnight or much longer) incubation at 4 °C to ensure complete immunoprecipitation (Maurer, 1971). Immunoprecipitates can be removed by centrifugation (e.g. 80 000*g* for 1 min), washed repeatedly with iso-osmotic saline and analysed for incorporated radioactivity.

Carrier enzyme is often required to ensure a reasonably sized precipitate. This may represent a significant limitation of the method. The amount of carrier enzyme needed for each immunoprecipitation varies considerably, depending on the required size of the precipitate, but may be between 5 and 50 μg. In most experiments many samples need to be analysed (e.g. 10–20) and therefore, since often small amounts of purified carrier enzyme are available (e.g. 1–5 mg), a restriction is set on the possible number of immunoprecipitations which can be carried out. Partially purified enzyme could be used as a carrier but this may aggravate problems of co-precipitation (see below).

Careful controls are essential for immunoprecipitation analyses. Two basic problems are non-specific precipitation and co-precipitation. Non-specific precipitation is due to non-specific interactions between the proteins in a tissue extract and immunoglobulins. Also, proteins may simply adhere to the walls of incubation or centrifuge tubes in the absence of control serum. These interactions can be measured with control serum. This is added to tissue extracts and the incubations processed exactly as described for antiserum. In this way the contribution of non-specifically precipitating material to an immunoprecipitate can be estimated.

Co-precipitation is a significant problem of immunoprecipitation analysis. Adventitious protein is often trapped in immunoprecipitates as they are formed. The degree of immunoprecipitate depends on the amount of carrier added and therefore the size of the immunoprecipitate. At least three approaches have been taken to overcome this problem: control treatments have been devised to correct for co-precipitation; methods have been designed to minimize co-precipitation; and techniques have been developed for further analysis of immunoprecipitates.

Two methods for the estimation of co-precipitation have been developed. Schimke *et al.* (1965) designed a control based on two immunoprecipitations from the same tissue extract. Carrier enzyme is first added to the extract, followed by antiserum, and immunoprecipitation occurs. After removal of the precipitate the same amount of carrier enzyme is added to the extract and a second immunoprecipitation is carried out. The principle of the method assumes that all of the antigen of interest and accompanying co-precipitants will be in the first immunoprecipitate, while only the co-precipitants will be in the second immunoprecipitate. Subtraction of radioactivity precipitated in the second case from that precipitated in the first case should give the radioactivity incorporated into the antigen.

A second method for the assessment of co-precipitation has been developed by Cho-Chung and Pitot (1968). Carrier (a) is added to a radiolabelled tissue extract and twice the amount of carrier (b) is added to another equal volume of the same tissue extract. If no co-precipitation occurs then the radioactivity precipitated should be identical in both cases. If co-precipitation occurs then radioactivity incorporated into the antigen of interest is given by

$$A-(B-A)$$

where $A$ = radioactivity in precipitate ($a$)
$B$ = radioactivity in precipitate ($b$)
$B-A$ = co-precipitated radioactivity

Attempts have been made by several authors to minimize the extent of co-precipitation. Since co-precipitation occurs by non-specific trapping, similar co-precipitants should be precipitated with immunoprecipitates to any antigen. This has led to the development of pre-immunoprecipitation techniques. Tissue extracts are treated with appropriate quantities of an antigen and antiserum (e.g. ovalbumin–antiovalbumin serum) before treatment with carrier enzyme and antiserum to the antigen of interest.

An example of this type of analysis is shown in Figs 4.14 and 4.15.

Other methods for reducing co-precipitation include washing immunoprecipitates with detergents (Hopgood *et al.*, 1973) and centrifuging immunoprecipitates through a 1 M-sucrose cushion (cf. Figs 4.15 and 4.16). It is hoped that co-precipitants may be removed from the immunoprecipitates by this procedure (Rhoads *et al.*, 1973). Further improvement may also be obtained by using affinity-purified antibodies and partially purified antigen preparations.

The best way of assessing co-precipitation is by analysing immunoprecipitates by polyacrylamide gel electrophoresis in the presence of sodium dodecyl sulphate and 2-mercaptoethanol (e.g. Hopgood *et al.*, 1973). This approach should be used in conjunction with a technique designed to reduce co-precipitation together with extensive washing procedures.

This procedure will allow the measurement of radioactivity incorporated into the subunit(s) of an enzyme or protein and can clearly identify co-precipitants of different molecular mass. Therefore, radioactivity incorporated into the antigen of interest can be measured. It is advisable to compare the radioactivity profile from precipitates obtained with control serum and

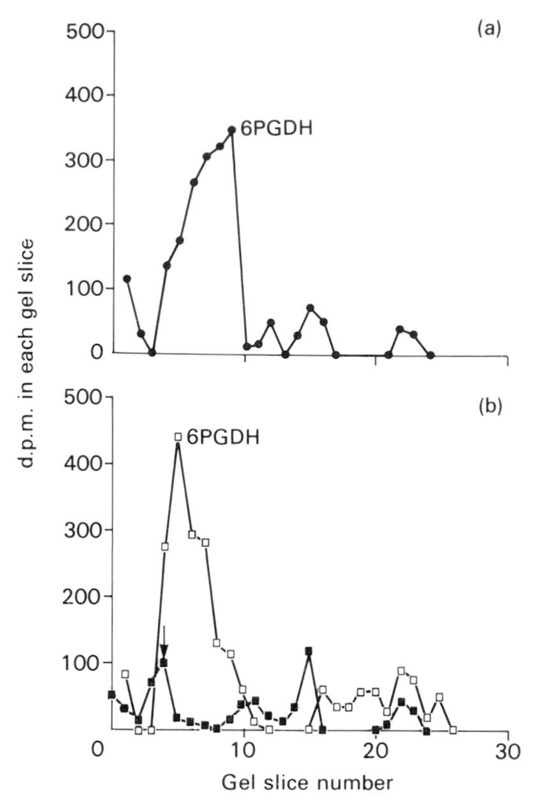


**Fig. 4.14** Analysis of immunoprecipitates by electrophoresis on polyacrylamide gels in the presence of sodium dodecyl sulphate. L-[4,5-$^3$H]leucine (1 mCi) was injected *in vivo* into a duct of the mammary gland of a 2 day *post partum* rabbit. After 4–6 h, the gland was excised and a particle-free supernatant prepared ($6 \times 10^6 g_{av}$ for 1 min). (a) Immunoprecipitation of 6-phosphogluconate dehydrogenase (6-PGDH) from samples (200 $\mu$l) of this supernatant was carried out with the addition of 20 $\mu$g of partially purified enzyme as carrier. Incubation was carried out for 30 min at 30 °C and overnight at 4 °C. To ensure complete precipitation of the enzyme, the volume of antiserum used was twice that required to precipitate all of the enzyme present. (b) Yeast alcohol dehydrogenase (25 $\mu$g) and its antiserum (200 $\mu$l) were added to a sample (200 $\mu$l) of particle-free supernatant for preliminary immunoprecipitation. The sample was incubated for 30 min at 37 °C and 30 min at 4 °C. Subsequently, 6-phosphogluconate dehydrogenase was immunoprecipitated as described above. After incubation, the immunoprecipitates were sedimented by centrifugation and boiled in electrophoresis sample buffer for 2 min. After electrophoresis, the gels were cut into 2 mm slices, digested with hydrogen peroxide and radioactively measured. Electrophoresis of immunoprecipitate of 6-phosphogluconate dehydrogenase (●), alcohol dehydrogenase (■) and 6-phosphogluconate dehydrogenase after preliminary immunoprecipitation (□).

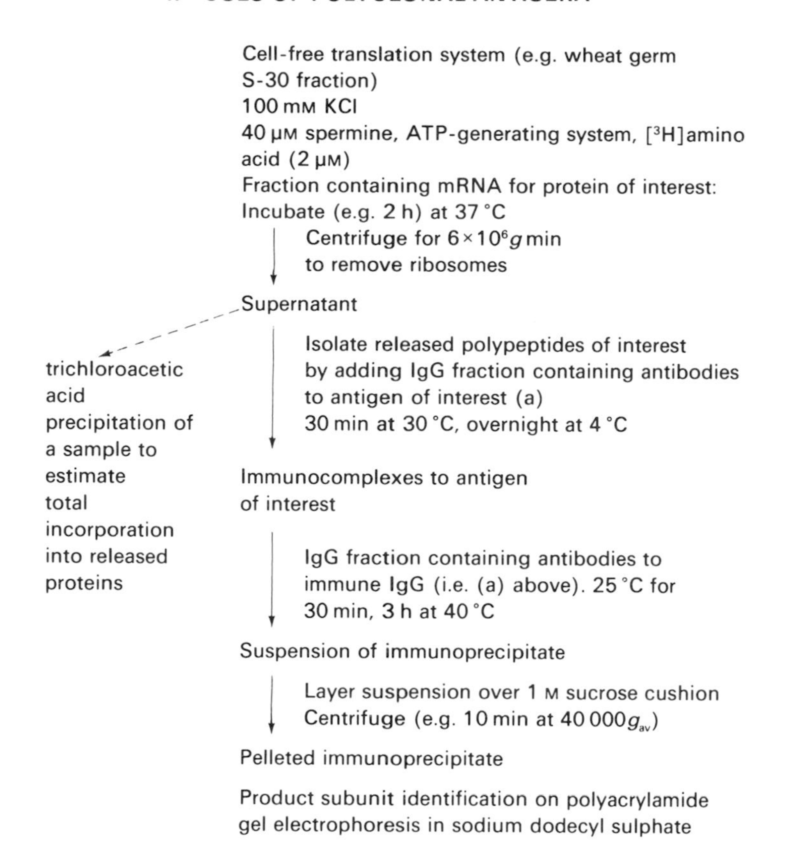


**Fig. 4.15** Immunoprecipitation of polypeptides in cell-free translation systems.

antiserum and even to analyse precipitates given with different amounts of carrier antigen. The method will not distinguish radioactivity in subunits of equal size derived from different proteins.

Another problem which can be encountered with immunoprecipitation is low incorporation of radioactivity into an antigen, e.g. by pulse-radiolabelled amino acid *in vivo* into a slowly synthesized protein. Furthermore, the antigen may have several subunits (e.g. cytochrome oxidase). In these conditions low radioactivity in the separated peptides may be expected. The immunoprecipitate may often be predominantly immunoglobulin, e.g. 90% immunoglobulin in immunoprecipitates of cytochrome oxidase (Walker and Mayer, unpublished observations). These problems mean that the radioactivity in any band on a polyacrylamide gel may be so low as to be undetectable. In such cases another technique for immunoisolation of an antigen (e.g.

The technique depends on the immunoreactivity of polysome-bound nascent polypeptide chains of the antigen of interest. A typical procedure is shown below.

Polysomes

(e.g. in 50 mM Tris-HCl buffer, pH 7.5, containing
150 mM NaCl, 5 mM $MgCl_2$, 1 mg/ml heparin)

↓ incubate with IgG from non-immunized animal to reduce non-specific binding of antibodies to ribosomes

Treated Polysomes

↓ Add detergent (e.g. Triton X-100; final concentration 0.5% w/v). Add appropriate amount of antibodies (purified by immunoadsorbent chromatography) Incubate 20–30 min

Antibody–Polysome Complex

↓ Add IgG fraction containing antibodies to antibodies of interest. Incubate 60 min Centrifuge 10 min at 40 000$g_{av}$

Immunoprecipitated Polysomes

↓ Resuspend in starting buffer, layer on 1 M sucrose cushion. Centrifuge 10 min at 40 000$g_{av}$ Repeat 3–4 times

Immunoprecipitated Polysomes

↓ Specific mRNA extraction in phenol-sodium dodecyl sulphate

Specific mRNA to enzyme or protein
of interest

**Fig. 4.16** Immunochemical isolation of polysomes translating mRNA of interest.

immunoadsorption) must be used in order that only antigen and not predominantly immunoglobulin can be loaded on a polyacrylamide gel.

Several methods based on immunoprecipitation have been successfully used to isolate antigens from a wide range of tissue extracts and for a variety of purposes. The variety of these approaches shows the extreme usefulness of immunochemical techniques to biological scientists.

Antisera have been used to isolate newly synthesized polypeptides from *in vitro* protein translation systems (Fig. 4.15: Rosen *et al.*, 1975). This procedure (Fig. 4.15) relies on a double-antibody technique to immunoprecipitate the antigen of interest, and centrifugation of the immunoprecipitate through a sucrose cushion to remove contaminating proteins.

Antisera have also been used to immunoprecipitate polysomes (Fig. 4.16) containing mRNA for abundant proteins such as ovalbumin (Palacios *et al.*, 1972) and immunoglobulin (Schechter, 1973). For mRNAs of lesser abundance, immunoaffinity columns (Schutz *et al.*, 1977) and Protein A-Sepharose columns (Shapiro and Young, 1981) are more successful, especially when monoclonal antibodies are used. Monoclonal antibodies to the human HLA-DR antigen have been used in conjunction with Protein A-Sepharose to purify the corresponding mRNA 2000–3000-fold (Korman *et al.*, 1982). The purified mRNA can then be used to prepare a cDNA probe for screening a total cDNA library. Alternatively, it may be used to generate a double-stranded cDNA clone.

Studies on rapid covalent enzyme modification such as may occur *in vivo* during acute regulation of some metabolic pathway can be very effectively carried out immunochemically. For example, an antiserum to acetyl-CoA carboxylase from rabbit mammary gland (raised in sheep) has been used to rapidly immunoisolate the enzyme from fat cells prepared from rat epididymal fat-pads. The antiserum was used to precipitate the enzyme from cell-homogenate supernatants (prepared by centrifugation for 100 000$g_{av}$ for 60 min. The antiserum was only incubated with the cell extracts for 30 min at 30 °C (which leads to a 90% loss of enzyme activity) and the immunocomplexes were sedimented by centrifugation at 80 000*g* for 30 min. Over 80% of purified [$^{14}$C]biotin-labelled enzyme which was added to such supernatants could be sedimented by these procedures. The short incubation of antibodies with tissue extract followed by rapid sedimentation of immune complexes by high-speed centrifugation is of obvious value to those interested in acute enzyme regulation by covalent modification (e.g. by phosphorylation or glycosylation), since the enzyme of interest can rapidly be isolated with minimal alteration of the modified protein. This is a prerequisite when examining physiological change in the extent of covalent modification.

The extent of phosphorylation of acetyl-CoA carboxylase in fat cells has been examined by the method described above. The results are shown in Fig. 4.17. Clearly, the phosphorylated enzyme is immunoprecipitated by antiserum and not by control serum. The other peaks in the sedimented material are due primarily to phosphorylated proteins of particulate origin (e.g. pyruvate dehydrogenase).

An alternative to rapid sedimentation of immune complexes by centrifugation (total isolation time approximately 1 h) is immunoadsorbent (immunoaffinity) chromatography (see below). For example, an immunoadsorbent to pyruvate dehydrogenase completely binds the enzyme from rat liver in 15 min at room temperature (Fig. 4.21; Burgess, Russell and Mayer, unpublished observations) while an immunoadsorbent to fatty acid synthetase from mammary gland binds the enzyme in extracts of mammary tissue within 5 min (Paskin and Mayer, unpublished observations). With some develop-

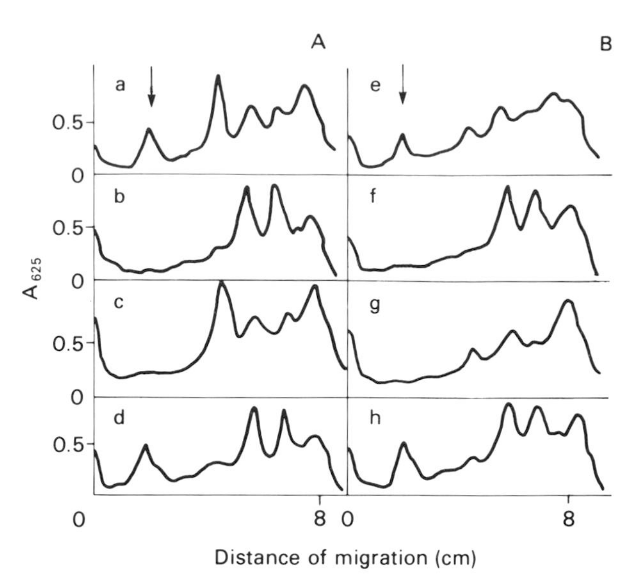


**Fig. 4.17** Densitometric traces of radioautographs demonstrating the specific precipitation of $^{32}$P-labelled protein from a 10 000$g$ supernatant of fat cells after incubation with antiserum to acetyl-CoA carboxylase. Fat cells were incubated for 75 min with $^{32}$P, with (A) or without (B) insulin (10 m.i.u./ml) added for the last 15 min. After centrifugation of the whole-cell extracts at 10 000$g$ for 1 min, the supernatants were incubated for 30 min at 30 °C with either antiserum to acetyl-CoA carboxylase (c, d, g and h) or control serum (a, b, e and f) (20 $\mu$l/ml in each case). Samples were then centrifuged at 80 000$g$ for 30 min at 4 °C and the proteins from the supernatant (a, c, e and g) and pellet (b, d, f and h) fractions separated on adjacent tracks by SDS/polyacrylamide-slab-gel electrophoresis (5% gel). The dye front (Bromophenol Blue) was allowed to migrate 10 cm. The arrow indicates the position of the subunit of acetyl-CoA carboxylase.

ment, immunoaffinity chromatography can be carried out much more quickly than the preparation of cell extracts, which therefore becomes limiting in this type of study.

Immunoprecipitation techniques coupled with polyacrylamide gel electrophoresis in the presence of sodium dodecyl sulphate can be used to considerable advantage when the antigen of interest is a glycoprotein. The heavy subunit of immunoglobulin can serve as internal standard for carbohydrate identification since it is glycosylated. Therefore, carbohydrate staining (e.g. with Periodic-acid-Schiffs reagent) or carbohydrate labelling with lectins (e.g. $^{125}$I or fluorescein-labelled lectins) can be carried out with some confidence (Cahill and Morris, 1979). Iodinated Protein A, which binds to the $F_c$ portion of immunoglobulins, can also be used effectively (Burridge, 1978).

### 4.3.2. Immunoadsorption methods

Immunoaffinity (immunoadsorption) techniques may prove to be the ultimate methods of choice with antibodies. A considerable number of approaches have been taken, some of which are described below.

Several immunoadsorption methods have been developed for the isolation of antibodies (Chapter 3) or antigens with immobilized immunoadsorbent (Kristiansen, 1976). Immunoadsorbents may be prepared in several ways, including direct polymerization (Ternynck and Avrameas, 1976) and coupling of antigens or antibodies to insoluble supports, e.g. Sepharose, polyacrylamide (Ternynck and Avrameas, 1976) or nylon (Edelman and Rutishauser, 1974). Immunoglobulins from non-immune sera should be coupled to Sepharose for use in control experiments. The immunoadsorbents can be conveniently packed into Pasteur pipettes, syringe barrels, or small glass columns, or may be used in batches. Clearly, the binding and elution conditions for each antigen–antibody system must be studied. The use of Protein A-bearing *Staphylococcus aureus* as immunoadsorbent is described in detail in Section 7.5.3.

Purified immunoglobulin has been routinely linked to Sepharose (Fig. 4.17) by cyanogen bromide procedures (Porath *et al.*, 1973) and to polyacrylamide or nylon by bifunctional reagents, e.g. glutaraldehyde (Edelman and Rutishauser, 1974; Ternynck and Avrameas, 1976). Practical details of these procedures are given in Chapter 7.

In general, the coupling of relatively high concentrations of antibodies is preferred if the immunoadsorbent is to be used to isolate an antigen from a tissue extract. However, coupling of increasing amounts of antibodies may result in a progressive increase in the number of inactivated antibodies due to factors including unfavourable orientation of the antibodies on the solid support (Kristiansen, 1976).

Non-specific binding is a common problem with all immunoadsorbents, and polyacrylamide has been recommended instead of Sepharose immunoadsorbents for isolation of detergent-solubilized membrane antigens (Haustein and Warr, 1976).

Columns of adsorbents prepared from non-immune immunoglobulins may be used in series with the immunoadsorbent columns to remove non-specific binding species. Subsequently, the non-specific components and antigen of interest can be eluted and analysed by polyacrylamide gel electrophoresis in the presence of sodium dodecyl sulphate. Prefiltration through Sepharose alone can be sufficient to remove some contaminants (Fig. 4.18). Alternatively, a preparation containing antigen may be passed through non-immune IgG-Sepharose columns either before or after elution from the immunoadsorbent in order to remove non-specific binding species.

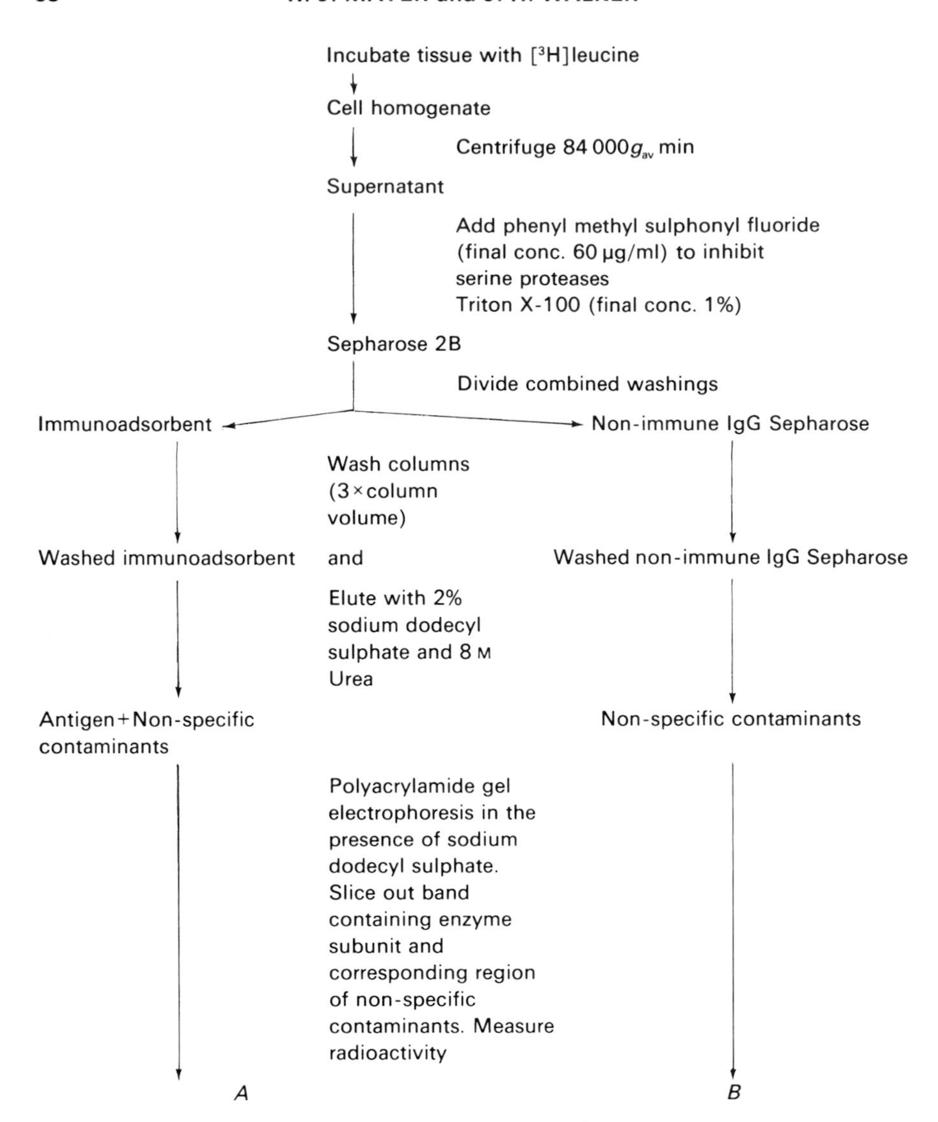


**Fig. 4.18** Isolation of fatty acid synthetase by immunoadsorbent chromatography.

The antigen of interest may be lost to some extent by binding to the non-immune immunoglobulin supports. Non-specific binding may be considerably reduced by the use of immunoadsorbents prepared with immunoaffinity-purified antibodies to the antigen of interest. An example of the use of immunoaffinity antibodies to casein in immunoaffinity chromatography is shown in Fig. 4.19.

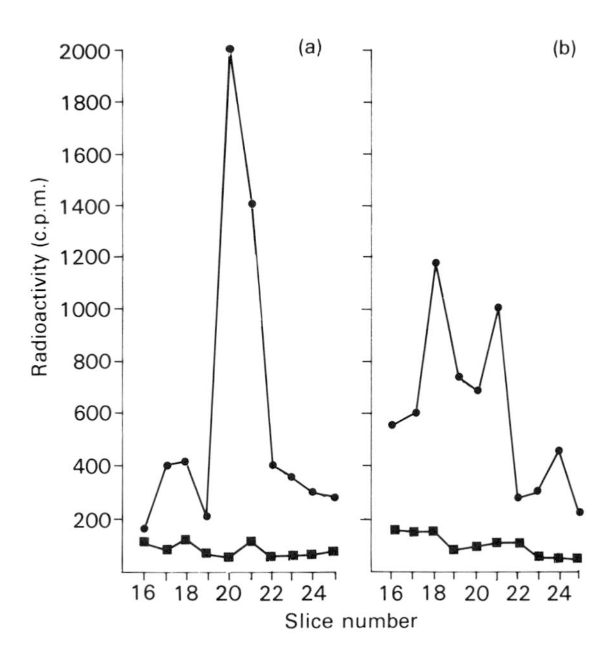


**Fig. 4.19** Polyacrylamide gel electrophoresis of radiolabelled casein isolated by immunoadsorbent chromatography and immunoprecipitation. (a) Samples (0.06–0.4 ml) of particle-free supernatant fractions obtained from mammary explants cultured for 24 h with hormones were applied to either (i) a Sepharose anti-casein column (3 ml, ●) or (ii) a Sepharose control serum IgG column (3 ml, ■) and left in contact with the column overnight at 4 °C. Radiolabelled casein was eluted from the columns with 2% sodium dodecyl sulphate containing 8 M urea, dialysed against water, freeze-dried and taken up in sample buffer containing 50 $\mu$g of casein (carrier) for electrophoresis. The position of the radioactive peak corresponds to the $R_f$ of purified recombined casein. (b) Samples (0.06–0.4 ml) of particle-free supernatant fractions were mixed with either (i) 1 ml of anti-casein IgG (●) or (ii) 1 ml of control serum IgG (■), and the mixture was incubated for 48 h at 4 °C. The immunoprecipitate was washed and processed for electrophoresis as described above.

The casein immunoadsorbent was used in conditions where non-specific binding was minimized so that it became the method of choice rather than immunoprecipitation where co-precipitation may complicate interpretation of the data. This is clearly shown with the casein–antibody system in Fig. 4.19, where analysis of material eluted from immunoadsorbent (Fig. 4.19a by polyacrylamide gel electrophoresis in the presence of sodium dodecyl sulphate gives a much clearer picture of the radiolabel incorporation into the subunits of the antigen than analysis of immunoprecipitates (Fig. 4.19b). The radiolabel incorporation into casein after immunoprecipitation and electrophoretic analysis cannot be accurately measured in spite of the fact that the specific anti-casein antibodies, prepared by immunoadsorption (Al-Sarraj *et al.*, 1978), were used for the procedure.

The kinetics for optimal binding of an antigen to an immunoadsorbent must be determined with respect to time taken for antigen binding and with respect to the pH and ionic strength of the tissue extract. This is especially important for detergent-solubilized membrane antigens, which may show complex binding characteristics with immunoadsorbents.

As a general rule, it appears that the binding of soluble antigens (i.e. enzymes or proteins) to immunoadsorbents is more rapid than the binding of detergent-solubilized membrane antigens. Further, the binding of detergent-solubilized antigens can be considerably influenced by detergent type and concentration. This might be expected in view of the micellar composition of detergent solutions, and equilibria which may exist between different micellar forms of an antigen, which may expose more or less antigenic determinants.

The binding of purified fatty acid synthetase to an immunoadsorbent is complete in 5–10 min at room temperature (Fig 4.20). Acetyl-CoA carboxylase binds to its immunoadsorbent in a similarly short time. The binding of enzymes in detergent-solubilized extracts of mitochondrial preparations show very different binding characteristics to their respective immunoadsorbents (Fig. 4.21). The binding of cytochrome oxidase is completed over a

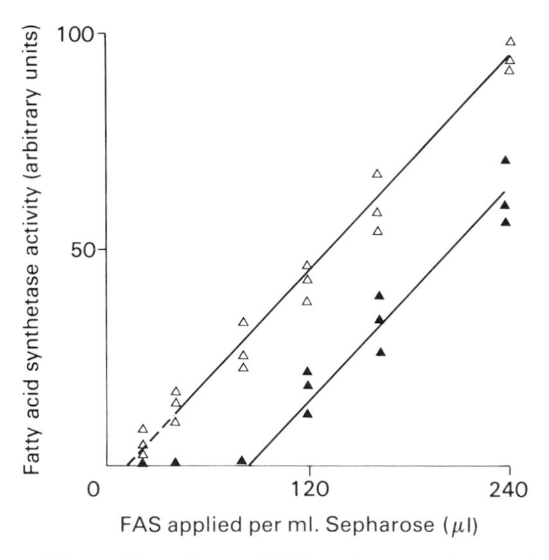


**Fig. 4.20** Binding of fatty acid synthetase (FAS) activity to columns of non-immune IgG-Sepharose or (anti-FAS)-IgG-Sepharose. Purified FAS in 0.25 M phosphate buffer, pH 7.0, was applied to 1.0 ml columns of either non-immune IgG-Sepharose or (anti-FAS)-IgG-Sepharose. After 5–10 min, the columns were eluted with 3 ml of the same buffer, and aliquots of the eluate were assayed for FAS activity. Activity binding to non-immune IgG-Sepharose columns (△); activity binding to (anti-FAS)-IgG-Sepharose columns (▲). The results with three series of columns are shown.

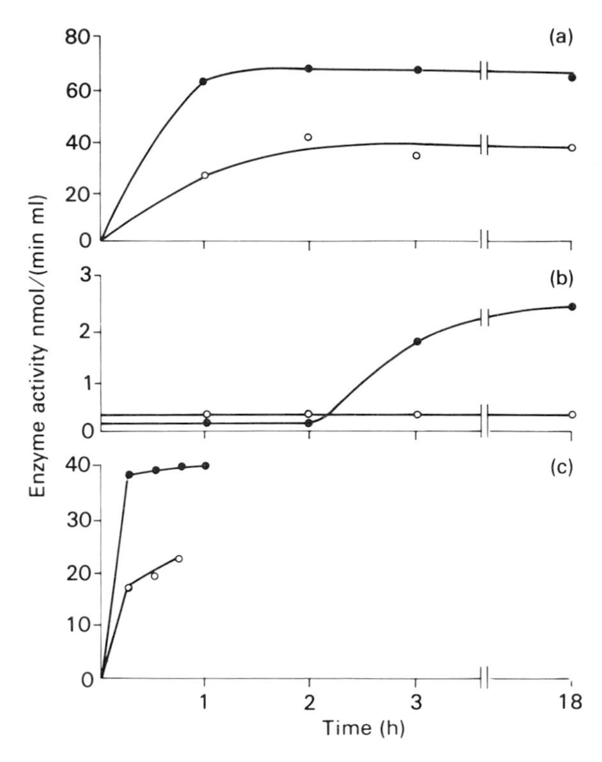


**Fig. 4.21** Binding characteristics of immunoadsorbents with antigens in detergent solution. Mitochondrial preparations were solubilized in 20 mM sodium phosphate buffer, pH 7.4, containing 1% (w/v) sodium cholate. The detergent-solubilized material was preincubated with Sepharose 4B for 30 min at room temperature before application to the immunoadsorbents. The Sepharose 4B was washed with three volumes of the buffer and the combined washings were used for enzyme immunoisolation. Samples of the combined washings were applied to Sepharose 4B immunoadsorbents (prepared by the cyanogen bromide method) containing antibodies to cytochrome oxidase (a), monoamine oxidase (b) and pyruvate dehydrogenase (c) and incubated at room temperature. The binding of the immunoadsorbent for cytochrome oxidase was measured with antibody excess, whereas binding to the other immunoadsorbents was estimated with antigen excess. Immunoglobulin G containing antibodies of interest was used to prepare immunoadsorbents to cytochrome oxidase and pyruvate dehydrogenase, whereas purified antibodies (prepared by mitochondrial adsorption; Fig. 3.12) were used to prepare the immunoadsorbent to monoamine oxidase. Both immunoadsorbent (●) and non-immune IgG-Sepharose (○) used for the isolation of cytochrome oxidase, monoamine oxidase, and pyruvate dehydrogenase, contained 10 mg, 0.2 mg and 10 mg of bound protein per millilitre of Sepharose respectively.

period of 2 hours, that of monoamine oxidase is only completed after overnight incubation, while pyruvate dehydrogenase binding is complete after 15–30 min. Pyruvate dehydrogenase is called a "soluble" mitochondrial enzyme although it is very difficult to remove from mitochondrial preparations without detergent. The behaviour of this enzyme in detergent (cholate) is very similar to that of other soluble enzymes, i.e. rapid binding to immunoadsorbent (Fig. 4.21c).

Cytochrome oxidase binding from a cholate solution (Fig. 4.21a) is similar to that seen when the enzyme is dissolved in Triton X-100 (Walker and Mayer, 1977). If the enzyme is left in contact with the immunoadsorbent for 2.5 h at room temperature, the binding characteristics are identical to those shown by soluble enzymes in much shorter periods of time, e.g. 10–15 min (cf. Figs 4.22 and 4.20). Monoamine oxidase shows much slower binding to its immunoadsorbent than does cytochrome oxidase (Fig. 4.21b) although both enzymes were solubilized by the same detergent under identical conditions. The only difference was that purified antibodies to monoamine oxidase were used for the preparation of its immunoadsorbent (Fig. 3.12).

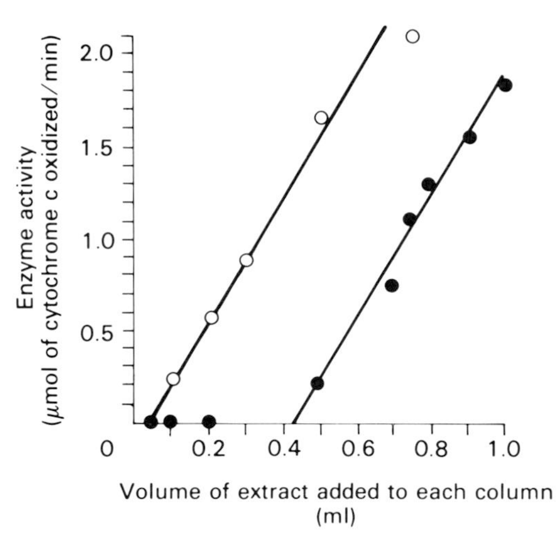


**Fig. 4.22** Binding of cytochrome oxidase to an immunoadsorbent after prolonged contact. Samples of a mitochondrial Triton X-100 (1%, v/v) extract were left in contact with the columns for 2.5 h at room temperature. Enzyme activity washed off non-immune IgG-Sepharose column (○), enzyme activity washed off antiserum-Sepharose column (●).

Non-specific binding of the enzymes to non-immune IgG-Sepharose is apparently a function of the concentration of coupled protein, since both the non-immune IgG-Sepharose preparations used for cytochrome oxidase and pyruvate dehydrogenase had 10 mg protein/ml and bound considerable enzyme activity non-specifically, whereas the non-immune IgG-Sepharose used with monoamine oxidase has 0.2 mg protein/ml and bound no enzyme activity.

Many elution systems have been described for immunoadsorbents (Ruoslahti, 1976) in which extremes of pH or ionic strength have been often used. These conditions alone are not always sufficient to elute an antigen of interest. Complete elution of acetyl-CoA carboxylase, fatty acid synthetase and cytochrome oxidase from their respective immunoadsorbents can only be achieved with 8 M urea containing 2% (w/v) sodium dodecyl sulphate (Table 4.1). Fortunately, when measuring incorporation of radioactive amino acids into protein antigens it is not necessary to preserve the conformation of the antigen on elution from the immunoadsorbent.

**Table 4.1** Elution of antigens from immunoadsorbents. (Immunoadsorbents were successfully eluted with the eluants.)

| Immunoadsorbent for | Acetyl-CoA carboxylase | Fatty acid synthetase | Cytochrome oxidase |
|---|---|---|---|
| | | Antigen eluted (%) | |
| Eluant | | | |
| 2.5 M $MgCl_2$ | — | — | — |
| 0.2 M glycine-HCl, pH 2.8 | — | 7.8 | — |
| 8 M urea | approx. 100 | 27 | 20 |
| 8 M urea +2% (w/v) sodium dodecyl sulphate | — | 75.3 | 74 |

Elution of immunoadsorbents with urea–sodium dodecyl sulphate brings about a varying degree of loss and binding capacity of the immunoadsorbent. Immunoadsorbents to casein and fatty acid synthetase lose 5–10% of their binding capacity after a single elution with the denaturing solvents. However, an immunoadsorbent to monoamine oxidase loses approximately 50% of binding capacity and an immunoadsorbent to cytochrome oxidase loses approximately 30% of binding capacity during each elution. The best solvents are obviously those which lead to the minimum loss of binding capacity, but which quantitatively elute an antigen.

The use of electrophoretic elution of antigens from immunoadsorbents (Dean *et al.*, 1977) seems to have great potential as a gentle but effective elution technique which does not require denaturing solvents. A convenient apparatus which can be used for this procedure is supplied by ISCO (ISCO Model 1750; sample concentrator, ISCO, Nebraska, USA; Allington *et al*., 1978). This apparatus can be used not only to elute an antigen but also to concentrate it at the same time.

Immunoadsorbents can be conveniently used to isolate antigens from biological fluids or tissue extracts. The conditions for immunoadsorption and elution must be carefully checked. If these conditions are known then rapid quantitative isolation of an antigen from a tissue extract can be carried out.

If an antigen is to be isolated from a tissue extract without much denaturation then the immunoadsorbent should be prepared with an antiserum containing antibodies with low affinities to the antigen. Under these conditions it might be expected that some compromise may be reached whereby an antigen would bind to an immunoadsorbent and be eluted from the immunoadsorbent in relatively mild conditions.

Finally, it should be noted that immunoadsorbents can be very sensitive to antigen modification. Modification of fatty acid synthetase or cytochrome oxidase by mild acetylation or iodination results in the production of species with complex binding and elution properties (Walker, Paskin and Mayer, unpublished observations).

Table 4.2 outlines the relative advantages and difficulties of immunoprecipitation methods and immunoaffinity chromatography for isolating antigens. In general terms, immunoprecipitation methods are the most convenient for isolating radiolabelled antigens from extracts of tissues or cells in culture. It is, however, essential that experiments are performed to control for non-specific contamination. Immunoaffinity chromatography is particularly useful for isolating antigen free from bound antibody. One final exciting use of antibodies in immunoisolation procedures is for the purification of cells and subcellular fractions.

The subfractionation of B and T lymphocytes by means of anti-IgG serum is sufficiently well characterized to have found its way into textbooks of practical immunology (e.g. Hudson and Hay, 1976). Similar subfraction schemes should be possible, in principle, for all cell types which possess characteristic antigenic determinants. Similarly, the subfractionation of subcellular organelles by means of antisera to specific surface antigens on each organelle should also be feasible. This is especially important in cases where heterogeneous populations of the same type of organelle are present (e.g. synaptic vesicles containing different transmitters) or where a particular type of subcellular organelle is present in a variety of forms (e.g. the plasma membrane, which on homogenization gives rise to sheets and vesicle). An immunological procedure for the isolation of plasma membranes has been developed (Luzio *et al*., 1976), and by the use of a double-antibody technique

**Table 4.2** Immunoprecipitation and immunoadsorption chromatography: a comparison.

| Point | | Immunoprecipitation[a] | Immunoadsorption chromatography |
|---|---|---|---|
| 1.(a) | Non-specific contamination | Co-precipitation | Non-specific adsorption |
| (b) | Elimination of non-specifically bound protein | Washing procedures[b] and use of immunoaffinity purified antibodies to antigen | Washing and use of specific antibodies to antigen[b] |
| (c) | Which method has most contamination | — | — |
| 2. | Carrier enzyme | Usually needed | Not needed usually |
| 3. | Protein in immunoisolated material | Often mostly immunoglobulin | Mostly antigen of interest |
| 4.(a) | Polyacrylamide gel electrophoresis in sodium dodecyl sulphate | Must be carried out | Must be carried out |
| (b) | Material applied to gel | Mostly immunoglobulin | Mostly antigen |
| 5. | Simple or complex antigen subunit composition on polyacrylamide gels | Causes problems of interpretation if low radiolabel incorporation (see 4b) | Causes less problems of interpretation of radiolabel incorporation |
| 6. | Quantity of antibodies needed | Large quantities can be needed<br>Depends on amount of carrier used | Low binding capacity and loss of binding capacity can lead to large antibody requirement |
| 7. | Rapidity of immunoisolation | Can be made rapid by ultracentrifugation of immunocomplexes | Extremely rapid with non-membrane antigens; complex slow binding characteristics with detergent-solubilized membrane antigens |

[a]Double-antibody methods are more routinely used than single-antibody methods.
[b]e.g. Washing with non-ionic detergent mixtures.

enabled specific purification of those membrane vesicles with "outside-out" conformation. Without the use of a double-antibody technique the authors could not obtain a purification equivalent to that obtained by conventional subcellular fractionation techniques.

Similarly, cholinergic synaptosomes have been isolated from mixed transmitter brain synaptosomes using specific antibodies to a cholinergic-specific ganglioside and a cellulose immunoadsorbent. The specificity of these lipid-binding antibodies was originally demonstrated by complement-mediated lysis of cholinergic synaptosomes (Richardson *et al.*, 1982; Fig. 4.23), which showed specific lysis of the cholinergic subpopulation of the synaptosomes. More recently, immunoblotting of lipid fractions have been used to characterize these antibodies (Ferretti and Borroni, 1986). Synaptic vesicles (Matthew *et al.*, 1981) and clathrin-coated vesicles (Merisko *et al.*, 1982) have also been isolated using immobilized Protein A and monoclonal or polyclonal antibodies respectively.

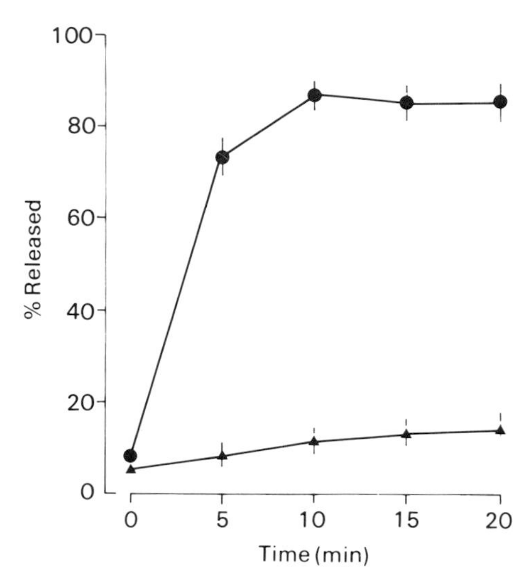


**Fig. 4.23** Complement-mediated lysis of guinea-pig cortical synaptosomes. Release of choline acetyltransferase (●) and lactate dehydrogenase (▲) from synaptosomes derived from 100 mg of guinea-pig cortex. Total activities (per 100 mg tissue): choline acetyltransferase, $210 \pm 21$ nmol/h (9); lactate dehydrogenase, $70.1 \pm 1$ $\mu$mol/h (9).

## 4.4. Methods for Characterizing Antigens

In this section we summarize methods which can be used to characterize the molecular properties of an antigen. Further details of the methods are given in Chapter 7.

### 4.4.1. Determination of antigen molecular weight

The native molecular weight of an antigen may be determined by gel filtration or sedimentation centrifugation using a specific antiserum to assay fractions from the separation by means of dot blots (Towbin *et al.*, 1979) or ELISA procedures (Voller *et al.*, 1979) or immunoelectrophoretic analysis (Svendsen, 1973). If a monospecific antiserum is not available, then column fractions may be analysed on SDS gels, transferred on to nitrocellulose, and then subjected to immune detection. In this case the blots will show the subunits of antigens with different molecular weights eluting from the column with different retention times. If samples are denatured with SDS in the absence of reducing agent, the presence of disulphide-linked subunits may easily be demonstrated on the immune blots as a shift in the antigens' apparent molecular weights. The method using gel filtration and SDS gels may also be applied to subcellular fractions as in the case of synapsin I where antigens of $M_r$ 86 000 and 80 000 were shown to copurify with synaptic vesicles from mammalian brain (Huttner *et al.*, 1983) during gel filtration on glass bead columns.

Multi-subunit antigens may be demonstrated after crossed immunoelectrophoresis of radiolabelled antigens against polyspecific antisera. Individual immunoprecipitation lines may be excised from the gel and analysed by SDS gel electrophoresis (Norrild *et al.*, 1977). The individual subunits may then be visualized by autoradiography, fluorography or scintillation counting of individual gel slices. If monospecific antisera are available, then the radiolabelled antigen may be immunoprecipitated from solution and the immunoprecipitate analysed by gel electrophoresis (Ross *et al.*, 1978).

### 4.4.2. Determination of antigen isoelectric point

Antigens may be separated by isoelectric focusing under non-denaturing or denaturing conditions in agarose or polyacrylamide gels and then transferred electrophoretically on to nitrocellulose paper and subsequently reacted with antiserum. Two-dimensional gels may also be probed with specific antisera to identify antigen isoelectric point (Changeux, 1973; Sudhof *et al.*, 1984). Gels which have been stained with Coomassie brilliant blue, de-stained and photographed may subsequently be treated with 25% isopropanol/10% acetic acid, to remove most of the Coomassie attached to protein. Then, after equilibrating the gel in running buffer containing SDS, the proteins may be electrophoretically transferred on to nitrocellulose and processed to detect antigenic components. Faint traces of Coomassie dye remain on the blot and aid in the comparison of antigen position relative to the major proteins in the fraction.

### 4.4.3. Determination of covalent modification of antigens

Glycosylated antigens may be demonstrated by using lectin columns to subfractionate extracts containing antigens (Buckley and Kelly, 1985). The column flow-through and the fraction bound and eluted with specific sugar may then be analysed by immunoblotting. Alternatively, cells in culture may be labelled with radioactive sugars, and antigens may be isolated by immunoprecipitation and analysed by gel electrophoresis. Clearly, this approach may be used to look for many covalent modifications, including myrystilation and, in particular, phosphorylation. In the case of proteins phosphorylated on tyrosine residues, antibodies may be prepared to phosphotyrosine (Comoglio *et al.*, 1984; Ek and Heldin, 1984) and used to immunoprecipitate protein substrates for tyrosine kinases. The antibodies to phosphotyrosine may be prepared by immunizing rabbits with phosphotyrosine coupled to an antigenic protein with carbodiimide. The antibodies obtained may also be used to localize antigens with immunofluorescence on cells in culture.

### 4.4.4. Determination of the intrinsic or extrinsic nature of membrane proteins

Charge-shift electrophoresis of membrane proteins (Helenius and Simons, 1977) distinguishes intrinsic membrane proteins from extrinsic membrane proteins or cytosolic proteins on the basis of the interaction of these proteins with charged and non-ionic detergents. Only the intrinsic membrane proteins bind both types of detergent and thus only they show an alteration in their electrophoretic mobilities. By using charge-shift electrophoresis as a first-dimension separation technique in crossed-immunoelectrophoresis it is possible to characterize membrane-protein antigens in terms of their amphiphilic nature (Jørgensen, 1977). Similarly, immune blots of charge-shift electrophoresis may be used to demonstrate changes in mobility of the antigen.

An alternative to charge-shift electrophoresis is to subfractionate membrane proteins with Triton X-114 (Bordier, 1981; Pryde, 1986). This non-ionic detergent solubilizes membranes at 4 °C, but at 30 °C it separates into an aqueous phase and a phase containing Triton X-114 and integral membrane proteins (details of the procedure are given in Chapter 7).

### 4.4.5 The vectorial orientation of antigens in membranes

Antigens can be assigned to the inner or outer surface of a membrane if intact and lysed cells or organelles are available. Thus methods based on adsorption of antibodies with intact or lysed cells (Bjerrum *et al.*, 1975) and intact or lysed synaptosomes (Jørgensen, 1976) have been devised. A comparison of the responses of antibodies adsorbed with intact and lysed cells or organelles should indicate where the antigen is found.

Alternatively, cell-surface antigens may be specifically labelled with dinitrophenyl moieties which may then be identified and isolated by means of antibodies to dinitrophenol and Protein A-bearing *Staphylococcus aureus* (see Section 7.5.3).

## 4.5. Immunohistochemistry

Immunohistochemical procedures now form one of the most important tools of cell biology and can often be used to rapidly provide new insights into many areas of research especially studies on the cytoskeleton and neurotransmitter pathways.

The localization of specific antigens in tissue sections and cells in cultures has also assumed great importance over the years, in particular because of the potential of the method for rapidly screening biopsy samples for human tumour diagnosis (Osborn and Weber, 1982a; Osborn *et al.*, 1984). The method is also of clinical value for scanning patients' sera for autoantibodies, and human autoantibodies have proven to be of great value in cell biology studies. Thus autoantibodies known by immunofluorescence to recognize the nuclear lamina have been used to identify cDNA clones expressing the lamina proteins 'lamins' (McKeon *et al.*, 1986). This led to the sequencing of these proteins and the demonstration that they possess considerable sequence homology with intermediate filament proteins. Important discoveries are being made with autoantibodies which recognize pericentriolar material. Only a single pair of centrioles is present in each cell and therefore the fortuitous occurrence of antibodies in certain patients has enabled specific components associated with centrioles to be identified (Valdivia and Brinkley, 1985) without the need for prior isolation of completely pure centrioles.

The remainder of this section outlines the different procedures available for locating antigens at the levels of resolution of the light and electron microscopes and is complemented by an extensive selection of recipes and protocols in Chapter 7.

### 4.5.1. Localization of antigens with the light microscope

#### 4.5.1.1. Tissue sections

The most usual way of locating an antigen in tissue is to prepare sections of the tissue, incubate the section with antibodies and then detect bound antibody with a labelled second antibody. The first essential step in this procedure is to fix the tissue to ensure that the structure of the tissue remains preserved well enough to allow the site of the antigen to be unambiguously defined. In general, two classes of fixatives exist—those which cross-link proteins (e.g. formaldehyde, glutaraldehyde, carbiodiimide), and those which precipitate proteins (e.g. acetone, methanol, ethanol). Optimal fixation of proteins is achieved with glutaraldehyde and of membranes is achieved by means of the lipophilic fixative osmium tetroxide. In immunohistochemical studies a frequently encountered problem is that it may be difficult to find the level of fixative which preserves cellular structure adequately without destroying all of the antigenic determinants in the tissue section. Optimal fixation of tissues is obtained by perfusion of animals with a Ringer solution followed by fixative (Priestley and Cuello, 1983).

However, adequate fixation is often achieved simply by rapidly excising small blocks of tissue and immersing in fixative. It is also true that for studies on frozen sections, fixation may not be necessary prior to sectioning for certain antigens. For example, intermediate filaments have been localized in brain tissue using 5 $\mu$m cryostat sections of unfixed tissue (Shaw *et al.*, 1981). In this case, however, cold-acetone extraction of the sections was performed prior to immunostaining, to permeabilize the membranes and to fix the tissue. For many antigens, especially soluble ones, this procedure would be unacceptable, but for some cytoskeletal and membrane components this method may give excellent results and has the advantage that results may be obtained very rapidly and with limited damage to antigenic determinants.

Cryostat sections are usually cut at 10 $\mu$m thickness although with some tissues (liver, kidney, brain) 5 $\mu$m sections can be obtained without difficulty. Sections of 10 $\mu$m may also be obtained with a Vibrotome (Oxford Instruments) which uses a vibrating razor blade to prepare sections from tissue immersed in buffer at 4 °C.

The alternative procedures for obtaining sections require the tissue to be embedded in a supporting medium which facilitates sectioning. The most usual of these embedding media are based on paraffin (e.g. Paraplast). These require that the fixed tissue is dehydrated through an ethanol series (10–100%) and then "cleared" to replace the ethanol with a solvent miscible with paraffin (e.g. xylene, oil of Wintergreen). The tissue is then impregnated with paraffin at around 57 °C (1 °C above the melting point of the paraffin). The tissue is then embedded in a block of paraffin which can then be sectioned and used for immunohistochemistry. Clearly, the treatment with fixative, organic solvents and prolonged heating may damage many antigenic determinants, but the ever-increasing sensitivity of antibody detection systems decreases the extent of this problem. An advantage of paraffin-embedded tissue is that it may be stored for prolonged periods and still provide excellent immunohistochemical results, although with immunofluorescence, problems of autofluorescence increase with storage.

Sections may also be obtained from samples of tissue embedded in Epon or other plastic media used for electron microscopy. In this case, semi-thin sections (0.5–1 μm) may be obtained easily and this decrease in section thickness considerably improves the resolution of antigens with immunofluorescence microscopy (e.g. DeCamilli *et al.*, 1983). As with paraffin sections the embedding medium must usually be removed from the section before immune staining can take place, and in the case of sections embedded in Epon this involves particularly drastic treatment with potassium hydroxide solution (see Chapter 7). Recently, however, plastic embedding media have been introduced which obviate some of these problems. In particular, Lowicryl resins have been developed which can be used for embedding tissues at low temperature. Lowicryl-embedded tissues can be sectioned both for light and electron microscopy and can be etched with sodium metaperiodate to expose more antigens, although this step is usually not necessary (Roth, 1984). The preservation of antigenic determinants is very high with Lowicryl-embedded tissue (Bendayan and Zollinger, 1983; Armbruster *et al.*, 1983) and it is even possible to detect enzyme activities, degrade lectin-binding sites with glycosidases (Roth, 1984) and hybridize cDNA probes (Binder *et al.*, 1986).

Sections are mounted on slides or cover slips. Several adhesives may be used to prevent sections detaching from the glass and are listed in Chapter 7. The normal protocol for identifying antigens in tissue sections involves (1) permeabilizing tissue section and blocking non-specific binding sites; (2) incubation with antibodies; (3) washing to remove unbound antibody; (4) detecting bound antibody; and (5) preparing the slide for viewing (Fig. 4.24).

Detailed consideration of these points (e.g. antibody concentrations, non-specific blockers, etc.) are given in Chapter 7 and we shall only discuss here certain points of general interest. The most important thing to consider is the detection system to use for identifying bound antibody. The simplest approach, providing that you have access to a fluorescence microscope, is to use second antibodies which have been conjugated with fluorescent dyes (Coons *et al.*, 1942; Haaijman, 1983). This method is especially useful for comparing the distribution of two antigens (e.g. Shaw *et al.*, 1981) since specific antisera from different species may be used to detect the antigens.

Specific second antibodies conjugated to rhodamine or fluorescein may then be used to locate the bound antibodies in a specific manner, and by viewing the specimen through appropriate filters one antigen will be localized by the red emission of rhodamine and the second antigen will be localized by the green light emitted from fluorescein. This method is not only very useful but also very pleasing aesthetically. Double labelling may also be used with fluorescently labelled lectins to identify specific sugars (Virtanen *et al.*, 1980) and also with fluorescently labelled toxins such as phalloidin, which binds to

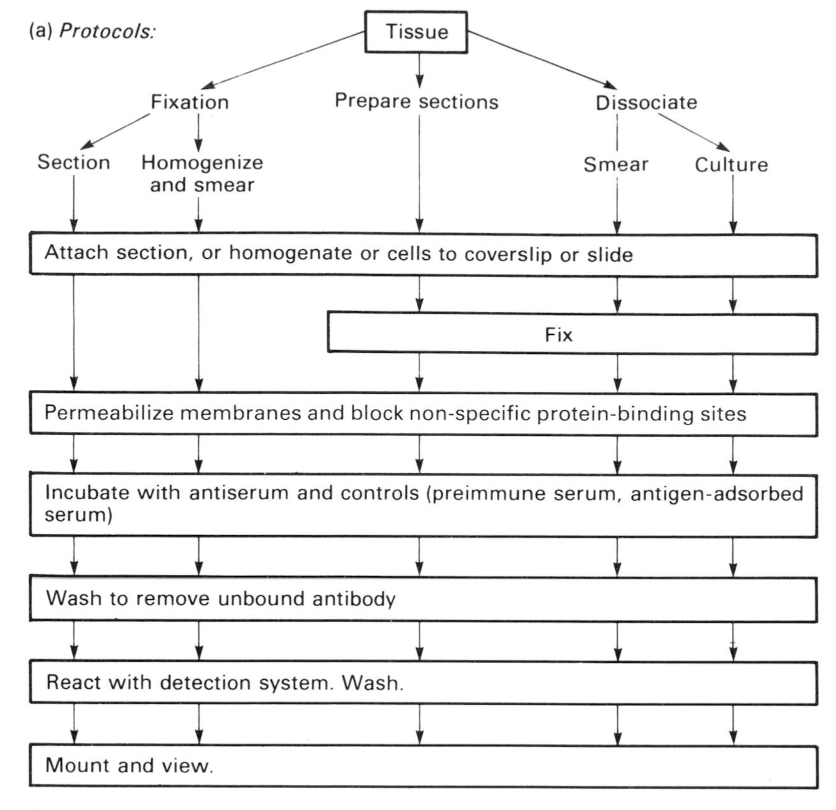


(b) *Detection systems:*

(i) Fluorescent-labelled second antibody
(ii) Enzyme-labelled second antibody
(iii) Bridge methods (e.g. anti-IgG and PAP)
(iv) Streptavidin methods
(v) Gold-conjugated second antibody

**Fig. 4.24** Immunohistochemistry.

F-actin (Wulf *et al.*, 1980), and alpha-bungarotoxin, which binds to the nicotinic acetylcholine receptor (Fig. 4.25; Ravdin and Axelrod, 1977).

Samples stained by immunofluorescent antibodies may be stored for prolonged periods of time (12 months or longer (Osborn and Weber, 1982b)) in semipermanent mounting media based on polyvinyl alcohol (e.g. Mowiol 4–88). Fading of fluorescence can be a problem (Fig. 4.25c) although addition of *p*-phenylenediamine considerably retards the loss of fluorescence (Platt and Michael, 1983). Normally, fluorescently labelled sections are compared with phase-contrast images of the same area of the slide, but it is also possible to perform immunofluorescence staining of sections previously stained with haematoxyoylin and eosin (Weinstein and Lechago, 1977; C. Vaillant, personal communication). This is especially valuable, since it enables a direct comparison of the immunofluorescence staining pattern with the conventional staining pattern found in standard histology textbooks. It is also possible to take photographs using a combination of low-intensity bright-field illumination and immunofluorescence emission.

The most usual alternative to fluorescently labelled antibodies is the use of peroxidase-conjugated second antibodies (or protein A) or the peroxidase–antiperoxidase complex, PAP (Sternberger, 1979; Taylor, 1976). The great advantage of these methods is that they provide a permanent record that can be counterstained with haematoxylin and eosin and that can be viewed with far less sophisticated microscopes than those required for immunofluorescence microscopy. The PAP method provides enormous amplification of the antibody signal and allows extremely high dilutions of first antibody to be used, which has the effect of increasing the effective specificity of the antiserum in favour of the major antigen. A similar method has been devised using soluble complexes formed between alkaline phosphatase and monoclonal antibodies to alkaline phosphatase (Cordell *et al.*, 1984).

Very sensitive peroxidase-linked methods are also available which utilize the very high affinity of avidin and streptavidin for biotin. Streptavidin is a 60 000 protein composed of four subunits (Chaiet and Wolf, 1964). Like the egg-white protein, avidin, it has a very high affinity ($K_d = 10^{-15}$ M) for biotin. Unlike egg-white avidin, streptavidin has minimal non-specific binding at physiological pH (Haeuptle *et al.*, 1983). Some manufacturers supply avidin which has been covalently modified to reduce non-specific binding. Sections are incubated in antiserum and then with biotin-labelled second antibodies. The biotin label is then recognized by one of several streptavidin detection systems. Most directly the streptavidin may be conjugated to peroxidase (or other enzymes). Alternatively, the streptavidin may form a bridge between biotin-conjugated second antibody and biotin-conjugated peroxidase. The most sensitive approach involves the use of soluble complexes prepared by incubation of streptavidin and biotinylated peroxidase. Kits for these staining procedures are available from Amersham and Vector Laboratories. The specific peroxidase activity associated with antigen is normally demonstrated by diaminobenzidine/$H_2O_2$ stain, which provides a highly insoluble brown reaction product which can be intensified with osmium tetroxide (black), or with cobalt (blue). Recently, a silver staining method for the intensification of the diaminobenzidine has been developed and marketed by Amersham. This greatly increases the sensitivity of the peroxidase-staining methods. Other stains for peroxidase activity include 4-chloronaphthol (blue) and aminoethylcarbazole (red) which cannot, however, be dehydrated and embedded in permanent mounting media.

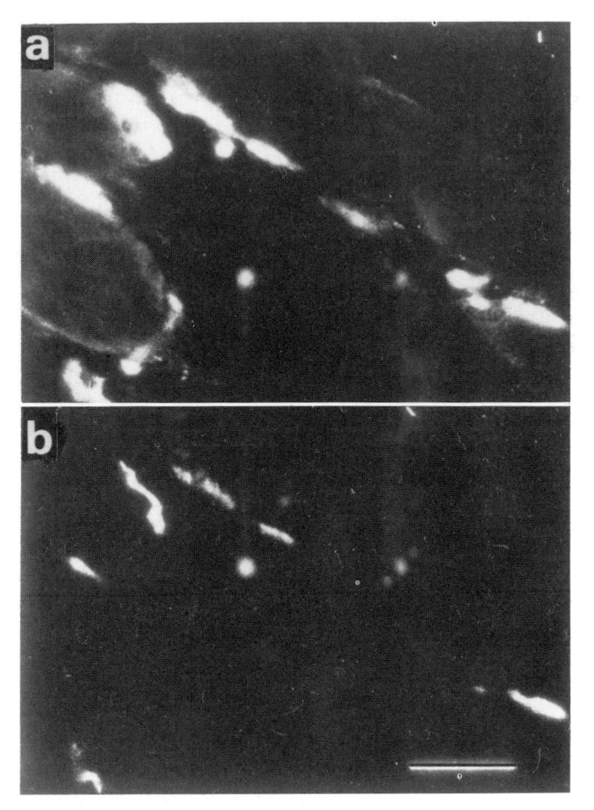


**Fig. 4.25** Immunofluorescence microscopy. Nerve endings have been identified in cryostat sections of rat diaphragm using (a) an antiserum specific to synaptic vesicles followed by FITC-labelled second antibody; and (b) rhodamine-conjugated $\alpha$-bungarotoxin. The scale bar represents 50 $\mu$m. Reproduced from Walker *et al.* (1985) with permission from the publisher. (c) Fading of immunofluorescence. Tissue stained with fluorescently labelled second antibodies was illuminated for 5 min using the $\times 40$ objective. Subsequently, the $\times 16$ objective was used to produce this photograph.

Endogenous peroxidase activity may occur in tissue sections and is often destroyed by preincubation in methanol and hydrogen peroxide.

A further development in immune-staining light microscopy is the development of reagents labelled with colloidal gold. Colloidal gold can be prepared with varying diameters, e.g. 15, 10, 6 and 3.5 nm (DeMey, 1984; Slot and Geuze, 1985). Immune reagents can then be prepared by adsorbing immunoglobulin (Faulk and Taylor, 1971) or Protein A (Roth, 1982) on to gold particles. At their simplest, immune gold-staining procedures are not particularly sensitive for light microscopy but they can be made very sensitive by intensifying the gold with silver precipitation (Danscher, 1981). Immuno-

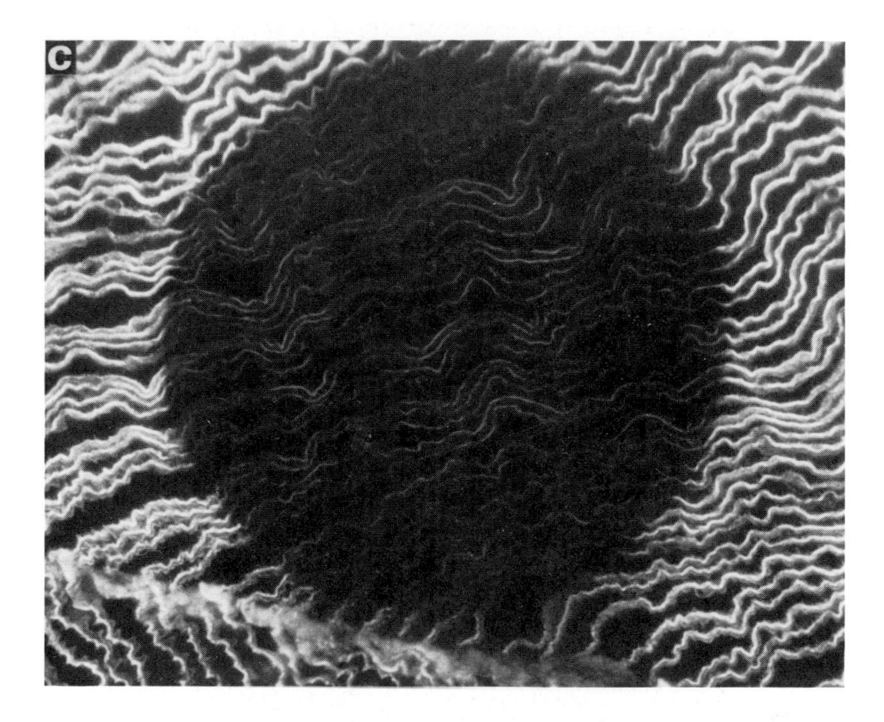


gold–silver staining methods (Holgate *et al.*, 1983; Hacker *et al.*, 1985) can increase sensitivity of staining such that antigens previously undetectable with the PAP method on paraffin sections are easily obtained (Springall *et al.*, 1984). It is also possible to use double-staining protocols in which one antigen is stained black with silver-intensified Protein A gold and then a second antigen is stained red with Protein A gold alone (Manningley and Roth, 1985). It is also of interest to note that gold-labelled streptavidin is also available (Amersham) linking the already sensitive biotin-labelled reagents to the even more sensitive gold/silver methods. Extensive details of gold-labelled reagents are also available from Janssen Pharmaceuticals, a firm which supplies these reagents and supports extensive research on their development.

### 4.5.1.2. Cells in culture

Cells which adopt a flattened extended morphology in cell culture are very useful for the localization of antigens whose subcellular location is unknown. Fibroblast-like (e.g. 3T3) and epithelial-like (e.g. HeLa) cell lines have well defined distributions of all the subcellular organelles (Fig. 4.27) and markers

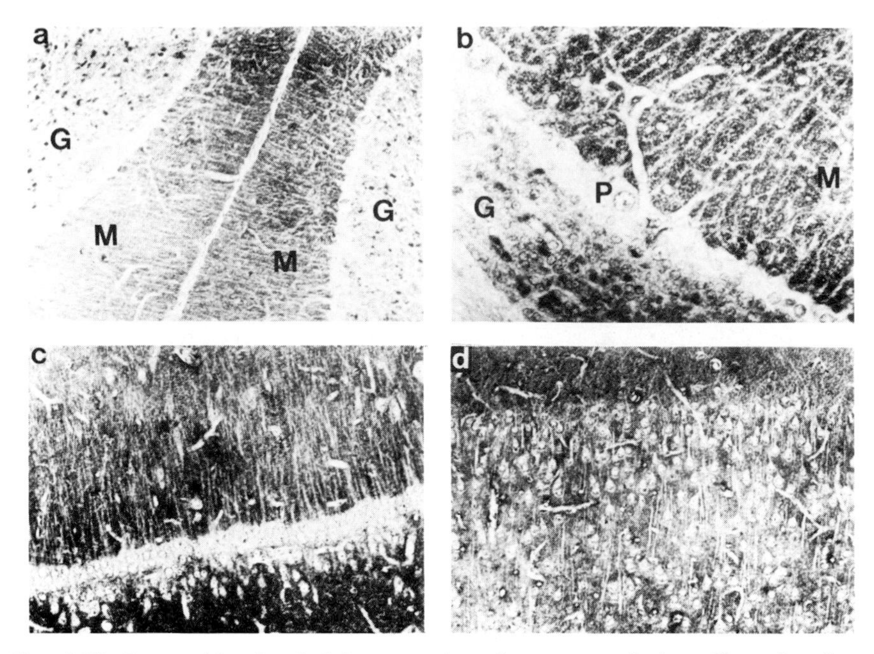


**Fig. 4.26** Immunohistochemical demonstration of a nerve-terminal specific antigen in rat brain. Paraffin sections (10 μm thick) of Bouin-fixed rat brain were stained by the indirect peroxidase method. Staining of cerebellum (a,b) was found in molecular (M) and granule (G) layers but was not associated with the Purkinje cells (P). In the hippocampus (c) staining was found around the pyramidal cell bodies which were not themselves stained. A similar staining pattern was also found in the cerebral cortex (d). Reproduced from Walker *et al.* 1985 with permission.

are available for these organelles (Table 4.3). Similarly, the patterns of staining for cytoskeletal proteins are also well established (Fig. 4.27). Clearly, the major advantage of cells in culture is that the cells are immediately accessible to changes in cellular function induced by drugs, hormones and viral infection.

As with tissue sections it is necessary to fix and permeabilize the cultured cells. The most usual procedures are treatment of the cells with acetone at −20 °C or fixation with formaldehyde followed by permeabilization with non-ionic detergent (usually Triton X-100). Superior staining of cytoskeletal structures may be obtained in some cases by first extracting the cells with detergent in a buffer that stabilizes the cytoskeleton and only then fixing the preparation (Osborn and Weber, 1982a,b).

**Fig. 4.27** Immunofluorescence staining patterns of cells in culture with antibodies to cytoskeletal proteins and cellular organelles. (a) actin, (b) tubulin, (c) plasma membranes, (d) mitochondria, (e) endoplasmic reticulum, (f) Golgi, (g) nuclei, (h) nucleoli, (i) centrosomes, (j) nuclear membrane. (k) is phase of (j).

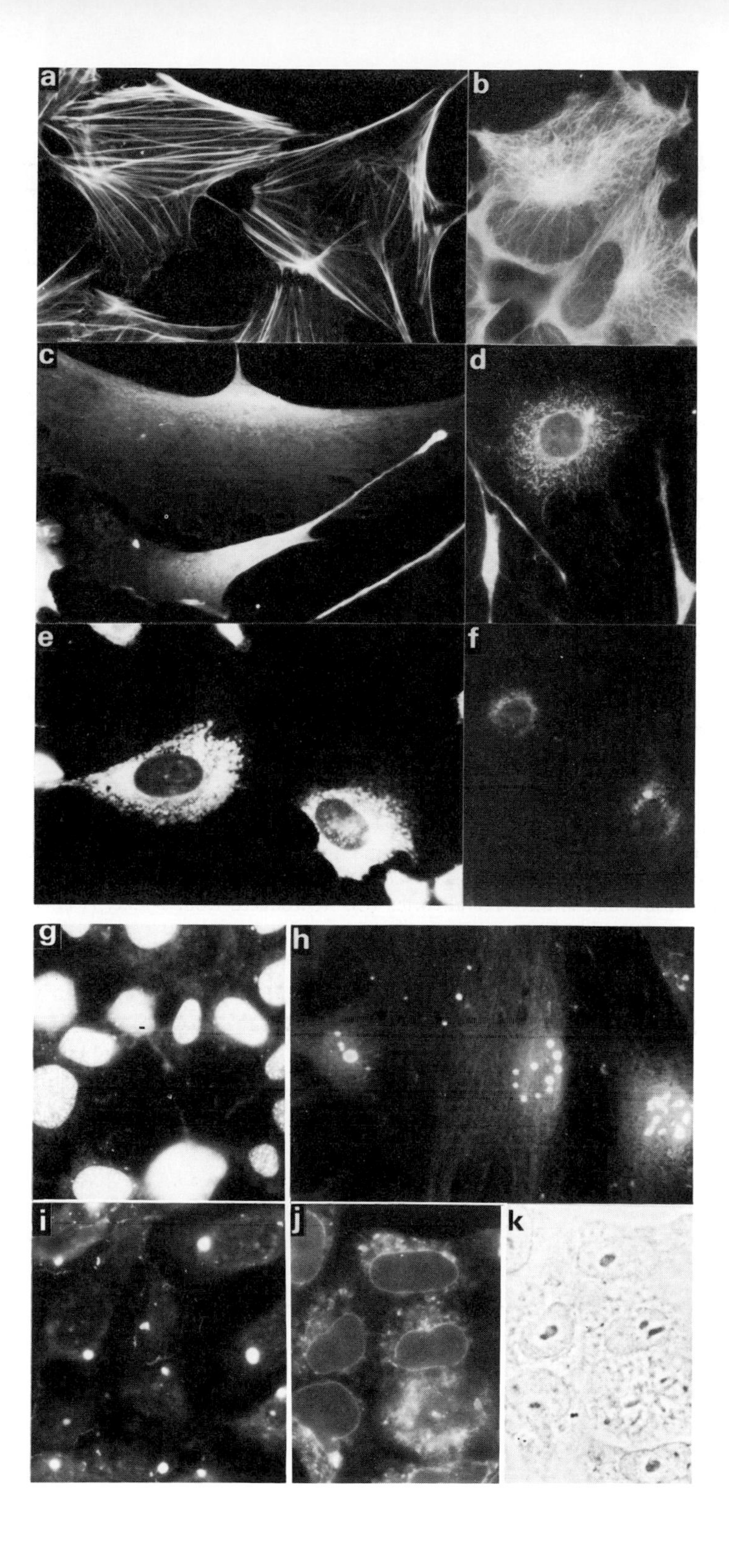
a
b
c
d
e
f
g
h
i
j
k

**Table 4.3** Markers for subcellular fractions in histochemistry

| Subcellular fraction | Marker | References |
|---|---|---|
| (1) Nuclei | (a) Bisbenzimide (Hoechst 33258) | Bainbridge and Macey (1983)<br>Hilwig and Gropp (1973) |
| | (b) Anti-nuclear antibodies | Pinnas *et al.*, (1973)<br>Sigma Tech. Bulletin 1000 |
| (2) Golgi | (a) Vital stain | Lipsky and Pagano (1985) |
| | (b) FITC-wheatgerm agglutinin | Virtanen *et al.* (1980) |
| (3) Mitochondria | (a) Vital stain, rhodamine 123 | Terasaki *et al.* (1984) |
| | (b) Antibodies | |
| (4) Lysosomes | (a) Adsorbed antiserum | Collot *et al.* (1984) |
| | (b) Antibodies to Cathepsin D | Poole, 1974 |
| (5) Endoplasmic reticulum | (a) Vital stain 3,3′-dihexyloxacarbocyanine iodide | Terasaki *et al.* (1984) |
| | (b) Specific antibodies | Louvard *et al.* (1982) |
| (6) Plasma mambrane | FITC-Concanavalin A | Nigg *et al.* (1983) |
| (7) Cytoskeleton | | |
| (a) Tubulin | (a) Specific antibodies | Weber *et al.* (1975) |
| | (b) Monoclonal antibodies | Wehland and Willingham, (1983) |
| (b) Actin | (a) Specific antibodies | Lazarides and Weber (1974) |
| | (b) FITC-phalloidin | Wulf *et al.* (1980) |
| | (c) DNase I–anti DNase I | Snabes *et al.* (1981) |
| (c) Myosin | Specific antibodies | Weber and Groschel-Stewart (1974) |
| (d) Intermediate filaments | Specific antibodies | Lehto *et al.* (1978)<br>Osborn and Weber (1983)<br>Small and Celis (1978) |
| (e) Clathrin | Specific antibodies | Anderson *et al.* (1978)<br>Bloom *et al.* (1980) |

The most usual procedure for viewing cultured cells is immunofluorescence, and an interesting point here is that this method enables structures below the resolving power of the microscope to be seen. For example, microtubules (25 nm in diameter) are too small to be resolved by lenses of the greatest resolving power ($\times 64/1.4$). However, individual microtubules can be clearly seen in immunofluorescence micrographs (Osborn and Weber, 1982b). The microscope produces an image of the light emitted from the microtubules but this image is $\simeq$ 200 nm diameter, i.e. at the limit of its resolving power—nevertheless, single microtubules are seen.

Although the bulk of work on cells in culture has been performed using immunofluorescence microscopy, there is clearly also much scope for the immunogold silver staining methods.

#### 4.5.1.3. Cell suspension and whole mounts

Suspensions of cells obtained from fixed homogenized tissues can provide valuable samples for immunohistochemical analysis. This has been especially true of epithelial cells from the small intestine (Bretscher and Weber, 1978). Similarly, myofibril preparations (Herman and Pollard, 1978) and muscle Z-discs (Granger and Lazarides, 1978, 1979) have proved useful for the location of cytoskeletal proteins. A further useful preparation has been described for the studies on the distribution of antigens at the cholinergic synapse (Walker *et al.*, 1985; Figs 4.28 and 4.29). In this case gentle homogenization of the electric organ of *Torpedo* results in the formation of large sheets of innervated membranes which can easily be studied for the localization of cytoskeletal proteins and nerve-terminal markers.

For very thin tissues, such as the mesentery, pericardium and the iris, whole mounts may be stained, especially for looking at neurotransmitter distributions (Costa and Furness, 1983).

### 4.5.2. Localization of antigens with the electron microscope

Antigens may be localized at the level of resolution of electron microscope with pre-embedding, post-embedding or negative-staining procedures.

#### 4.5.2.1. Pre-embedding procedures

These involve processing small sections of tissues obtained by cryostat

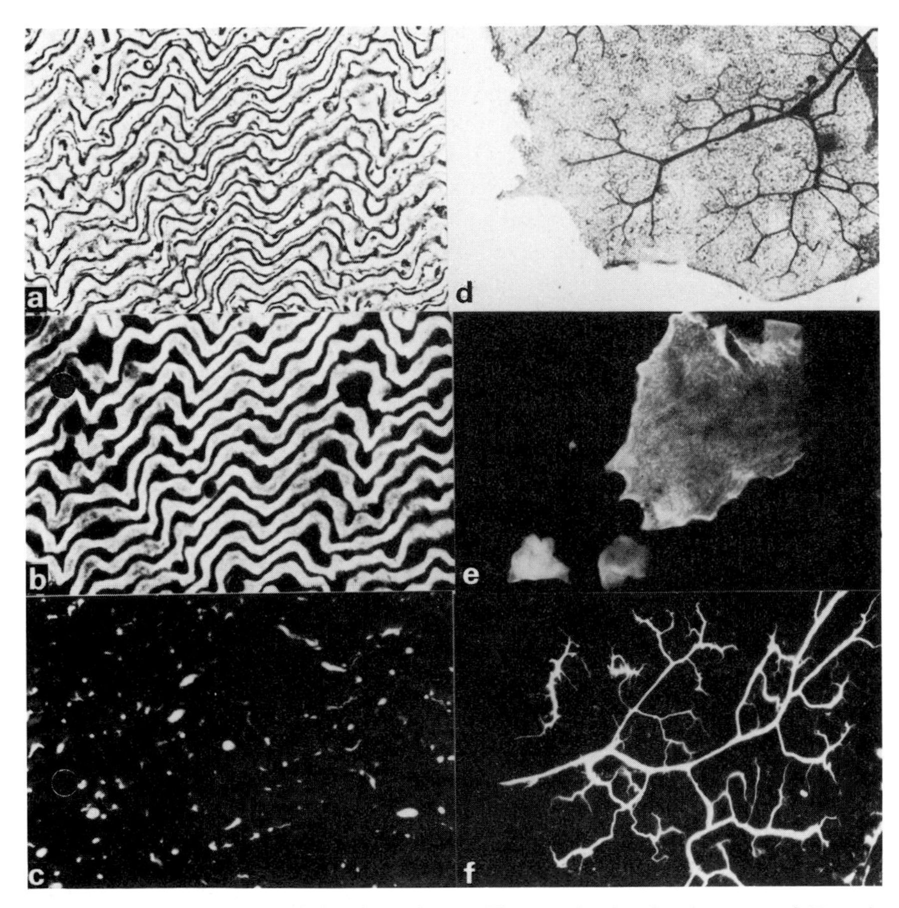


**Fig. 4.28** Demonstration of desmin and neurofilaments in the electric organ of *Torpedo*. Paraffin sections through the electrolyte columns are shown (a) in phase microscopy, (b) stained with the desmin antibody, and (c) frozen, taken in a similar orientation, stained with the antibody to the neurofilament L-polypeptide. In all these sections the dorsal surface is towards the top of the picture. The *en face* homogenate preparation is shown stained with Coomassie Blue (d), desmin antibody (e), and with antibody to the neurofilament L-polypeptide (f). a–c, 300 ×; d,e, 250 ×; f, 420 ×.

sectioning or vibratome sectioning, or alternatively cells in culture may be processed (Table 4.4). The major problem experienced in electron microscopic localization is that strong fixatives such as glutaraldehyde are usually necessary to maintain recognizable ultrastructure, and these tend to destroy the antigenic determinants you wish to localize. The fixation may also cross-link cytoplasm and thus seriously impede the access of antibodies to antigen, although this problem may be overcome to some extent by using $F_{ab}$

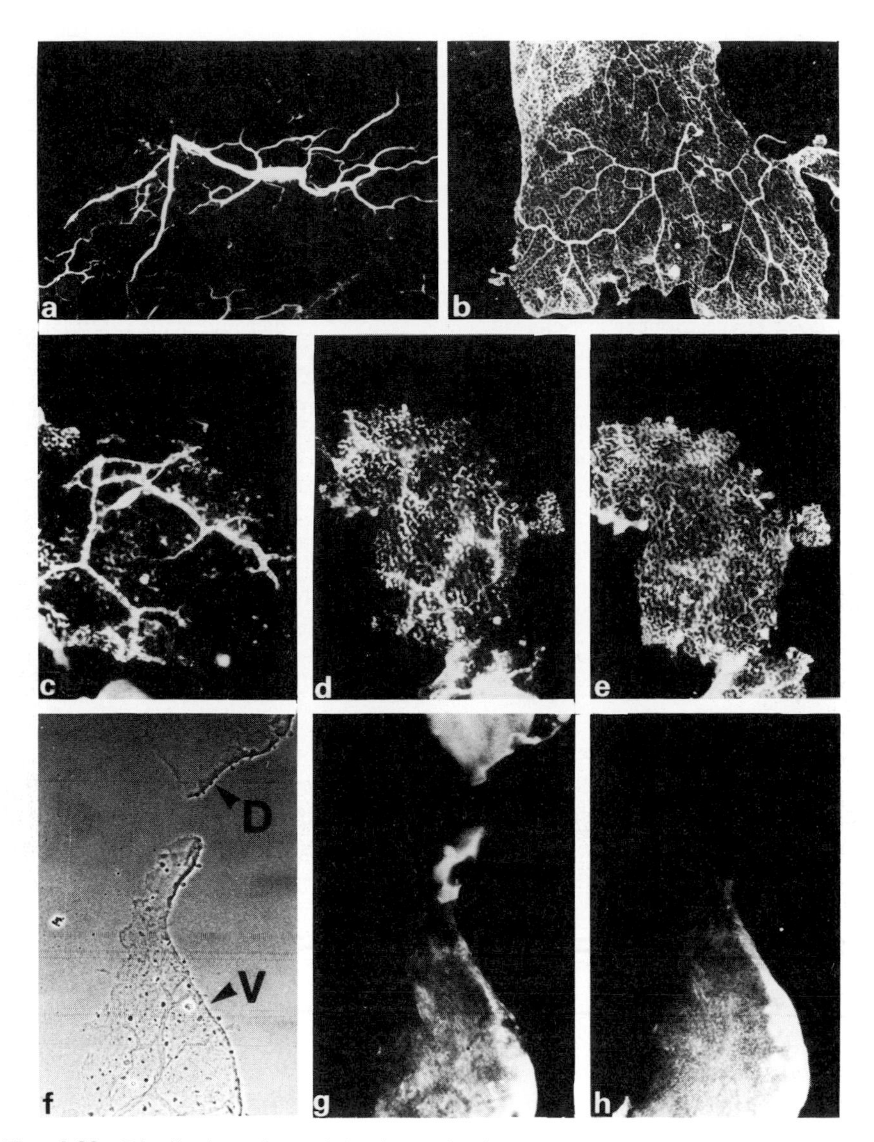


**Fig. 4.29** Distribution of cytoskeletal proteins in *en face* membrane preparations from *Torpedo* electric organ. (a) Tubulin. (b) Fodrin. (c) Actin. Note the similarity of the staining patterns seen for fodrin (b) and for actin using rhodamine-labelled phalloidin (c). (d,e). Double labelling with (d) phalloidin and (e) a vesicle antiserum to identify the nerve terminals. Note the similarity of the staining patterns. (f–h). Phase (f) and double label with desmin (g) and rhodamine-conjugated $\alpha$-bungarotoxin (h). Note that the dorsal (D) and ventral (V) membranes can be identified either from the phase micrograph, or by staining with $\alpha$-bungarotoxin. Desmin is found associated with both membranes. a, f–h, 300 ×; b, 180 ×; c–e, 280 ×.

**Table 4.4** Immunolocalization with the electron microscope

| | | Label | References |
|---|---|---|---|
| (1) | Pre-embedding methods | | |
| | Vibratome sections | IG | Lamberts and Goldsmith (1985) |
| | Cryostat sections | IG | Weidenmann and Franke (1985) |
| | Cryostat sections | IG | Kartenbeck *et al.* (1984) |
| | Tissue chopper sections | IP | Fiedler and Walker (1985) |
| | Synaptosomes in agarose | IF | DeCamilli *et al.* (1983) |
| | Cells in culture | IF | Webster *et al.* (1978) |
| | Cells in culture | IP | Willingham *et al.* (1978) |
| (2) | Post-embedding methods | | |
| | LR white sections | IP | Whitnall *et al.* (1985) |
| | Lowicryl K4M sections | IG | Valentino *et al.* (1985) |
| | Lowicryl K4M sections | PAG | Orci *et al.* (1985) |
| (3) | Ultracryosections | IG | Rindler *et al.* (1985) |
| | | PAG | Griffiths *et al.* (1982) |

Key: IG, immunogold; IP, immunoperoxidase; IF, immunoferritin; PAG, Protein A gold.

fragments linked to microperoxidase (low $M_r$ peroxidase generated by degradation of cytochrome C (Kraehenbuhl *et al.*, 1974). The second major problem is the need to permeabilize membranes in order to ensure access of antibodies to intracellular antigens. Unless tissue is well fixed, extraction with non-ionic detergent can seriously damage ultrastructure. One way around this is to use saponin, which selectively removes cholesterol from the membranes, allowing access of antibodies to the cell interior. Saponin must be included throughout the procedure if the membranes are to remain permeable (Willingham and Pastan, 1985). Furthermore, saponin does not permeabilize membranes sufficiently to allow large reagents such as colloidal gold-labelled antibodies to enter cells. However, excellent results have been obtained using immunoperoxidase staining of cells in culture permeabilized with saponin and fixed with water-soluble carbodiimides (Willingham *et al.*, 1978). Gold-, peroxidase-, and ferritin-labelled reagents may be used on cells in culture when membranes are disrupted by more rigorous means such as extraction with Triton X-100.

Sections of tissue 10–200 $\mu$m in thickness may also be processed with pre-embedding staining procedures (Table 4.4), although usually only the outermost part of the tissue is stained in a representative manner (Priestley and Cuello, 1983), which calls for careful interpretation of the results.

In all of the pre-embedding methods, the immunostained tissue must subsequently be processed for electron microscopic examination. Thus the tissue must be post-fixed with glutaraldehyde to maintain optimal preservation and to prevent movement of the immune stain during subsequent processing. The sample must then be dehydrated, embedded, sectioned and viewed in the electron microscope. A protocol is given in Chapter 7 and Table 4.4 lists various references in which these methods are detailed further.

An example of pre-embedding staining of thick tissue sections with an immunoperoxidase method is shown in Figs 4.30 and 4.31 (Fiedler and Walker, 1985). In Fig. 4.30, different samples of tissue have been processed with various antisera. In Fig. 4.30a an antiserum to a calcium-binding protein, calelectrin, which is found throughout the cells is studied. The same antiserum adsorbed with pure antigen no longer shows positive staining (Fig. 4.30b). The structure visible in (b) is due to the counterstaining of the Epon thin sections with uranyl acetate and lead citrate. An antiserum specific to synaptic vesicles was also used in this experiment (Fig. 4.30c) and exclusively stained nerve terminals in the tissue. At higher magnification (Fig. 4.31) calelectrin was seen to be associated with various cellular structures, including plasma membranes (Fig. 4.31a), mitochondria (Fig. 4.31c) and intracellular filaments (Figs 4.31a–d). If desired, sections may be viewed without

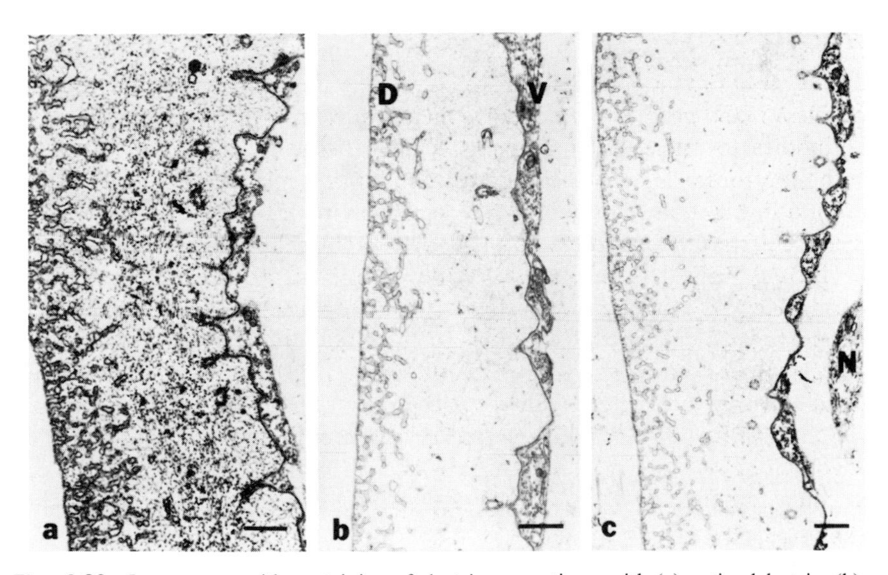


**Fig. 4.30** Immunoperoxidase staining of electric organ tissue with (a) anti-calelectrin, (b) a control serum prepared by adsorbing the anti-calelectrin antiserum with calelectrin and (c) an antiserum to *Torpedo* synaptic vesicles. Reaction product is also visible in the nerve (N). Ventral (V) and dorsal (D) membranes are also indicated. Bars 1 $\mu$m.

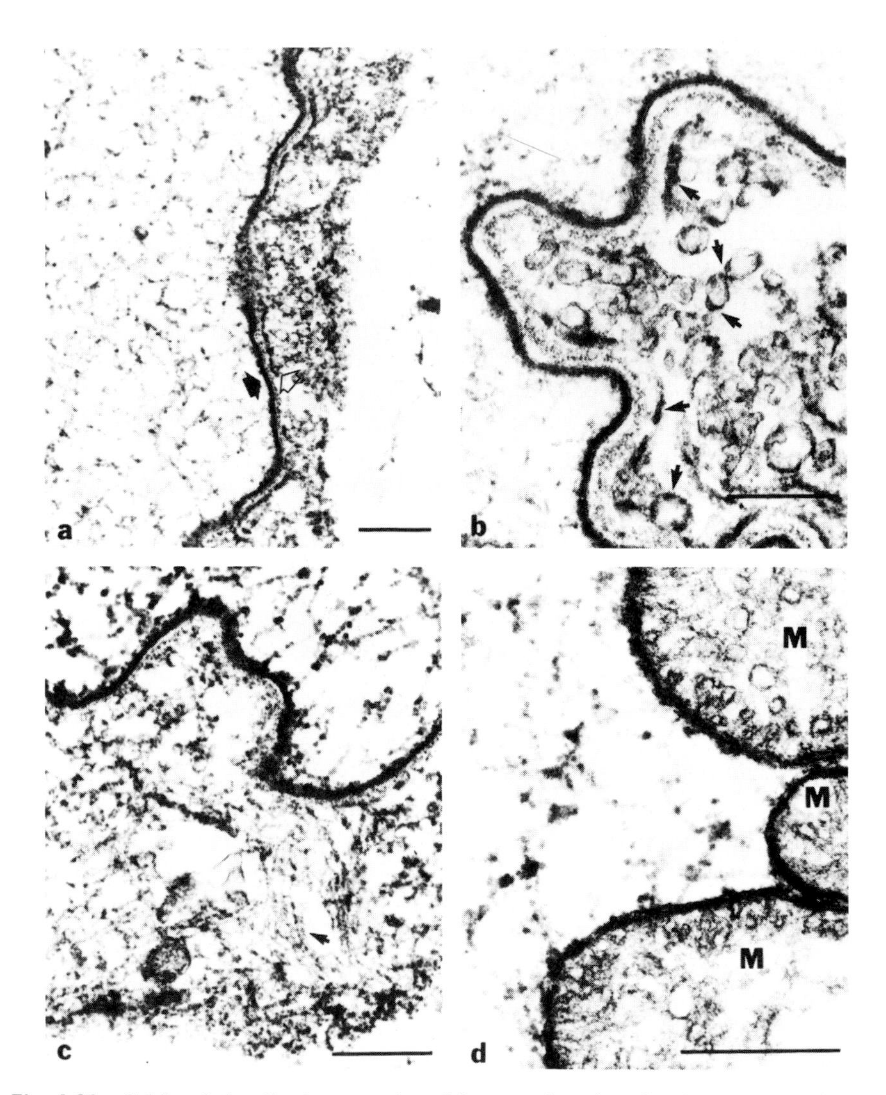


**Fig. 4.31** Calelectrin localization: overview of the synaptic region, showing strong staining on the receptor containing membrane (filled arrowhead) and weaker staining on the synaptosomal plasma membrane (open arrowhead). (b) Higher magnification of a nerve ending showing staining associated with synaptic vesicles and synaptosomal plasma membranes (arrow). (d) Peroxidase staining in the electrocyte associated with filaments and mitochondria (M). Bars (a, c, d) 0.5 $\mu$m, (b) 0.25 $\mu$m.

staining, and additionally after counterstaining with uranyl acetate and lead citrate. The immune staining stands out more clearly without counterstaining, but counterstained samples are far more convenient for viewing with the electron microscope.

### 4.5.2.2. Post-embedding procedures

Recently, considerable advances have been made with *post-embedding* immune-staining methods for electron microscopy. New hydrophilic resins such as Lowicryl, LR white and glycol methacrylate can be processed in such a way as to maintain antigenic sites in tissue samples. With Lowicryl K4M, tissue can be embedded at low temperatures (−35 °C) after dehydration, also at low temperature. Good ultrastructure and immunoreactivity are obtained on thin sections which can be directly viewed in the electron microscope (Roth *et al.*, 1981), LR white can also be used for immune staining with good results (Newman, 1984). References to post-embedding staining methods are given in Table 4.4 and procedures are given in Chapter 7 for embedding tissues in Lowicryl K4M and staining thin sections. Gold-labelled staining reagents are especially useful for these post-embedding staining methods and have been reviewed elsewhere (Beesley, 1985; DeMey, 1983). It is also of interest that biotin-labelled DNA probes have been hybridized to sections of Lowicryl K4M-embedded tissue to identify the sites of expression of specific genes at the level of resolution of both the light and electron microscopes (Binder *et al.*, 1986). Sensitivity equivalent to autoradiography was obtained using antibodies to biotin and Protein A gold for electron microscopy and silver enhancement for light microscopy.

A final procedure for post-embedding staining involves the use of ultra-thin cryosections (Tokuyasu, 1983). This method has been used with great success to localize antigens at the EM level (Geuze *et al.*, 1981; Slot and Geuze, 1983; Griffiths *et al.*, 1982; Rindler *et al.*, 1985; Keller *et al.*, 1984).

Finally, methods have also been devised which involve the staining of samples on negative-staining grids (Beesley, 1985). Samples containing, for example ciliated bacteria are adsorbed on to formvar- and carbon-coated grids. Specific antisera are then used to identify the location of specific antigens (the pili in this case). A similar method has been described in which affinity-purified antibodies are used to prepare antibody-coated, formvar grids. These "immunoaffinity" grids have been used to specifically isolate neurofilaments from crude homogenates of nerves (Willard *et al.*, 1980). Affinity antibodies could then be used to increase the filament diameters by immunobinding of specific antibodies, also showing that the antigen was distributed along the entire length of the neurofilament. Similar procedures have been used to decorate intermediate filaments in intact cytoskeletons of cells grown on formvar-coated gold grids (Henderson and Weber, 1981).

# 5
# Production and Use of Monoclonal Antibodies

## 5.1. General Introduction

The production and use of monoclonal antibodies constitutes a major revolution in immunochemistry, witnessed by the explosion in the number of laboratories engaged in this work. Only with the benefit of considerable hindsight will the full implications of the development of techniques to obtain continuous cell lines which produce homogeneous antibodies, by fusing mouse myeloma cells to spleen cells from immunized animals, be realized. Certainly, Köhler and Milstein (1975) have launched immunology on a dramatic new course of tremendous potential.

In this chapter we bring together the techniques for the production of monoclonal antibodies with the areas of intended application in biochemistry, cell biology and molecular biology. The majority of antigenic determinants (epitopes) of interest which will be considered are found on macromolecular antigens, particularly proteins or protein conjugates, although some attention will be given to nucleic acids as macromolecular antigens. For completion, the use of monoclonal antibodies to haptenic derivatives will also be considered.

The methods of production of monoclonal antibodies and the use of monoclonal antibodies in life sciences will be critically appraised in terms of advantages and disadvantages over polyclonal antibodies and practical difficulties in production and use.

Particular attention will be paid to the objectives and philosophy of researchers when embarking upon monoclonal antibody production and use. The growth in the production of monoclonal antibodies is almost exponential. Examination of the literature reveals that the principles behind many of the new exciting areas of application of monoclonal antibodies were established in earlier papers on polyclonal antibodies and that subsequently there has been an explosion of production of monoclonal antibodies for uses based on these principles throughout the life sciences. In this chapter we attempt to identify these areas of use so that the guiding principles in the production and use of monoclonal antibodies can be defined and analysed. There is clearly no point in this chapter in simply listing the thousands of papers describing monoclonal antibody production and use which are to be found in cumulative form in bibliographic texts on library shelves.

As mentioned earlier, this chapter will concentrate on proteins and protein conjugates, e.g. glycoproteins as macromolecular antigens for monoclonal antibody production. The specific macromolecular antigens of interest which have been considered are shown in Table 5.1. These macromolecular antigens are chosen for their interest to biochemists, and cell and molecular biologists. Monoclonal antibodies to epitopes on these macromolecular antigens have found applications in the areas described in Table 5.2.

The principles and objectives underlying the production and use of monoclonal antibodies to macromolecular antigens will be analysed in the

**Table 5.1** Monoclonal antibodies and macromolecular antigens

1. Enzymes and proteins
2. Membrane antigens (including receptors)
3. Membrane antigens (cell-surface markers)
4. Viruses and cell-transforming proteins
5. Chromosomes, genes and modified DNA
6. Clinical radioimmune assays
7. Cloning and screening of recombinant expression systems
8. Antibody microinjection

**Table 5.2** Uses of monoclonal antibodies

1. Protein structure
2. Immunoaffinity chromatography
3. Radioimmune assay
4. Histocompatibility testing
5. Microbial and parasitic disease
6. Viral disease
7. Differentiation antigens
8. Neurochemistry
9. Receptor studies
10. Cytoskeletal studies
11. Chromatin structure
12. Recombinant DNA research
13. Studies of the cell surface and surface–cytoskeletal interactions
14. Target-cell isolation from a cell mixture, e.g. by fluorescent activated cell sorter (FACS)
15. Assessment of intracellular antigen function, e.g. by microinjection of $M_{ab}$s
16. Studies of cell–cell interactions
17. Use of anti-idiotypic $M_{ab}$s to purify and analyse "intractable" cell antigens
18. Drug targeting
19. Clinical use as antitoxins

order shown in Table 5.1. For brevity and conciseness the uses of monoclonal antibodies to each group of macromolecular antigens will be presented in tabular form and textually analysed to an extent determined by the general usefulness of the objectives and pitfalls described in the articles.

Consideration of Table 5.2 will show the extent to which monoclonal antibodies have found application in the life sciences. Table 5.2 serves to draw together clearly identifiable uses of monoclonal antibodies, most of which will be described in detail later in the chapter. Inspection of Table 5.2 shows that there is something for everyone, which is an indication of the widespread application of these new homogeneous immunochemical reagents in the life sciences.

For simple convenience, this chapter will be divided into two parts: Section 5.2 will consider techniques in the production of monoclonal antibodies, and Section 5.3 will consider the uses of monoclonal antibodies (summarized in Tables 5.1 and 5.2).

There have already been several excellent reviews on the production of monoclonal antibodies (Yelton *et al.*, 1978; de St Groth and Scheidegger, 1980; Galfre and Milstein, 1981), their properties, and some applications (Staines and Lew, 1980; Edwards, 1981).

A monoclonal antibody will be abbreviated as $M_{ab}$ (plural $M_{ab}$s).

## 5.2. Production of Monoclonal Antibodies

### 5.2.1. Introduction

Methods for the preparation of $M_{ab}$s have been reviewed several times (de St Groth and Scheidegger, 1980; Göding, 1980; Oi and Herzenberg, 1980; Edwards, 1981; Galfre and Milstein, 1981; Mayer and Billett, 1985), and Kennett *et al.* (1981) have prepared an excellent book on the subject. This section assesses the methodology, including more recent developments and our own experiences with the technology, placing special emphasis on the routine methods used for screening antibodies, in particular antibodies to protein antigens, and considers some of the problems encountered.

### 5.2.2. Background

A $M_{ab}$ is the product of a single clone of B lymphoid cells; myelomas, tumours of an antibody-producing cell, thus secrete $M_{ab}$s. Köhler and Milstein (1975) devised a method of constructing hybrid myelomas which secreted antibodies to antigens of choice. In their technique a myeloma cell line is fused to spleen lymphocytes secreting useful antibodies (probably activated plasma cells) to produce permanent hybrid myeloma (hybridoma) lines secreting defined $M_{ab}$s. These hybridomas are thus a source of very specific antibodies, each antibody having precise binding properties, and since they can be grown in continuous culture and also as ascites tumours in animals, they provide an unlimited supply of antibodies. The crucial feature of this fusion is that most hybrids continue to produce all of the immunoglobulin polypeptide chains synthesized by each of the parent cell lines, i.e. expression of the immunoglobulin chains is codominant. In the technique the spleen cells are immortalized, but it does mean that if the myeloma line also expresses immunoglobulin chains the hybrid will express these also, so that a mixed molecular immunoglobulin species will be secreted. To overcome this problem, stock myeloma lines have been selected which have lost the ability to produce their own immunoglobulins, or at least the immunoglobulin heavy chain (see later). In addition, mutant myeloma lines are used; usually these lack hypoxanthine phosphoribosyl transferase, or thymidine kinase, enzymes required in one of the salvage pathways of nucleotide biosynthesis. These cells die in medium containing hypoxanthine, aminopterin and thymidine (HAT medium) (Littlefield, 1964) because aminopterin (a toxic analogue of folic acid) blocks the *de novo* biosynthesis of purines and pyrimidines, and the cells are not able to utilize the exogenous supply of hypoxanthine or thymidine (see Fig. 5.1) (Oi and Herzenberg, 1980). However, if

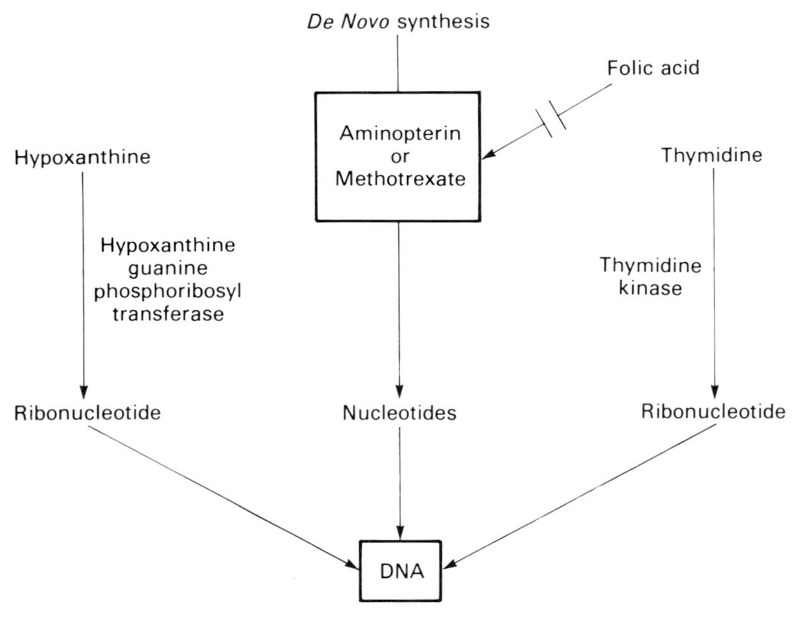


**Fig. 5.1** Basis of HAT selection.

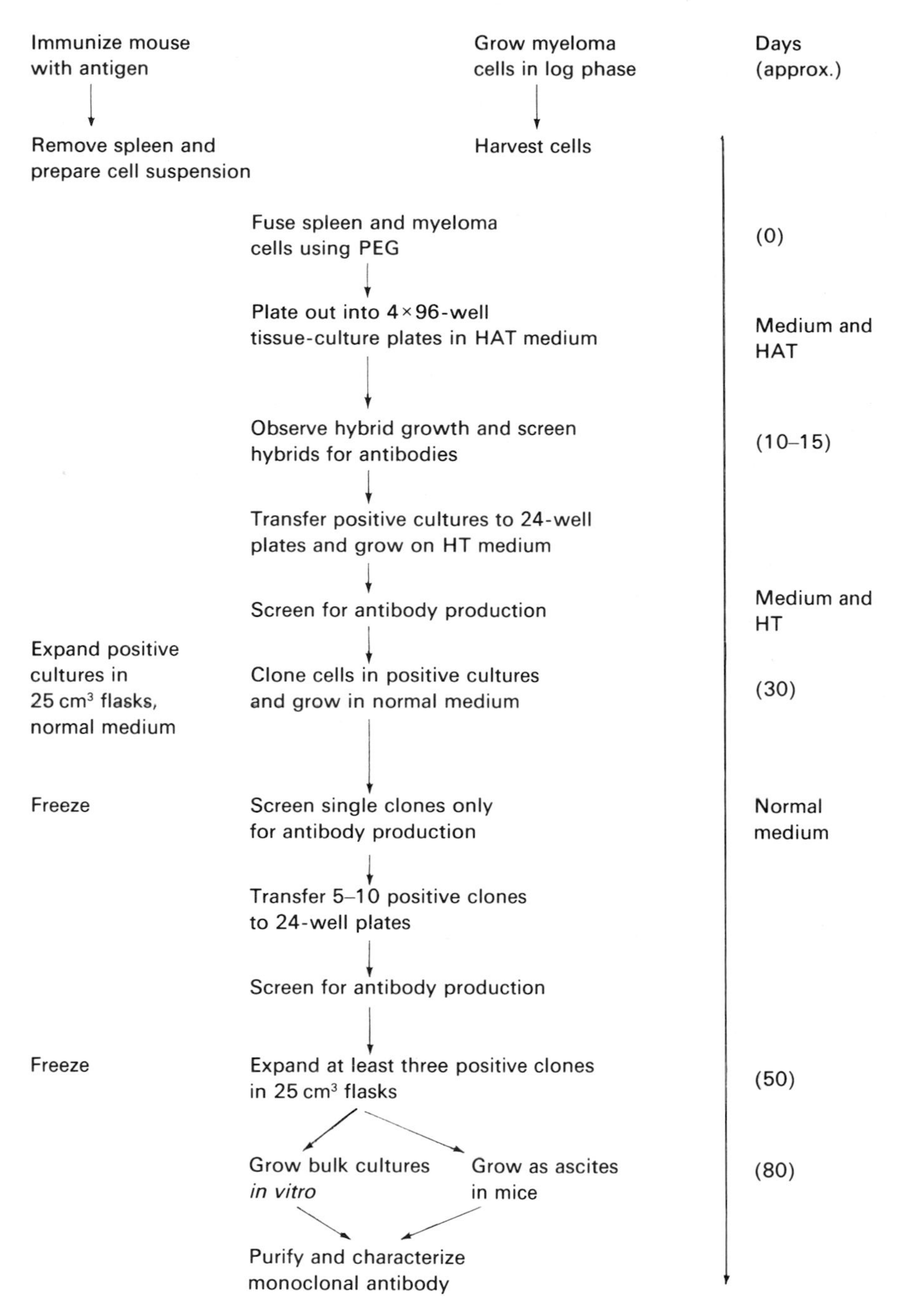


**Fig. 5.2** Production of monoclonal antibodies. *HAT = medium containing hypoxanthine, aminopterin and thymidine. HT = medium containing hypoxanthine and thymidine.

mutant myeloma cells fuse with spleen cells, which are wild-type, the resulting hybridomas can utilize hypoxanthine and thymidine and grow in HAT. There is no positive selection against unhybridized spleen cells but most of them die within a week.

The exact procedure used for the development of useful hybridomas varies between laboratories, but the general protocol is standard (see Fig. 5.2, showing production of mouse × mouse hybridomas). Most of these steps will be considered in more detail in the next section. Note, however, that the method is very time-consuming, involving approximately 3–4 months of continuous bench work from immunizing the animals to when stable cloned lines are established. Often it takes longer than this to make the antibodies, and then, of course, the antibodies need to be purified and characterized more fully. Indeed, a wide range of experience, materials and equipment is needed, and the task should not be taken on lightly.

### 5.2.3. Important properties, advantages and disadvantages of monoclonal antibodies

For convenience, some advantages and disadvantages of monoclonal antibodies are assembled in Table 5.3. Clearly, anticipation of these features of monoclonal antibodies may influence choice of a monoclonal or polyclonal route in antibody production.

**Table 5.3** Properties of monoclonal antibodies

1. Precise specificity
2. Unlimited availability
3. Wide range of binding affinities
4. Non-reaction in classical precipitation reactions with single $M_{ab}$s
5. Lack of complement activation by single $M_{ab}$
6. Monoclonal IgM can aggregate on storage
7. Many $M_{ab}$s to globular proteins are conformationally directed
8. Many $M_{ab}$s to membrane proteins detect epitopes in denatured antigen

### 5.2.4. Problems of production

Two main continuing problems plague the production of $M_{ab}$s. The first is the low yield of hybrids. Sometimes this is because the myeloma cell line has been subject to spontaneous genetic drift. However, not all lines are equivalent and, for example, some sublines of NS1 have been reported not to fuse adequately. Fresh cells should thus be obtained from a successful laboratory. Low hybrid yields can also be the result of using a poor immunogen. This problem can sometimes be overcome by changing the immunization technique or mouse strain.

The second problem, which is often more worrying, is the instability of hybrids, whether this leads to complete loss of hybrids or loss of antibody production. This is probably due to a number of factors. For example, if the antibody-forming hybrid is one of a number of hybrids in a well, it may be overgrown by non-secretors. To overcome this problem, after fusion one should aim to plate out cells so that single clones are isolated directly (de St Groth and Scheidegger, 1980).

Even if one has a cloned hybrid line, there may be sudden or gradual loss of antibody production. The reason for this is less clear, but it is no doubt due in part to the loss of heavy- or light-chain synthesis associated with chromosome loss and segregation of genes. Sometimes a subpopulation within an apparently stable clone is responsible and the only way to attempt recovery is by repeated cloning. It is worth noting that Yelton *et al.* (1978) have found subclones of apparently stable populations of antibody-forming hybrids which are heterogeneous with respect to the amount of immunoglobulin produced. They found that some subclones, the "maxi-producers", devoted a higher percentage of the protein synthetic activity to antibody than did the "mini-producers", and that the "maxi-producers" are more likely to die after a number of generations in culture. We have also found that very good secretors are more likely to die, and we suspect, in agreement with Yelton *et al.*, that the "maxi-producers" are terminally differentiated.

Another reason for loss of antibody production during the early stages may be that other cells in the "cocktail", for example macrophages (feeder cells) and myeloma hybrids, can suppress some antibody-secreting hybrids while facilitating the growth of other secreting hybrids. Finally, mycoplasma contamination can also cause cultures to die out gradually. These hybrids can be cleaned by injecting into animals, as for the production of ascites; mycoplasma-free cells are then recovered when the ascites is drained (Edwards, 1981).

### 5.2.5. Production protocol

#### 5.2.5.1. General materials

Methods used in the author's laboratory (RJM) are summarized below.

### 5.2.5.2. Choice of media

The most commonly used media are Dulbecco's Modified Eagle's Medium (DMM) and RPMI-1640, and some workers supplement the medium with glucose ($25.0\ mol/m^3$) (Kennett *et al.*, 1981, p. 364), sodium pyruvate ($0.5$–$1.0\ mol/m^3$) and glutamine ($2\ mol/m^3$) (Oi and Herzenberg, 1980). We find that with fresh media the addition of glutamine is not needed and, indeed, it inhibits the growth of the mouse myeloma line NS1. Addition of glucose may be beneficial at high cell densities. Tissue-grade water is needed and it is essential that only fresh media are used; we make fresh media every month. We find that media prepared directly from powder are superior to those bought as 1 × or 10 × concentrated solutions.

### 5.2.5.3. Choice of serum

Serum must be chosen carefully since it is so variable. Most workers use foetal calf serum (FCS) since it supports cell growth at low density, gives a high frequency of fusion and has a low immunoglobulin content, and thus does not interfere with screening assays. Heat inactivation is not usually necessary but it may be required, e.g. when assays involve complement lysis. Different batches of serum must be tested by measuring cloning efficiency. Myeloma cells growing in log phase are diluted to 10, 50 and 100 $cells/cm^3$ in medium containing test serum(s) or control serum and 10% and 100 $mm^3$ aliquots of each cell suspension are dispersed into replicate wells (say 24) of a 96-well microtitre plate. These are incubated for 1 week without disturbing in a 5% or 10% $CO_2$ humid incubator at 37 °C. Wells are then examined and the number of clones counted. A good serum lot will support over 90% cloning efficiencies of NS1, but sera yielding 70–80% cloning efficiencies are satisfactory.

For the maintenance of myeloma cells in log phase growth, serum is used at 10%; for the fusion, selection and cloning we recommend medium containing a minimum of 15% FCS; when hybridoma cells are well established they can be kept in 5% FCS.

Several companies selling serum have now recognized the problems of batch variation and are either issuing a "description" of their sera or selling special batches specifically for hybridoma work. Synthetic substitutes for serum have been developed and used by some laboratories (Chang *et al.*, 1980; de St Groth, see Wingerson, 1983), based on hormone or lipid mixtures, but they are used by only a very small group of workers, partly because the lipids are not commercially available. However, a product called Nu-serum has recently been put on the market by Collaborative Research, and one called HB101 by Flow Laboratories, and if, as is claimed, they can support hybridoma growth, they will simplify the purification of antibodies.

#### 5.2.5.4. Contamination control

A good sterile technique is obviously essential for this type of work, but sometimes cultures become contaminated with either bacteria, fungi or yeasts. In our laboratory, penicillin and streptomycin are routine additives to media; these are added at a final concentration of 50–100 units/$cm^3$. Kanamycin sulphate is sometimes added, but mainly it is held in reserve to control outbreaks of penicillin-streptomycin-resistant bacteria in important cultures. We have found that while antifungal agents (nystatin and fungizone) can control some forms of yeast, they are toxic to all our hybridoma cells. Yeasts are usually confined to isolated wells and do not spread, but once fungi have sporulated, the spores spread very rapidly both within cultures and within incubators. It is therefore essential to check important cultures as often as possible, preferably daily, and to destroy infected cultures immediately. We have no experience of mycoplasma infection but some at least are controlled by gentamycin. If mycoplasma are suspected, cultures can be analysed using commercially available kits based on either DNA-staining (Hoechst stain, e.g. from Flow) or cytotoxicity (Mycotect, BRL), or kits containing $M_{ab}$s to the five major mycoplasma species (Mycospec, BRL). Alternatively, cultivation tests, which are usually the most reliable, can be undertaken using either the mycoplasma isolation kit sold by Flow or in collaboration with an experienced mycoplasma laboratory (e.g. PMLS, Mycoplasma Reference Laboratory, Bowthorpe Road, Norwich NR2 3TX, UK).

#### 5.2.5.5. Plastics

Plastics vary in quality and we have found that some hybridoma cells dislike being moved from one brand to another. If this transfer occurs in the early stages, before cells are established, it can cause the complete loss of a culture.

### 5.2.6. Choice and maintenance of myeloma line

It is not known how similar the continuous cell line has to be to the spleen lymphocyte, the other parental cell, but to date only myeloma cells have been used. The myeloma stock is usually but not always (Tables 5.5–5.11) of the same species as the immunized animal. The most commonly used species is the mouse, partly because mutant rat myelomas were developed later, and partly because non-secreting rat myeloma mutants are still not available. All mouse myelomas are derived from the BALB/c strain (see Table 5.4) and they are all equal in general performance except Sp2/0, which is difficult to grow and gives unstable hybrids. Sp2/0 is a fast grower but cannot thrive at temperatures below 35 °C (de St Groth and Scheidegger, 1980).

The main advantage of using myeloma cells and spleen cells from the same species is that hybridomas can be grown as ascites in the animal. Interspecies hybridomas prepared between rat spleen cells and mouse myeloma cells are quite satisfactory in culture (see Tables 5.5–5.11), but often they cannot be grown in conventional rats or mice. Athymic or immunosuppressed animals (e.g. by total-body irradiation) thus have to be used to achieve successful growth of interspecies hybridomas *in vivo* (Kennett *et al.*, 1981). The rat × rat combination is newer, and derivation of the hybrids appears to be more difficult than with the mouse lines. However, it has been suggested that rat × rat hybrids are more stable than mouse × mouse hybridomas. For example, Lachmann *et al.* (1980) found that on cloning rat × rat hybrid cultures which were difficult to grow, 100% active clones were isolated. This may be because the percentage of growing hybrids expressing spleen immunoglobulins is 60% when mouse myeloma parental lines are used and over 90% with rat (Galfre and Milstein, 1981).

The other major consideration when choosing myeloma lines is which animal gives the best immune reaction; indeed this may be the determining factor.

The recently developed rat myeloma YC6/41 allows the production of bispecific monoclonal antibodies which can be used in immunohistochemistry.

The human myeloma lines available are noted in Table 5.4, for completeness. Unfortunately, human myelomas are difficult to grow in tissue culture and human hybridomas have low growth rates and poor immunoglobulin secretion (Sikora and Neville, 1982).

There is no doubt that the derivation of hybrids is only successful if the myeloma cells used for fusion have high viability (>90%). This means that cells must be kept in log-phase growth for at least 1 week prior to fusion. Mouse myeloma cells should be kept at a density of $3–8 \times 10^5$ cells/cm$^3$ by splitting every 2–3 days; we have found that they can be grown successfully in stationary suspension cultures. Our laboratory has no experience of rat myelomas but, according to Galfre and Milstein (1981), it may be essential to grow them in spinner cultures, and others (Gunn, personal communication) have confirmed this finding.

**Table 5.4** Myeloma lines used in hybridoma production

| Species | Line | Derived from | Immunoglobulin produced | Reference |
|---|---|---|---|---|
| Mouse | P3 × 63/Ag8 | P3K | MOPC 21 $IgG_1$ (k) | Köhler and Milstein (1975) |
| | P3NS1/1 Ag4.1 | P3K | k chains (not secreted) | Köhler and Milstein (1976) |
| | Y63/Ag8.653 | P3 × 63/Ag8 | None | Kearney *et al.* (1979) |
| | Sp 2/0 Ag14 | Hybrid Sp2[a] | None | Schulman *et al.* (1976) |
| | FO[b] | Sp 2/0 | None | de St Groth and Scheidegger (1980) |
| Rat | Y3 Ag.1.2.3 | R210.RC.Y3 | S210 k chain | Galfre *et al.* (1979) |
| | YC6/41(Y4)[c] | Y3 Ag.1.2.3 | | Milstein and Cuello (1983) |
| Human | LICR-LON-HMy2 | ARH 77 | IgG | Edwards *et al.* (1982) |
| | SKO 007 | | IgE | Kaplan and Olsson (1980) |
| | GM1500 6TG A12 | | IgG | Croce *et al.* (1980) |

[a] Sp2 is a hybrid myeloma prepared with × 63/Ag8 and a spleen cell from a BALB/c mouse immunized with red blood cells.

[b] Self-fused Sp 2/0.

[c] A clone which was recently derived from Y3 and spleen cells from a rat subjected to prolonged immunizations with horseradish peroxidase. This HAT-sensitive hybridoma secretes anti-peroxidase antibody.

### 5.2.7. Choice of animal/immunization

Most people have used either BALB/c mice or LOU rats as a source of spleen cells and, unless there are specific problems (see later), these are the best partners to the myeloma cells available. Naturally, the immune state of the animal from which the spleen is taken determines whether secreting hybrids are obtained. Since individual animals may respond differently, more than one individual should be tried; in addition, more than one immunization protocol should be tried. Workers do use other strains of animals; for example, Adolf *et al.* (1980) used outbred albino mice, and Choo *et al.* (1980) used Wistar rats.

It is not yet clear which type of spleen cell actually fuses with the myeloma cell to produce an antibody-secreting hybrid, but it is generally believed that blast and/or plasma cells are involved. It is the number of activated blast/plasma cells available for fusion that is important, and it appears that recently activated B cells are larger in size (Stahli *et al.*, 1980). This idea is further supported by the general experience that the best time for fusion is 3 or 4 days after boosting, when antigen-induced proliferation is strong (Göding, 1980), and not at the peak of antibody secretion, namely 7–8 days after boosting. There is some evidence that hyperimmunization may not be beneficial (Oi and Herzenberg, 1980) and if hyperimmune animals are used they should be rested (3–4 weeks) before receiving the final antigen boost prior to fusion.

In general, cell-surface antigens are extremely immunogenic when presented on intact cells and no adjuvants are needed. Most workers prime with between $1 \times 10^6$ and $1 \times 10^7$ cells intraperitoneally (i.p.) and then boost with the same dose 3 weeks afterwards; fusion is undertaken 3 days later (Bartlett *et al.*, 1981; Birrer *et al.*, 1981). Membrane preparations are often (Gee *et al.*, 1983) but not always (Brown *et al.*, 1983) used for immunization in the presence of adjuvant, using complete adjuvant for the prime immunization and incomplete adjuvant for the boost.

Young *et al.* (1979) were successful in producing $M_{ab}$s to the carbohydrate moiety of a soluble glycolipid, ganglio-*N*-triosylceramide (asialo $GM_2$) by immunizing mice with asialo $GM_2$ non-covalently adsorbed to naked *Salmonella minnesota*. Others (e.g. Eisenbarth, 1981) have generated $M_{ab}$s to the Forsmann antigen (a glycolipid-*N*-acetyl galactosaminosyl-(al-3)-*N*-acetyl-galactosaminosyl (Bl-l)-ceramide) by immunizing mice with viable cells.

### 5.2.8. Weakly immunogenic antigens

The antigens are considered operationally as "soluble" and "poorly soluble" macromolecular antigens.

### 5.2.8.1. Soluble proteins

Some soluble proteins can be weakly immunogenic in aqueous solution, and the addition of some sort of adjuvant is essential. Kearney *et al.* (1979), for example, used 100 μg of alum-precipitated nitrophenylated (NP) chicken globulin mixed with inactivated pertussis vaccine as antigen and, following a single immunization, obtained a high frequency of hybrids against the NP chicken globulin. Although several investigators have been successful with this type of protocol, a more complicated regimen is often needed for soluble proteins and, if possible, the final immunization should be given intravenously (i.v.) (Galfre and Milstein, 1981). An impressive immunization method for soluble proteins has been used by Stahli *et al.* (1980), and others have now followed his recommendation (Kramer *et al.*, 1980; Borrebaeck and Etzler, 1981). The basis of Stahli *et al.*'s method is the inclusion of several immunizations with high doses of antigen in saline on each of the last 4 days before fusion, one at least being i.v. This protocol appears to generate a large number of activated blast/plasma cells, as shown by an increase in size of the lymphocytes. A suitable protocol for soluble proteins would thus involve one or two i.p. injections of 10–60 μg antigen (perhaps more for rats) emulsified in complete Freund's adjuvant or precipitated with alum and supplemented with $10^9$ heat-killed *B. pertussis* organisms at monthly intervals followed by i.v. and/or i.p. injections of 50–500 μg antigen in saline each day for 4 days before fusion.

### 5.2.8.2. Poorly soluble antigens

Weakly immunogenic macromolecules are not restricted to soluble antigens. Indeed, many macromolecules which are difficult to solubilize are also weakly immunogenic. Different strategies have been adopted to produce $M_{ab}$s to weakly immunogenic macromolecules. Two conceptually different approaches are worth describing.

*In vitro* immunization, i.e. presentation of antigen *in vitro* to a mixed-cell preparation containing spleen cells, thymocytes and macrophages, followed by fusion of cells with myeloma cells, can result in $M_{ab}$s to highly conserved or weakly immunogenic determinants (Pardue *et al.*, 1983). It is claimed that very little antigen is required, i.e. ng quantities (Luben *et al.*, 1982), although considerably more protein (1–50 μg) of protein species conserved through evolution may be required (van Ness *et al.*, 1984). It is worth noting that *in vitro* immunization may well be the only procedure to eventually generate hybridomas secreting human antibodies.

There are several problems with the *in vitro* immunization technique: the most significant is competition of introduced antigen with serum components in the rabbit or foetal calf serum required to support B cell division. Serum-free *in vitro* stimulation systems have been described, e.g. Iscove and Melchers (1978), but these systems rely on serum protein replacements which are themselves immunogenic and generally support B cell proliferation less well than serum-containing media. A final problem concerns the presentation of insoluble antigens to the *in vitro* stimulation reaction in a form which can be recognized by lymphocytes or monocytes. Obviously, presentation in cytotoxic solvents must be avoided: insoluble precipitates which cannot readily interact with the majority of cellular species in the stimulation reaction have been used. Finally, suppressor T cell activity in the *in vitro* immunization system should be avoided if such T cell activity is present in the stimulation reaction (Click *et al.*, 1972).

van Ness *et al.* (1984) have recently described procedures which may overcome some of the problems indicated above and result in the production of $M_{ab}$s to weakly immunogenic antigens. The general strategy is outlined in Fig. 5.3. There are several critical features of the method. The antigen(s) of interest (a chromosomal scaffold protein (SC-1); non-histone chromosomal proteins; and vimentin) were coupled individually to fumed silica to allow the best presentation of the insoluble poorly immunogenic species to cellular elements in the *in vitro* immunization. Both soluble and insoluble antigens are easily irreversibly bound to silica, giving antigen-coated silica which is easy to disperse, e.g. by sonication, and induces lymphocyte aggregation, enhancing cell–cell contact and antigen binding. Silica itself may act as a macrophage attractant or activator, triggering the possible macrophage requirement in the T cell-mediated response. Silica may also be mitogenic.

The coupling of antigen to silica was achieved as follows. Proteins were subjected to polyacrylamide gel electrophoresis in the presence of sodium dodecyl sulphate and mercaptoethanol. Gel slices containing antigens of interest were crushed and mixed with sodium dodecyl sulphate (2%) and 2-mercaptoethanol (1%) in 5 mM Tris-HCl, pH 7.2. Crushed slices were left for 24–48 h and then heated at 100 °C for 2 min. The filtered supernatants were used with the silica. Fumed $SiO_2$ (10–1000 $\mu$g) was suspended in water by sonication and added to the solubilized proteins for 12–24 h at room temperature. The antigen-coated beads were recovered by centrifugation (10 000*g* for 5 min), washed with distilled water and sterilized by irradiation or autoclaving. More than 90% of all $^{125}$I-labelled non-histone proteins or vimentin bind to the silica.

The period of culture of spleen cells (Fig. 5.3) with antigen-coated beads and macrophages (induction period) defines the subsequent immune response of *in vitro* immunization. Induction occurs in the absence of competition by serum components. Only the most immunogenic antigens appear to elicit an immune response in the presence of foetal calf serum. A high cell

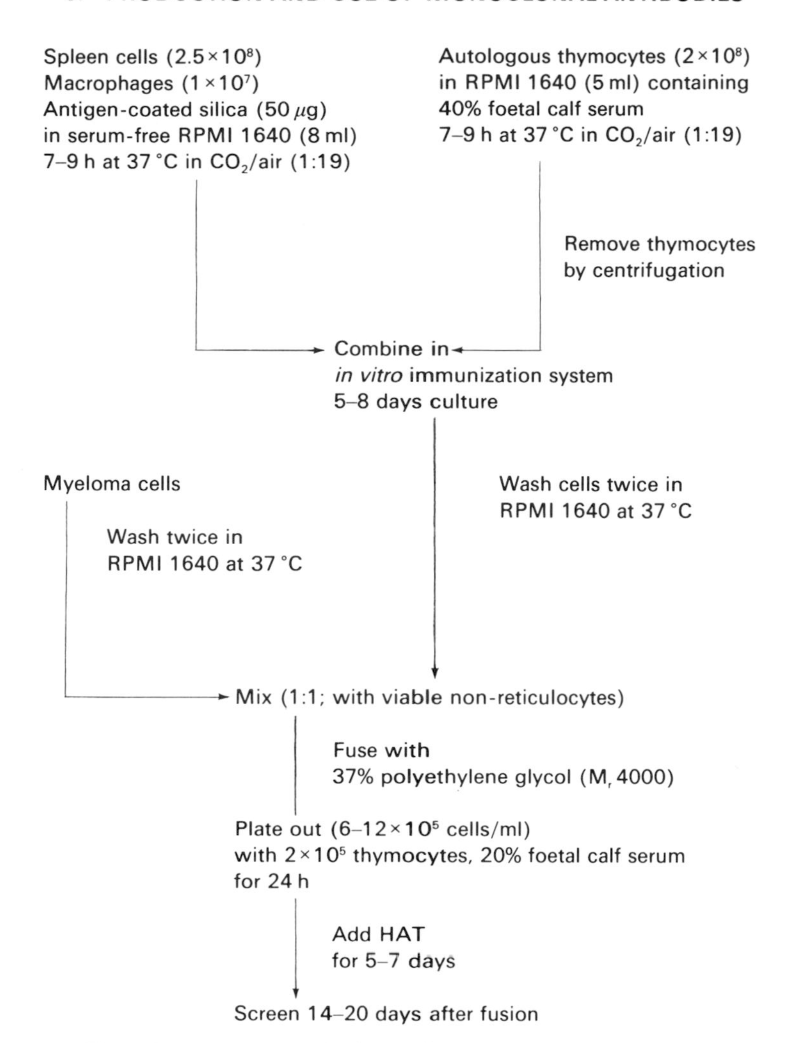


**Fig. 5.3** *In vitro* immunization and monoclonal antibody production.

concentration may be required for lymphocyte stimulation but a narrow range of cell concentrations appears not to be critical. Perhaps the silica-induced cell clustering is responsible.

A mixture of anti-suppressor T cell drugs and non-specific growth factors was included during the stimulation period, e.g. cimetidine, hydrocortisone, dimethylhydrazine, mercaptoethanol, glutathionine, nucleosides and hypoxanthine; this may improve the immunization response. Thymocyte-conditioned medium (Fig. 5.3) is an absolute requirement.

Microgram quantities of poorly immunogenic antigens are required in these procedures.

Finally, the immune response elicited *in vitro* is a primary response which gives rise to hybridomas producing antibodies of the IgM subclass. IgM antibodies can present problems for certain immunochemical studies, e.g. low ascites titres, lack of binding to Protein A, labelling problems, lability on freezing and thawing, inefficient antigen precipitation, and penetration and diffusion *in vivo*. These problems have led to a search for an alternative strategy to produce IgG antibodies to weakly immunogenic antigens.

Frosch *et al.* (1985) have produced a $M_{ab}$ to the weakly immunogenic bacterial polysaccharide antigens, i.e. meningococcus group B and *E. coli* K1 polysaccharides (homopolymers of $\alpha(2\rightarrow 8)$ linked units of *N*-acetylneuraminic acid) by immunization of the immunologically hyperreactive NZB mouse followed by hybridoma production. The strategy produced a high-titre $IgG_{2a}$ $M_{ab}$ that is specific for the K1 and meningococcus B capsules. NZB mice were obtained from the Jackson Laboratory.

### 5.2.9. Other immunization problems

A problem often encountered is the presence of an immunodominant component in the antigen preparation; this may even be a very minor contaminant in a soluble "purified" antigen preparation. Timothy Springer, working on cell-surface differentiation in mouse, has devised a method of overcoming this problem, namely using one or more $M_{ab}$s to the immunodominant antigen(s) to purify the antigen preparation prior to immunization (see Kennett *et al.*, 1981, p. 185). One $M_{ab}$ is covalently linked to Sepharose and a solubilized-antigen preparation is fractionated on it, allowing the removal of the immunodominant antigen. Indeed, it may be possible to simply mix the antigen preparation with sufficient $M_{ab}$ to mask the immunodominant region and to inject this mixture.

Although drug-marked human myelomas are now available, technical problems with the production of human $M_{ab}$s still exist, including low fusion frequency, low antibody secretion and poor cloning efficiency. Preselection of lymphocyte-secreting antibodies of the desired specificity would facilitate the establishment of useful human hybridomas, but this is not an easy task, especially if the antigen has low potency or where immunization occurred at a remote time. However, one ideal approach, supported by Zurawski *et al.* (in Kennett *et al.*, 1981, p. 19) is to use $M_{ab}$s to activated B cells, for example a $M_{ab}$ to a human B cell differentiation antigen, to separate subpopulations of B lymphocytes, perhaps using the fluorescence-activated cell sorter. Improving the fusion frequency would also be beneficial and a new method of fusion has recently been suggested by Bischoff *et al.* (1982).

It was recently reported (Sikora and Neville, 1982) that D. Crawford has had some success in producing human $M_{ab}$s to influenza and antirhesus D antibodies with a method used by others (Zurawski *et al.*, in Kennett *et al.*, 1981, p. 19) over the past few years. He transformed peripheral blood lymphocytes from donors by infecting the Epstein–Barr virus (EBV), stimulated them *in vitro* with antigen, and, by early cloning, could select lymphocytes which secreted at least as much immunoglobulin as the best human hybridoma systems available. The utility of this approach may be limited since all transformed cells reported thus far stop producing antibody with time. However, several groups are now working on the possibility of initial EBV transformation and *in vitro* stimulation followed by immortalization of selected cells by somatic cell fusion in a hybridoma system.

### 5.2.10. Anti-idiotypic antibodies

The clonal theory of humoral antibody production *in vivo* is supported by the practical immortalization of single antibody-producing cells in hybridomas. Each $M_{ab}$-producing clone is derived from a single cell genetically programmed to synthesize and secrete a single antibody which is unique in terms of sequence and structural conformation in the complementarity binding region. The specific sequences define the idiotype of the immunoglobulin molecule as defined by the *in vitro* production of anti-idiotypic antibodies to the specific sequences. In turn, such an anti-idiotypic antibody has its own idiotype which can further generate an anti-anti-idiotype and so on. This concept is embodied in the idiotypic network theory of humoral antibody production (Jerne, 1974).

Cell-surface receptors for hormones, neurotransmitters and other ligands share several common features with antibodies, e.g. functional binding domains for ligands (cf. epitope), interaction together with a signal transduction element (cf. $F_c$ tail), and a binding site which can distinguish related ligands (cf. epitopes) in terms of binding affinities.

Such analogies are confirmed by the fact that anti-idiotypic antibodies directed against the "active site" of anti-hormone immunoglobulins may interact with the hormone receptors. This "cross-reactivity" has been explained by the ability of anti-idiotypic antibodies to mimic a hormone by acting as "internal images" of the immunogen (Jerne, 1974). These relationships are shown diagrammatically in Fig. 5.4. In this simplified model the antibodies to a ligand (L), raised to haptenically derivatized L, are *idiotypes*.

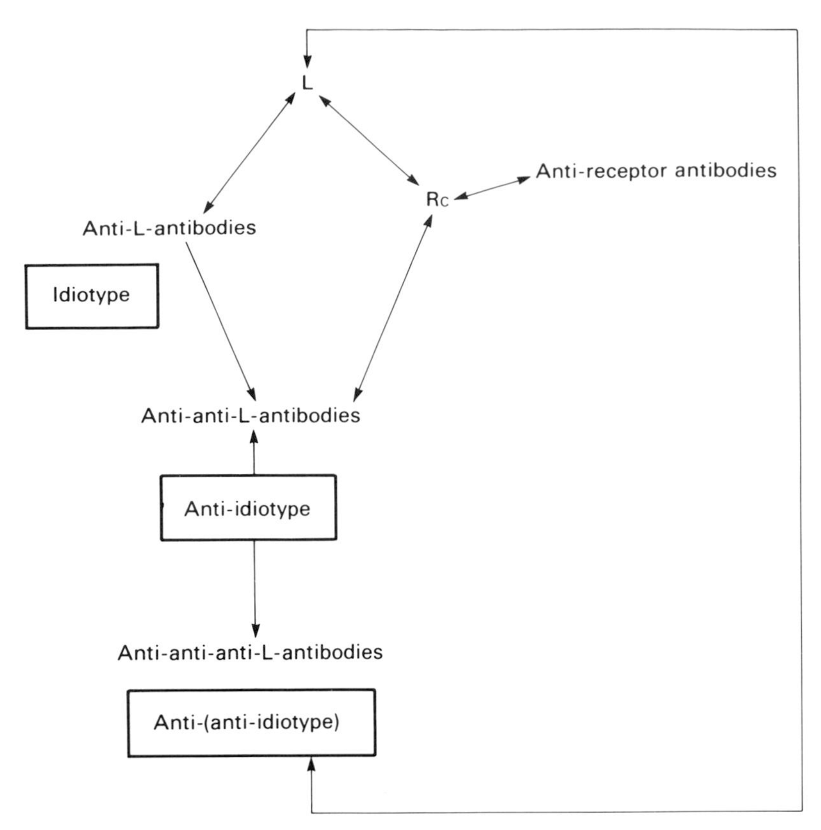


**Fig. 5.4** Anti-idiotypic network.

The antibodies produced *in vivo* to the idiotypes, i.e. anti-anti-L, are *anti-idiotypes*. In turn, further anti-(anti-idiotypes) etc. can be raised. The *in vivo* consequence of such an immunological network is the regulation of all antibody production, presumably by antigen–antibody binding and elimination from the body. The practical consequence is that the anti-L antibody and the receptor may bear binding sites of some similarity, i.e. internal images of the ligand. Therefore, the anti-anti-L antibodies may bear internal images for the idiotype of the anti-L antibodies and the receptor binding site, i.e. the anti-idiotypic antibodies may bind to the receptor in the region of the ligand binding site. Similarly, the anti-anti idiotype may have an internal image of the ligand L, and bind the ligand. Such relationships have been observed (Schreiber *et al.*, 1980; Sege and Peterson, 1978). One consequence of the multiple anti-idiotypic network is the amount a single (monoclonal) antibody *in vivo* will oscillate and eventually be damped by the anti-idiotypic responses, i.e. an anti-idiotypic response may be transitory.

One major practical and exploitable consequence of the idiotypic response is that antibodies to receptors can be produced. Normally, ligand receptors are found in very small amounts in cell surfaces, making their purification long and tedious. Anti-idiotypic antibodies which react with receptors therefore provide a means of purifying and characterizing a receptor, e.g. before microsequencing, oligodeoxynucleotide synthesis and probing recombinant DNA libraries.

Polyclonal and monoclonal anti-idiotypic antibodies have been raised which react with receptors (e.g. Strosberg, 1983; Guillet *et al.*, 1985; Amit *et al.*, 1986). Polyclonal anti-idiotype antibodies must be purified to remove anti-isotope antibodies, e.g. on IgG-affinity matrix, and possible anti-ligand antibodies on ligand-affinity matrix. Affinity-purified polyclonal and monoclonal antibodies are assayed by specific binding to the anti-ligand antibody and binding to cells containing the ligand receptor (see below). Anti-idiotypic antibodies are defined: (a) by the inhibition of ligand binding to the anti-ligand antibody in a dose-dependent fashion; (b) by the inhibition of ligand binding to membrane bound receptor, possibly non-competitively, which indicates binding of anti-idiotypic antibodies at the receptor at a site distinct from the ligand binding site; (c) by the anti-idiotype being isolatable on an idiotype-affinity matrix; (d) by the fact that anti-idiotypes *may* activate or inhibit basal or ligand-stimulated cellular effects, e.g. stimulation of adenyl cyclase; (e) by the anti-idiotype production *in vivo* being transient. It should be emphasized that idiotype structure is complex in that anti-idiotypic antibodies to anti-ligand antibodies may be generated which recognize precisely the ligand binding site, so blocking ligand binding directly, or which recognize an adjacent site, and which therefore may or may not block ligand binding sterically (see (b) above). Guillet *et al.* (1985) raised anti-idiotypic $M_{ab}$s to a monoclonal anti-alprenolol. Six anti-idiotypes were produced (defined by the reaction of [$^3$H]leucine radiolabelled anti-alprenolol with anti-idiotype ascites on nitrocellulose plates). Three anti-idiotypic $M_{ab}$s recognize $\beta$-adrenergic receptors. One $M_{ab}$ (IgM) binds to digitonin-solubilized receptors, and can therefore be used to immunoisolate $\beta_2$-adrenergic receptors (subunit 55 kD). The anti-idiotypic IgM stimulates receptor-bearing cells by activating adenyl cyclase. This activation is inhibited by propranolol.

Anti-idiotypic antibodies may have several uses besides the isolation of receptors, e.g. eliminating tumour cells or virally infected cells, or mimicking drug structure. However, the potential of this approach in the life sciences is arguably much larger, since if analogues of any substrate for a biologically active macromolecule can be produced and rendered haptenic, the possibility exists that $M_{ab}$s can be produced to anti-hapten antibodies which recognize the substrate binding site of the biologically active macromolecule and will bind effectively to this site. With luck, therefore, intractable biologically active macromolecules, for example membrane proteins such as cytochrome $P_{450}$s, or currently uncharacterized biologically active macromolecules, such as those having carcinogen metabolite binding sites, can be characterized and isolated.

### 5.2.11. Fusion, selection and maintenance of hybrids

Fusion methods are well documented (de St Groth and Scheidegger, 1980; Oi and Herzenberg, 1980; Galfre and Milstein, 1981; Kennett, in Kennett *et al.*, 1981, p. 365) and they involve the use of polyethylene glycol (PEG) as a fusing agent. Different authors use PEG of different molecular weights, e.g. PEG 1000 (Fraser and Venter, 1980; Frackelton and Rotman, 1980; Milne *et al.*, 1981), PEG 1500 (Galfre and Milstein, 1981; Kennett, in Kennett *et al.*, 1981), PEG 4000 (Beisiegel *et al.*, 1981), and different concentrations of PEG varying from 25% (Hoffman *et al.*, 1980) to the more common concentration of 50%. Dimethylsulphoxide (DMSO) is sometimes added to the PEG (e.g. Hoffman *et al.*, 1980; Borrebaeck and Etzler, 1981) to improve fusion, but de St Groth and Scheidegger (1980) found that the effect was minimal.

When preparing spleen cells, some workers, including us, prepare a single cell suspension simply by teasing pieces of spleen in medium using blunt forceps and removing the debris by a low-speed centrifugation for 5 min. Contaminating erythrocytes can be removed by lysis in ammonium chloride (Kennett, in Kennett *et al.*, 1981, p. 365; Milne *et al.*, 1981) but this is not really necessary. A conventional fusion mixture consists of a 10:1 ratio of spleen to myeloma cells (e.g. $10^8$ spleen cells plus $10^7$ myeloma cells) but some use a 5:1 ratio (e.g. Beisiegel *et al.*, 1981), some a 2:1 ratio (e.g. Frackelton and Rotman, 1980) and some a 1:1 ratio (e.g. Fraser and Venter, 1980).

It seems that the choice of fusion protocol need not be too critical, since all of them have been successful.

In our laboratory we use the method of Galfre and Milstein (1981), but myeloma cells are grown in RPMI medium and we find that our sample of PEG (BDH Chemicals Ltd, Poole, Dorset, UK, mol. wt 1500) lowers the pH of the medium and we adjust it to approximately pH 7.6 with NaOH. We find that, unless the pH is alkaline, fusion rates are very low. DMSO (5%) is added to the molten PEG, with the medium, so that if the PEG is diluted a little too quickly during the fusion procedure the effect is less harmful (de St Groth and Scheidegger, 1980). In addition, the final cell pellet is resuspended directly into HAT medium and is distributed into at least four 96-well microtitre plates, rather than 24-well (1 $cm^3$ wells) plates, as described by Kennett (in Kennett *et al.*, 1981, p. 367). Indeed, cells should be distributed into as many wells as possible so that clonal hybrids are produced, optimizing the frequency of surviving hybrids. Only about 10% of the wells should contain hybrids (de St Groth and Scheidegger, 1980), but we often find that, using our procedure, there are hybrids in over 90% of the wells.

Hybrid survival is also greatly improved by plating the cells on a feeder cell layer. This consists of non-dividing cells which provide a favourable environment for the hybrids (Göding, 1980). Macrophages (peritoneal, rat or mouse), irradiated thymocytes (mouse or rat) and irradiated 3T3 cells have been used at approximately $10^4$ cells/microtitre well and $10^5$ cells/$cm^3$ well (e.g. St Groth and Scheidegger, 1980; Greene *et al.*, 1980; Lachmann *et al.*, 1980).

A different method of fusing cells has been used in the production of human hybridomas by Bischoff *et al.* (1982). This involves electrofusion, a method pioneered by U. Zimmerman, and circumvents the use of HAT-sensitive myeloma cells. Fusion is observed under the microscope so that the resulting hybridoma cells can be identified and removed from the chamber using a micromanipulator. It claims to be particularly useful for human hybridoma production since, in certain electric fields, human lymphocytes do not fuse with each other. The method has yet to be tested effectively. Apparatus for this procedure is now commercially available.

### 5.2.12. Screening methods

Screening methods should be well worked out before a fusion is undertaken and should be fast, reliable, sensitive and relevant. Since $M_{ab}$s react with a single epitope they can only cross-link antigens into dimers and they generally do not work in the Ouchterlony double-diffusion assay, they do not immunoprecipitate their antigens, and they cannot be used in complement fixation and cytotoxicity assays. In addition, the concentration of useful immunoglobulin in culture filtrates is very low, being approximately 1–25 $\mu$g/$cm^3$ compared to 0.1 mg/$cm^3$ in hyperimmune serum (Edwards, 1981).

A point worth remembering is that we only get the $M_{ab}$s that we look for, and that, depending on the type of antigen and immunization schedule, less than 1% or more than 90% of the culture may produce antibodies. From our own experience in the laboratory we feel that extreme care should be taken over choosing a screening assay, ensuring that only useful hybrids are kept, and that selection should be undertaken as early as possible. The screen should be geared to pick out antibodies which can be used in the laboratory experiments. For example, if they are to be used for immunohistochemistry, marking special cells or special structures, they should be tested against cells or sections. It takes much more time to characterize an antibody than to perform another fusion. If specialized assays are too expensive or laborious, it may be worth performing a preliminary rapid test for antibody secretion *per se*, or streamlining the assays by pooling wells along rows and columns for the preliminary screening (Brown *et al.*, 1980).

Mechanical sampling devices are commercially available, ranging from 4-, 8- or 12-channel multi-dispenser pipettes to a 96-well sample-transferring system (Schneider and Eisenbarth, 1979).

Various methods are used to screen culture filtrates, but binding assays are particularly suitable for $M_{ab}$ work, because they are simple and sensitive while being applicable, in principle, to all classes of antibodies and to most antigens. Other assays will also be considered below.

### 5.2.12.1. Indirect trace binding solid-phase assays

In this type of assay an antigen needs to be bound to a solid support. Cell-surface antigens, of course, are naturally insoluble, while others are rendered insoluble by binding to a variety of surfaces, including plastic, filter paper, Sepharose or nitrocellulose paper. Tissue-culture filtrates are added to the bound antigen, the free antibody is removed, and the $M_{ab}$ bound measured by binding of a second, labelled antibody (see Fig. 5.5 for the principle of the method). The second antibody can be labelled using a radiolabel, an enzyme or a fluorescent probe; alternatively, labelled Protein A is used, either instead of a second antibody or in combination with it. The choice between the various labels is largely a matter of personal preference and the availability of measuring equipment (gamma counters, microtitration plate readers, etc.).

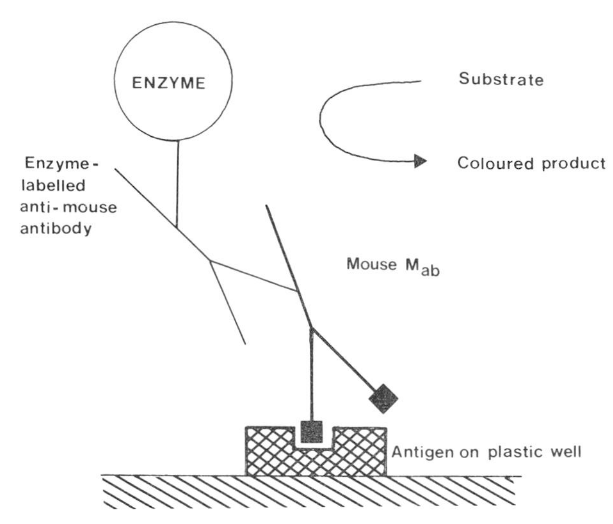


**Fig. 5.5** Indirect solid-phase (ELISA) for antibody production.

Antibodies to all sorts of antigens can be detected using this method, including proteins, e.g. ovalbumin (Bjercke *et al.*, 1981), pyruvate kinase (Hance *et al.*, 1980) (see Table 5.5), DNA (Haugen *et al.*, 1981), and lipoprotein (Patton *et al.*, 1982; Brockhaus *et al.*, 1982). The most common support is plastic. Flexible polyvinyl chloride (PVC) microtitre plates with U-shaped bottoms are the most popular. To prevent wastage of antigen, one has to ensure that maximum antigen binding is achieved, and sometimes different batches of plates show different binding capacity. In addition, different antigens bind optimally at different pH values; we routinely use pH 7.4. Many antigens adsorb well and tightly to the plastic, e.g. low-density lipoprotein (Patton *et al.*, 1982), but some do not, e.g. alkaline phosphatase (Slaughter *et al.*, 1981; Kennett *et al.*, 1981, p. 101). This problem can be circumvented by precoating the plates with either a specific polyclonal anti-antigen antibody (Slaughter *et al.*, 1981; Stocker and Heusser, 1979) or polylysine. The antiserum dilution used to precoat the plastic will need to be adjusted to give maximum antigen binding. When polylysine (1–5 mg/ 100 $cm^3$) is used to precoat the wells, the protein antigen can be fixed to it using glutaraldehyde (0.25% final concentration), and this is a particularly useful technique if one is looking for antibodies which will eventually react with fixed sections or cells, or which are inside cells.

After binding of the antigen to the PVC plates, non-specific adsorption must be prevented by blocking with a protein such as BSA (1%), gelatin (0.1%) or ovalbumin (1%). When polylysine is used it is advisable to block firstly with glycine (100 mM in 0.1% BSA—blocking is more effective at alkaline pH) for 30 min and then with BSA for 4 h at room temperature. A no-antigen blank should always be included since different $M_{ab}$s bind differently to plastic.

The concentration of antigen used to coat wells varies a lot; it can be as little as 10 ng/ml when preparations are fairly pure (e.g. Haugen *et al.*, 1981), and indeed it is better not to use a high concentration unless one is specifically looking for low-affinity antibodies.

If the antigen is bound to filter paper discs or Sepharose, activated with cyanogen bromide, one can specifically bind only the protein antigen in a mixture, and not glycolipids and carbohydrates, so that the assay can be made specific to protein antigens (McKay, 1980; Newman *et al.*, 1981).

An antigen can also be bound to nitrocellulose paper, and Hawkes *et al.* (1982a,b) have used a dot immunobinding assay to detect antibodies against synaptic plasma membranes. The antigen is dispensed on to small squares of nitrocellulose filter paper, allowed to dry to promote binding, and incubated with test antibody. Hawkes *et al.* detected specific antibody with horseradish-peroxidase-conjugate rabbit anti-mouse Ig, but radiolabelled second antibody could also be used.

If radiolabelled second antibody is used to detect bound $M_{ab}$ it is usually labelled with $^{125}I$ using conventional methods (e.g. see Galfre and Milstein, 1981), to a specific activity of approximately 2–20 $\mu Ci/\mu g$, diluting before storing in buffer containing 0.1% bovine serum albumin (BSA). $^{125}I$-labelled antibodies to mouse Fab (both $F(ab^1)_2$ fragments and intact antibodies) can also now be bought.

After incubation with labelled second antibody and extensive washing of the wells, the wells can be cut out (e.g. using a wire cutter, available from Biotec) and counted or subjected to autoradiography. Autoradiography is very simple and the plates are exposed to pre-flashed X-ray film in a sandwich between two intensifying screens (Young *et al.*, 1979; Weiler and Zenk, 1981).

An increasing number of enzyme-linked (ELISA) and fluorescein-linked species-specific second antibody reagents are coming on the market and they are particularly appealing since they exclude radioactive hazards, are fairly cheap and are stable (Amersham, Dakopatts, BRL and NEN each produce a wide range of these reagents). Enzymic reagents provide a quick visual result (often all that is required) when making positive/negative decisions when screening culture filtrates. Both $\beta$-galactosidase- and horseradish-peroxidase-linked antibodies are generally offered, and since animal tissues contain only very low levels of $\beta$-galactosidase activity this is the preferred enzyme when animal cells are used. Any peroxidase present in tissue sections can be blocked, and since its small size allows good penetration, peroxidase is often favoured for tissue sections. A more expensive reagent has recently been developed by BRL and Amersham involving a biotinylated secondary antibody which detected 2 ng amounts of antibody. It is a sandwich system in which the biotinylated secondary antibody is first added followed by a streptavidin bridge, biotinylated horseradish peroxidase and finally the substrate *O*-phenylenediamine. The streptavidin binds biotin with extremely high affinity and acts as the bridging molecule.

Another class of more sensitive ELISAs are termed ultrasensitive enzymic radioimmunoassays and they can be 100–1000-fold more sensitive than normal ELISA or radioimmune assay. In this case, bound second antibody–enzyme conjugate is measured using a radioactive substrate; usually $^{3}H$- or $^{14}C$-labelled substrates are used, which are less hazardous than $^{125}I$ (Hsu I-C. *et al.*, 1981; Hsu, S. *et al.*, 1981). However, they may be more inconvenient than other ELISA assays because of the problem of separating the product from the substrate, and the dependence on a liquid scintillation counter.

$^{125}I$-labelled Protein A can also be used instead of a secondary antibody, but it is worth noting that it does not bind to the $F_c$ region of all classes (notably IgM) or even to some IgG subclasses of mouse and rat immunoglobulins (e.g. it binds only weakly to mouse $IgG_1$ and not at all to rat $IgG_3$) (Surolia *et al.*, 1982).

Special problems may be encountered in the ELISA method with antigens which are insoluble or poorly soluble in aqueous solutions. Poor water solubility means that uniform coating of the wells in microtitre plates is difficult. Furthermore, with integral membrane proteins which are solubilized in non-ionic detergents, the surfactant properties of the detergent might diminish the interaction between detergent-solubilized antigen and the plastic surface. Several successful attempts have been made to overcome these problems.

Integral membrane proteins solubilized in non-ionic detergent, e.g. Triton X-100, can be conveniently bound to plastic for ELISA screening by additions of the histochemical fixative Bouin's fluid, followed by brief centrifugation and washing. In this way up to 2 $\mu$g/well of cytochrome oxidase could be bound to each well for $M_{ab}$ screening (Noteboom *et al.*, 1984).

Another problem which may increasingly occur as recombinant DNA technology develops is the production of $M_{ab}$s to bacterially expressed antigens, which are frequently insoluble. The production of insoluble aggregates containing cloned gene products in *E. coli* may be a physiological response to the production of large amounts of protein or the physicochemical consequence of the production of large amounts of fusion gene products. Such aggregates can be solubilized in sodium dodecyl sulphate (and urea). Therefore, proteins (or their subunits after gel electrophoresis) in such detergent/urea mixtures may be the material for immunization and ELISA screening. One interesting example concerns the screening of hybridoma products for antibodies to sequences of the oncogenes v-*erb*-B and v-*myb*. The sequences are expressed in fusion proteins in *E. coli*. The objective of the work was to obtain $M_{ab}$s to the oncogene protein sequences which could be used to detect and characterize the oncogene or proto-oncogene products in cells (cf. Chapter 6). The method developed for the ELISA detection procedure (Evan, 1984) is outlined in Fig. 5.6. The principle of the method is to cross-link the antigen in sodium dodecyl sulphate to the plastic wells with glutaraldehyde for use in the ELISA screening procedure. Approximately 0.1 $\mu$g of antigen is fixed to each microtitre plate well. Since the antigen is cross-linked to the plastic, the microtitre plates can be washed (with low pH) and re-utilized (up to five times).

#### 5.2.12.2. Immunoprecipitation

When one is interested in a specific protein or hapten, and particularly if a precipitating antibody is needed, the culture filtrate can be tested for its ability to precipitate labelled protein on the addition of a secondary antibody or Protein A (single $M_{ab}$s will not precipitate proteins). This is more

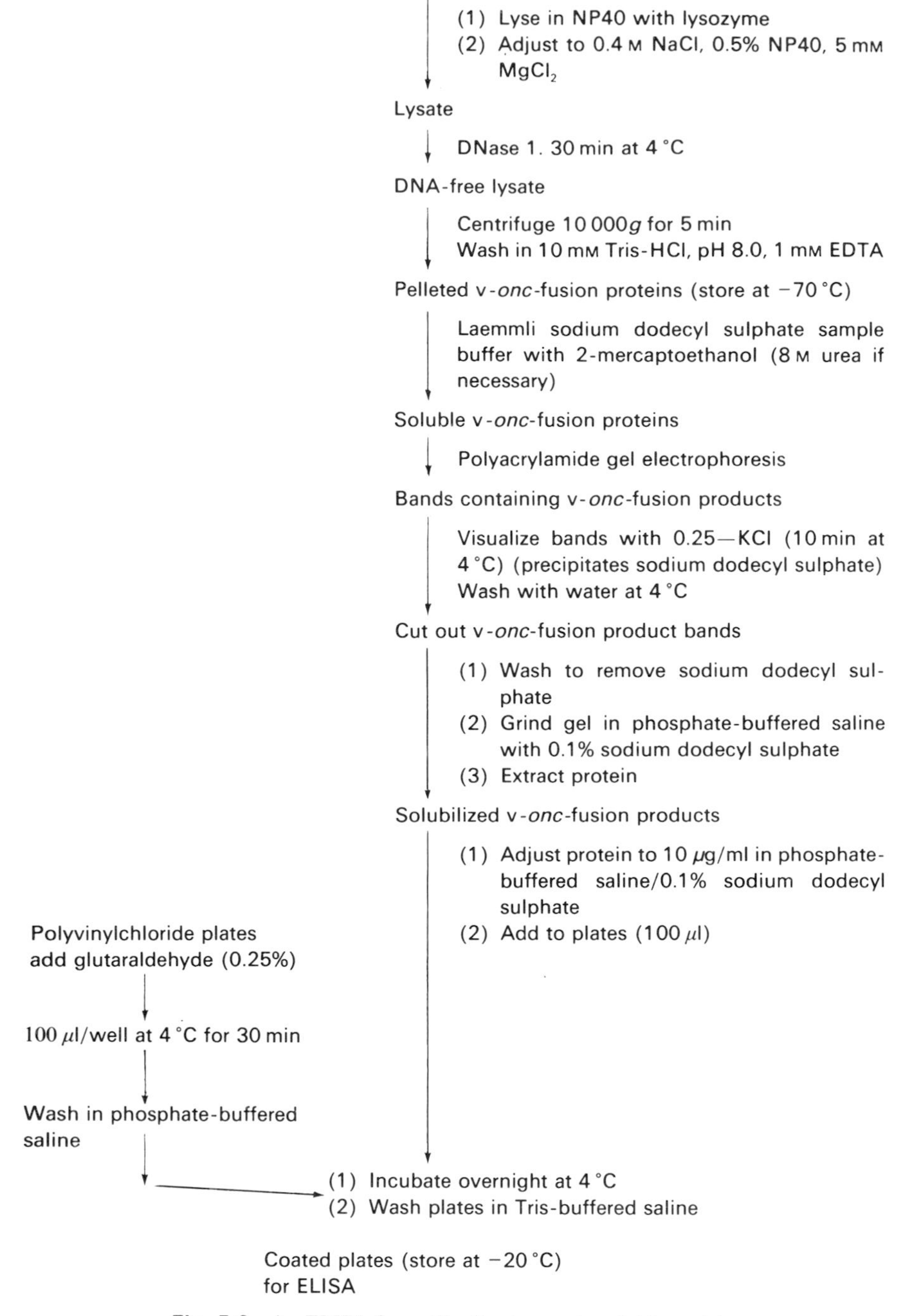


**Fig. 5.6** An ELISA for antibodies to poorly soluble proteins.

laborious than the solid-phase binding assay, but it can be streamlined by pooling culture filtrates. Unless the immunoprecipitate is to be analysed on a polyacrylamide gel, pure antigen is needed, and the antigen (or BSA-linked hapten) is usually iodinated (Galfre and Milstein, 1981) or tritiated (Berzofsky *et al.*, 1980). Hybridoma culture filtrates are incubated for 1–3 h at room temperature or overnight at 4 °C with the labelled antigen ($10^4$–$10^5$ cpm), usually in the presence of non-immune mouse/rat serum, and the immune complex precipitated with a second antibody, with PEG or with formalin-fixed *Staphylococcus aureus* cells (NEN) or Protein A-coated beads (immunobeads, Biorad) and counted. The second antibody can be added free or bound to Sepharose, agarose or Protein A on cells or beads. By coating the Protein A with an anti-mouse immunoglobulin, we make it class-independent. The immunoprecipitate can also be analysed by gel electrophoresis and fluorography.

This type of analysis can be tedious in screening procedures, but can serve to define useful properties of a $M_{ab}$, i.e. the ability to bind to an antigen in the presence of protein denaturants such as sodium dodecyl sulphate (Fig. 5.7, Mayer, unpublished observations).

Other examples of these methods are given in Table 5.5. One should note, however, that some $M_{ab}$s will not immunoprecipitate their protein antigens. This is a problem that we and others (Noonan *et al.*, 1981) have encountered, and it can be very distressing and annoying when attempting to identify a protein antigen (see Section 5.2.17.1). However, if one is specifically looking for a precipitating antibody this should be the screening method of choice.

We have found that detergent extracts of membranes tend to adsorb non-specifically to *S. aureus* cells, despite adding 1% BSA and detergent (e.g. 0.5% Triton) during incubation with the immune complex. We have found that this problem can be reduced to workable levels by repeated (4 × ) pre-washing of the cells in phosphate-buffered saline (PBS) containing 0.1% Triton, 0.1% sodium dodecyl sulphate (SDS) and 0.1% BSA.

### 5.2.12.3. Immunohistochemistry (see also Chapter 7)

The indirect staining of antigens in tissue sections and in isolated cells using indirect immunocytochemical and immunofluorescence techniques can be used very successfully to pick out antibodies. It is particularly useful for screening antibodies that mark particular cells, organs (Haspel *et al.*, 1983) or special structures, e.g. myofibrillar components of skeletal muscle (Lin, 1981). Whether frozen or paraffin sections are used depends to a large extent on the antigen of interest, and some antigens are irreversibly denatured either by the fixation or the dehydrating steps. Cryostat sections are usually fixed with acetone (− 10 °C, 5 min) (Osborn, 1981) but there is no general rule.

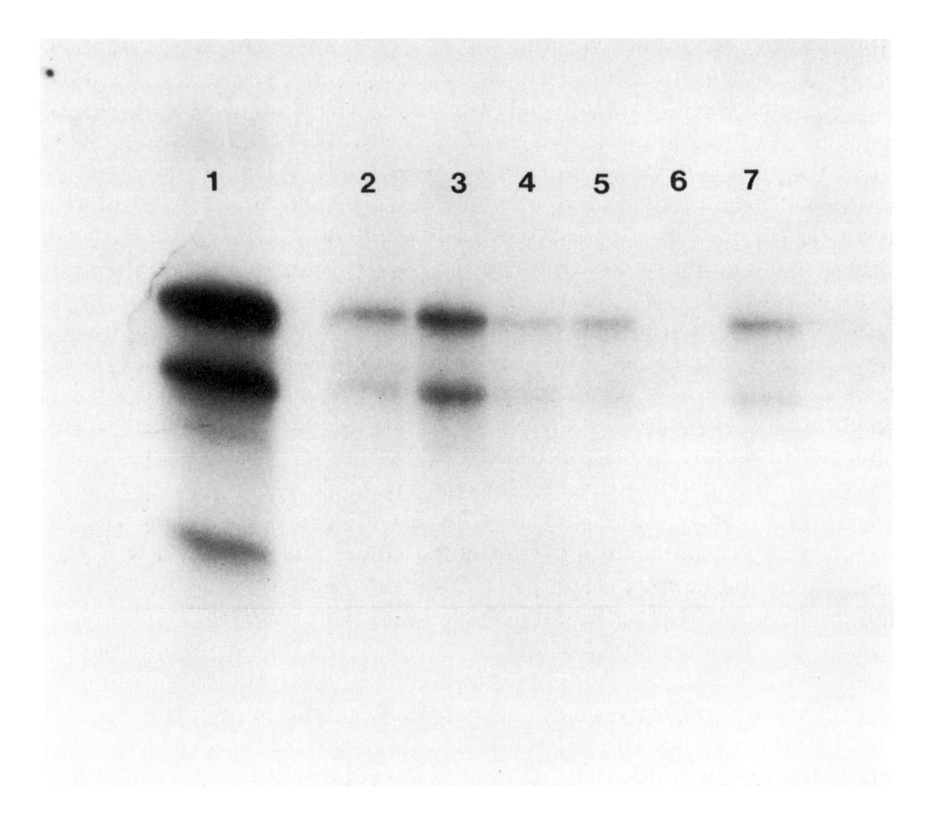


**Fig. 5.7** Immunoprecipitation of casein with monoclonal antibody. Reductively methylated [$^3$H]casein (track 1) was immunoprecipitated by incubation with monoclonal anti-casein antibody overnight at 4 °C on a rotary shaker, followed by incubation with sheep anti-mouse IgG overnight at 4 °C. Immunoprecipitation was carried out in the absence (track 7) and presence (tracks 2–5) of increasing concentrations of sodium dodecyl sulphate (0.05, 0.1, 0.25 and 0.25% respectively). No bands are seen in the absence of monoclonal antibody (track 6). Extraction buffers for immunoprecipitations contained 0.1% (w/v) sodium dodecyl sulphate. Fluorograms are of polyacrylamide (12.5%) gels run in the presence of sodium dodecyl sulphate and 2-mercaptoethanol.

Indirect immunofluorescence is becoming a somewhat less popular technique than immunocytochemical methods, mainly because of the reproducible and very sensitive signals given by Sternberger's peroxidase-antiperoxidase (PAP) method (Sternberger *et al.*, 1970).

Reasons why immunofluorescence assays of tissue sections with many culture supernatants present problems are: (a) no counterstain can be used to visualize overall cellular structures in the sections; (b) FITC conjugates can fade on examination or storage; (c) visualization of epitopes can be very time-consuming. Immunohistological screening of tissue sections during screening or in evaluation of subsequent $M_{ab}$s can be very useful with the peroxidase technique (Naiem *et al.*, 1982). The use of multi-test slides permits assessment of 100 samples in less than 3 h. Non-specific background labelling can be minimized. The semiquantitative assessment of immunoperoxidase staining of tissue sections can give much more information concerning cell types stained and shared epitopes which are useful in histopathology. Since many $M_{ab}$s against protein macromolecules are to conformational epitopes (Chapter 6) the immunohistological screening method with cryostat sections (cf. paraffin sections) may detect extra $M_{ab}$s to "intact" epitopes. However, $M_{ab}$s detected in this way may have limited usefulness in biomedical science (except for immunohistochemistry) relative to those detected by other techniques, e.g. immunoprecipitation or Western blots. The general principles of the method with cryostat (Naiem *et al.*, 1982) and paraffin sections (Allsop *et al.*, 1986) are shown in Fig. 5.8 (also see Chapter 6, Fig. 6.2). Immunocytochemistry with $M_{ab}$s can also be usefully performed by autoradiographic techniques at the light microscope or electron microscope level with internally radiolabelled $M_{ab}$s, e.g. with antibodies radiolabelled in hybridoma cells with [$^{3}$H]amino acids (Cuello *et al.*, 1982).

#### 5.2.12.4. Biological assays

These assays are not commonly used, but in some cases there is no alternative. They include, for example, assays which directly monitor the effects of $M_{ab}$s on the activity of enzymes (Choo *et al.*, 1980), inhibition of antiviral activity and nucleic acid synthesis of interferon (Secher and Burke, 1980; Hochkeppel *et al.*, 1981), and receptor-binding inhibition assays (Trowbridge and Lopez, 1982). The important characteristic of these methods is that they are very specific and do not require the purification of antigen. However, they are generally not as sensitive as other methods, they may need a large source of antigen, and, because of their exquisite specificity, they may not detect some antibodies directed to epitopes not involved in the biological activity.

#### 5.2.12.5. Fluorescent-activated cell sorting (FACS)

This is a method which can be used for the direct detection of antibody-producing cells. It can also be used for the identification of cells bearing the antigen recognized by a $M_{ab}$ when attempting to characterize the specificity of a $M_{ab}$.

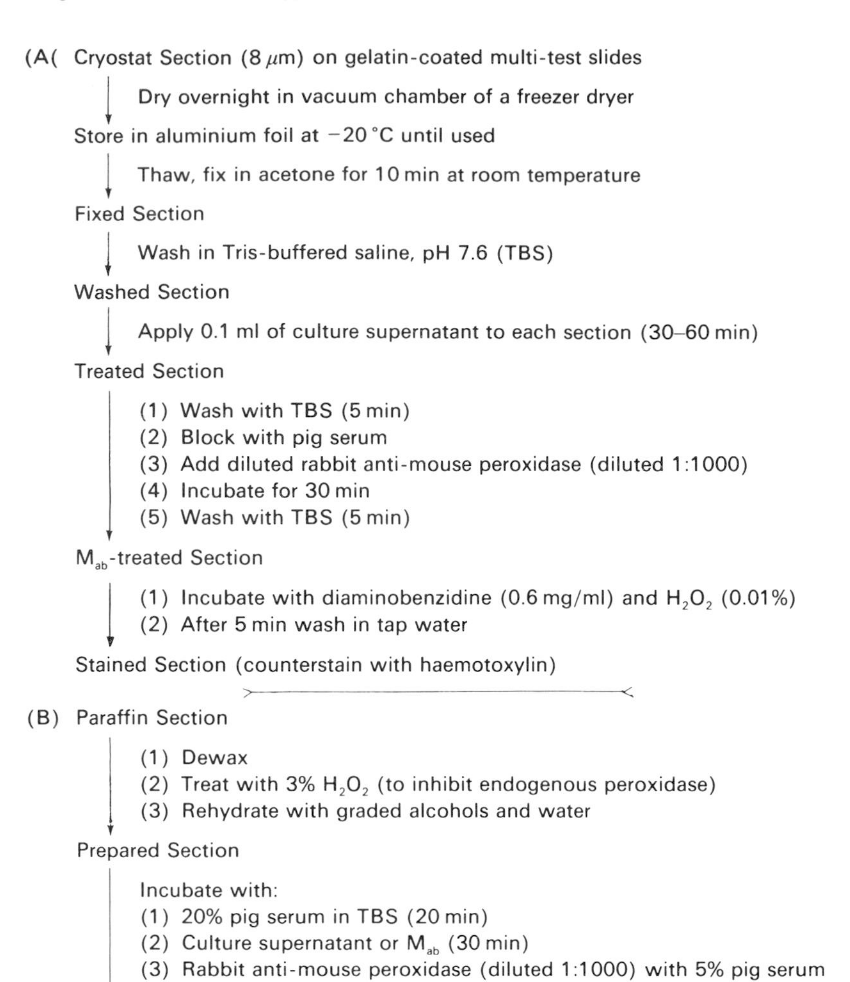


**Fig. 5.8** Immunoperoxidase staining of cryostat (A) and paraffin (B) tissue sections with monoclonal antibodies.

Antibody-secreting cells, under appropriate conditions, bind a certain amount of antigen. It has been possible to detect cells secreting specific antibodies by attaching the antigen to fluorescent microspheres. For example, Schwartz *et al.* (1981) coupled 200 μg affinity-purified asialoglycoprotein receptor to highly fluorescent (green) latex spheres (0.77 μm average diameter) using 1-ethyl-3-(3-dimethylaminopropyl)carbodiimide. Hybridoma cultures were incubated with these beads and sorted under sterile conditions in a fluorescence-activated cell sorter. For general information on FACS analysis see Herzenberg and Herzenberg (1978).

#### 5.2.12.6. Other assays

These include haemagglutination and lytic assays, both of which are well documented by Galfre and Milstein (1981). The former depends on the ability of an antibody to agglutinate red cells carrying the specific antigen, while the latter is based on lysis of cells by antibody and complement. The disadvantages of haemagglutination assays are the inhibitory effects due to excess antibody and quantitative inaccuracy; in practice these assays miss a number of antibody-secreting clones. Since only some classes of antibody fix complement and since the lytic assay may depend on antigen density and distribution, only certain $M_{ab}$s will be detected using the lytic method. However, they can be useful, particularly since a replica-plating method of analysis has been introduced recently which allows the assay of 1000–2000 microcultures in a 3 h period (Bankert, 1982).

### 5.2.13. Cloning

Most workers agree that initial cultures should be cloned as soon as possible, even if derived from a single hybrid cell, since chromosome losses leading to specific heavy- or light-chain deletions are particularly frequent in early stages of growth. This can be done either in semi-solid agarose or by limiting dilution (e.g. Kennett, in Kennett *et al.*, 1981, pp. 373–375). We use dilution cloning, however, since agar gives a lower cloning efficiency and this may select for vigorous, but unwanted hybrids, and since colonies of cells picked from agar are not always pure clones (Edwards, 1981; see also Galfre and Milstein, 1981). A FACS with a cloning attachment is very useful for the limiting dilution method.

### 5.2.14. Storage of cultures

Positive clones are grown up and stored by freezing (Kennett, in Kennett *et al.*, 1981). Only healthy cultures growing in mid-log phase should be frozen, otherwise they will not grow on thawing. When thawing out cells it is advisable to use a feeder layer and to try growing cells in 1 $cm^3$ wells in addition to flasks.

### 5.2.15. Large-scale production

Large amounts of $M_{ab}$ can be produced either *in vitro*, or *in vivo* by culturing cells as ascitic tumours. The method used depends on how much antibody is needed and on the facilities available. The concentration of $M_{ab}$ in culture is 5–25 $\mu g/cm^3$, perhaps up to 100 $\mu g/cm^3$, while in ascitic fluid 0.5–5.0 $mg/cm^3$ is usually obtained. Although the animal serum is often 100–1000 times the concentration of the culture filtrate, this is not always the case, depending on the immunoglobulins involved.

Culture filtrate contains antibody, usually together with FCS (foetal calf serum), but no irrelevant mouse or rat immunoglobulin. In cultures used to produce antibody, the concentration of FCS is usually kept low (2.5%) or completely removed for 2 days prior to harvesting time. Only cultures growing vigorously in 5% FCS are used for this purpose, and the cells are left for 1 day after the cells have reached the stationary phase of growth before the supernatant is collected by centrifugation. The important factor for antibody production is to grow cells slowly at maximum density; this is best achieved in spinner cultures. If the serum-free medium now on the market proves capable of sustaining good clonal growth of hybridomas, the problem of serum contamination will be eliminated.

The more common method of producing large amounts of $M_{ab}$ is *in vivo*, despite the fact that ascites fluid contains a variable, but usually small, amount of immunoglobulin impurities from the animal due to haemorrhage. The derivation of tumours is fairly straightforward and the high concentrations of $M_{ab}$ in the ascitic fluid is very useful. Accounts of how to grow intraspecies (mouse × mouse, rat × rat) and interspecies (mouse × rat) hybridomas in animals are given by Galfre and Milstein (1981) and Kennett (in Kennett *et al.*, 1981, p. 403). Cell lines can be established in mice as follows. Approximately $0.5 \times 10^7$ cells in medium without serum are injected i.p. into mice which were given an i.p. injection of 0.5 $cm^3$ pristane (2,6,10,14-tetramethylpentadecane) 3 weeks to 3 months previously. After 8–14 days, 0.3 $cm^3$ of the ascitic fluid is collected using a sterile hypodermic syringe and transferred (i.p.) into another pristane-treated mouse. After a further 8–14 days, 0.3 $cm^3$ of the ascitic fluid is removed and transferred into a non-pristane-treated mouse. When this fluid is ready to collect it is now diluted 1:10 with culture medium and transferred into another non-pristane-treated mouse. This is repeated and the line is now established. It is important to remember, however, that cells should not be subjected to multiple passages in culture or in mice without periodic cloning.

Culture filtrates are remarkably stable and can be stored at 4 °C with the addition of 0.1% azide for several months; they should not be frozen and thawed more than once if possible. Ascites is fairly stable in general and can be stored frozen at −20 °C for a few months, and longer at −70 °C. It should be stored in small aliquots so that it is not subject to repeated thawing and freezing. Unfortunately, some IgM antibodies tend to aggregate on storage and only fresh samples can be used.

## 5.2.16. Assessment of antibodies

### 5.2.16.1. Purification of antibody

The ease of purification depends on the species and class of antibody and whether the antibody has been produced *in vitro* or *in vivo*. Whatever the method used, the preparation should first be centrifuged to remove cells and debris (culture medium) or fibrin deposits (ascites), and the immunoglobulin isolated by precipitating with ammonium sulphate (final concentration 50%) and dialysed against PBS containing sodium azide (0.01–0.1%). For many purposes this is all that is required. Some antibodies are unstable to this salt treatment and they have to be concentrated using ultrafiltration devices.

Several conventional methods of purifying immunoglobulins can be used with $M_{ab}$s, including ion-exchange chromatography and gel filtration (Sephadex G-100 and Sepharose 6B). DEAE-ion-exchange chromatography does not give good recovery of IgM, which binds very tightly to the column, but can be used for mouse and rat IgG from culture and ascites (see Göding (1980) for detailed consideration of method). However, it does not remove contaminating protease and nuclease activities from antibodies. Another common method used is affinity chromatography on Protein A-Sepharose CL-4B. As indicated previously, the problem with this method is that not all immunoglobulins bind to Protein A. Rat and mouse IgM and mouse $IgG_1$ bind poorly, if at all, while murine $IgG_{2a}$, $IgG_{2b}$ and $IgG_3$ bind strongly. Rat $IgG_{2c}$ binds strongly, but there is some dispute as to whether rat $IgG_1$ binds (Göding, 1980; Surolia *et al.*, 1982; see also Oi and Herzenberg, 1980).

A new one-step purification has recently been published (Bruck *et al.*, 1982) for mouse monoclonal IgG from ascites. This uses DEAE affi-gel, a beaded cross-linked agarose bearing covalently linked cibacron blue F3GA and diethyl-aminoethyl groups, a dye with differential affinity for several serum proteins (Bio-Rad 153–7307). At pH 7.2, immunoglobulins (at least $IgG_1$ and $IgG_2$) were eluted at a lower ionic strength (30–50 mM NaCl) than proteases (100 mM NaCl) and were devoid of ribonucleases. The method does not work well with mouse polyclonal immunoglobulins from serum, due to the high albumin/immunoglobulin ratio and the diffuse isoelectric point of serum IgG, but can be used successfully in conjunction with a salt-fractionation step.

For monoclonal IgM antibodies, purification by gel filtration, e.g. Bio-Gel A-1.5 M agarose (Greene *at al.*, 1980), Sepharose 6B (Blythman *et al.*, 1981) and affinity chromatography on anti-Ig columns can be used (Göding, 1980; Greene *et al.*, 1980), agarose being the common material employed.

IgM can also be purified in a single step from ascites fluid by high-performance liquid chromatography with the Bio-Rad HPHT MAPS™ system. The chromatography system can also be used to purify IgM from other fluids, e.g. tissue-culture media and serum. The high-performance liquid chromatography system uses high-performance hydroxyapatite (HPHT) columns. The HPHT MAPS system can also be used to purify IgG.

High-performance liquid chromatography can also be used for the rapid purification of IgGs from ascites fluids. Transferrin and albumin are the two major contaminants of immunoglobulins in ascites fluids. Good recoveries (80–90%) of IgGs separated from contaminating proteins have been achieved by sequential chromatography (high-performance liquid chromatography) of ascites by anion-exchange chromatography (Mono Q, Pharmacia) and gel-permeation chromatography (Bio-Sil TSK 250, Bio-Rad) at neutral pH (Burchiel *et al.*, 1983). IgGs can be separated from each other ($IgG_1$, $IgG_{2a}$, $IgG_{2b}$) and from transferrin and albumin by high-performance liquid chromatography on a Spherogel TSK DEAE-5PW ion-exchange column (Deschamps *et al.*, 1986).

Conditions for a typical separation of IgG on Mono Q by high-performance liquid chromatography are shown in Fig. 5.9. The procedure works

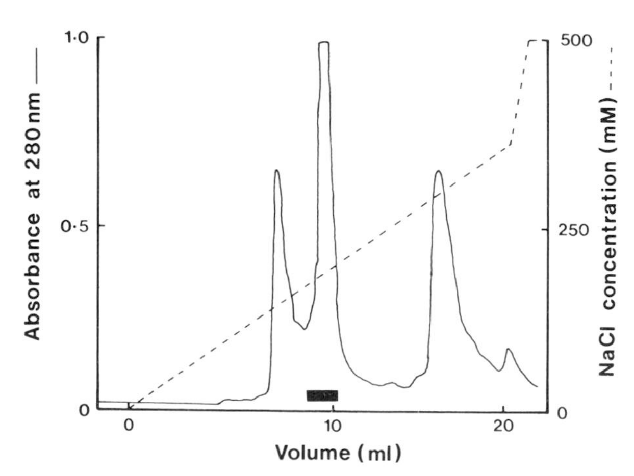


**Fig. 5.9** Purification of monoclonal antibodies to plaque core protein by high-performance liquid chromatography on Mono Q anion exchanger. Ascites fluid was passed through a 0.22 μm filter and samples (100 μl) were injected on to a Mono Q column. The column was eluted at 1 ml/min with a 20 min gradient of 0–375 mM NaCl in 20 mM Tris-HCl, pH 8.0. Bar corresponds to anti-plaque core protein immunoreactivity.

well under these analytical conditions, i.e. small sample volumes. For larger volumes of ascites fluid, e.g. 5–10 ml, the ascites should be filtered through glass wool (to remove lipid aggregates), and then the immunoglobulins should be precipitated with 50% saturated ammonium sulphate. The immunoglobulins should be dissolved in phosphate-buffered saline, dialysed against 20 mM Tris-HCl, pH 7.6, containing 0.025 M NaCl. Samples containing up to 8 mg of protein may be loaded on to the Mono Q column.

In situations where cell-borne Fc receptors are a problem it is useful to use $F(ab^1)_2$ fragments of monoclonal immunoglobulins. In addition $F(ab^1)_2$ fragments may be preferred in some systems where penetration of labelled antibody into cells is required. Care must be taken when preparing mouse $F(ab^1)_2$ fragments, since mouse IgG tends to precipitate from solution at the acidic pH required for pepsin action (Mishell and Shiigi, 1980). However, $F(ab^1)_2$ can be prepared by the method of Lamoyi and Nisonoff (1983). A second Sephadex G-100 fractionation can be used to separate $F(ab^1)_2$ from undigested IgG, Fc and small cleavage fragments (Burchiel *et al.*, 1984). On the other hand, it is fairly easy to prepare $F(ab^1)_2$ fragments from mouse–rat immunoglobulins (Springer, 1981).

### 5.2.16.2. Characterization of antibody

The monoclonal origins of the antibodies must be confirmed by demonstrating the production of homogeneous antibody molecules. Chain composition can be determined by various gel analyses. If the myeloma cell used is a non-secretor, only heavy and light chains of the spleen cell parent will be produced; if the myeloma is NS1, the MOPC-21 light chain of NS1 origin can also be produced, resulting in the production and secretion of mixed molecules. Several types of gel analyses have been used, including polyacrylamide isoelectric focusing, e.g. 5%, pH range 4–10 (Köhler and Milstein, 1975; Williamson, 1978; Fraser and Venter, 1980), reducing and non-reducing SDS electrophoresis (Lennon *et al.*, 1980; Gee *et al.*, 1983) and electrophoresis on cellulose acetate membranes (Young *et al.*, 1979). The antibody can be internally labelled at high specific activity using radioactive amino acid precursors (Göding, 1980; Galfre and Milstein, 1981) before being subjected to fractionation and the detection of Ig chains by autoradiography.

The class and subclass is often determined by Ouchterlony analysis using antisera specific for IgM or IgG (H + L), for IgG subclasses or for kappa or lambda light chains (e.g. Beisiegel *et al.*, 1981; Bjercke *et al.*, 1981). One should note again that this assay does not always work with $M_{ab}$s. Recently, Towbin *et al.* (1982) have used a dot immunoassay for antibody typing. Class-specific anti-mouse (or rat) antibodies were diluted and 1 $\mu$l aliquots applied as dots to nitrocellulose, incubated with hybridoma supernatants, and binding monitored using a peroxidase-coupled anti-mouse or anti-rat immunoglobulin. The secondary antibody could, of course, be radiolabelled instead. An alternative method is to use the trace binding solid-phase assay for screening culture filtrates, but with class-specific second antibodies (e.g. Berzofsky *et al.*, 1980).

### 5.2.16.3. Antigen identification

Ideally the identity of the antigen needs only to be confirmed, and the amount of work needed for a complete identification depends greatly on the success of the screening assays. Some of the techniques are briefly discussed below.

### 5.2.16.4. Antigen distribution

This normally involves immunocytochemistry (Chaper 7) and is used to demonstrate cell and tissue specificity and species specificity. It can be very informative, especially if the antigen is completely unknown. Most immunocytochemistry is undertaken using labelled second antibody rather than labelled $M_{ab}$; it is now possible to use ultrathin sections of tissues/cells and electron microscopy to look at the very fine detail, employing a second antibody with an electron-dense label, e.g. ferritin, iron-dextran (Imposil) or gold (see Osborn, 1981). In addition, it is possible to determine the cell specificity to a $M_{ab}$ by FACS analysis. In this case single-cell mixtures are incubated with the test $M_{ab}$ followed by a fluorescein-labelled second antibody prior to sorting (e.g. Girardet *et al.*, 1983).

### 5.2.16.5. Chemical nature of antigen

The preparation which reacts with the $M_{ab}$ can be subjected to digestion with proteolytic enzymes (trypsin, pronase) or carbohydrases (neuramidase $\alpha/\beta$-galactosidases, $\alpha$-glucosidase, $\alpha$-mannosidase) (Nudelman *et al.*, 1980; Hochkeppel *et al.*, 1981), or to organic solvent extraction, in an attempt to determine the chemical nature of the antigen (Girardet *et al.*, 1983; Nudelman *et al.*, 1980). Protein antigens can then be identified by immunoprecipitation or immunoblotting, and lipids by, for example, thin-layer chromatographic analysis (see below). Competition-binding assays and immunoaffinity chromatography can be tried for all types of antigen.

### 5.2.16.6. Immunoprecipitation

If the epitope recognized is on a protein it may be possible to immunoprecipitate the protein and identify it by gel electrophoresis (see Fig. 5.7 and screening methods, Section 5.2.12, for details). The antigen preparation is often labelled, usually with $^{125}I$ or $^{3}H$, after isolation, but sometimes it can be backbone-labelled prior to isolation; for example, if it is a preparation from a small animal, the animal can be injected with a radioactive amino acid precursor (Walsh *et al.*, 1981). If the antigen containing the epitope is a predominant protein in the preparation, labelling may not be necessary and the protein band can be stained with Coomassie blue or a similar stain (Gee *et al.*, 1983); of course, the immunoglobulin chains (mouse/rat $M_{ab}$ and second antibody, if used) will also be stained in this case.

### 5.2.16.7. Immunoblotting

Proteins can be fractionated by polyacrylamide gel electrophoresis, electroblotted to nitrocellulose membrane (Western blotting), reacted with the test $M_{ab}$ and then, if this binds to a particular band, it can be detected with $^{125}I$-labelled, fluorescein-labelled or peroxidase-labelled anti-mouse immunoglobulin (e.g. Kramer *et al.*, 1980; Gee *et al.*, 1983); for the Western blotting technique see Towbin *et al.* (1979) and Gershoni and Palade (1983).

One of the problems with this technique is that immunoglobulins tend to adsorb non-specifically to the nitrocellulose paper. This means that protein blots need to be blocked by preincubating in buffers containing a blocking agent, e.g. BSA foetal calf serum, horse serum, or milk powder extract, for 3–24 h before addition of $M_{ab}$, and washed extensively in buffers containing 3% BSA prior to and after the addition of second antibody. The other problem is that some $M_{ab}$s will not recognize their antigens following the SDS treatment they are subjected to during electrophoresis.

### 5.2.16.8. Thin-layer chromatographic (TLC) analysis of lipid antigens

Lipids extracted by organic solvents can be subfractionated by methods similar to that of Christie (1973) prior to analysis by TLC. For example, Girardet *et al.* (1983) have recently extracted glycolipids from the aqueous and the organic phases of an ether–butanol partition and subjected them to TLC analysis in chloroform/methanol/aqueous $CaCl_2$. The TLC plates were then incubated with test $M_{ab}$ followed by a second incubation with $^{125}I$-labelled rabbit anti-mouse $F(ab^1)_2$ antibody, washed extensively, dried and exposed to X-ray film.

### 5.2.16.9. Competition trace binding solid-phase assays

These assays are undertaken as described in Section 5.2.12.1 but the $M_{ab}$s are incubated in the presence of serial dilutions of inhibitors (proteins, lipids, carbohydrates, etc.) in the antigen-coated wells. Inhibition of $M_{ab}$ binding to the antigen-coated wells indicates that the incubation contains the epitope recognized by the $M_{ab}$ (e.g. Brockhaus, 1982). In addition, this type of assay can be used to determine whether $M_{ab}$s react with the same or different antigens (Fisher and Brown, 1980). Saturation binding levels of single $M_{ab}$s and $M_{ab}$s in pairs are incubated in antigen-coated wells, and if two antibodies with different specificities are used in the mixture a higher level of second antibody will be bound to the wells. Alternatively, a labelled "standard" $M_{ab}$ is used, and the ability of other $M_{ab}$s to inhibit binding to antigen-coated plates is measured (Breschkin *et al.*, 1981).

Competition trace binding assays can also be used to detect hapten conjugated to a protein which is bound to the microtitre wells. Fig. 5.10 shows a competition trace binding assay for a hapten in which the competition for antibody is between a conjugate attached to the plate and the "free" hapten in solution. Although the data is from a polyclonal serum (Middleton, Billett and Mayer, unpublished) the same principle can be applied with a $M_{ab}$. The antibody affinity is reasonable, allowing use at a dilution of 1:20 000, which permits the routine detection of approximately 1 ng of micromolecular free hapten in competition with plate-bound hapten–protein conjugate.

### 5.2.16.10. Affinity chromatography

This is a method often used with protein antigens and it should allow one to identify protein antigens present at low concentrations. The $M_{ab}$ is covalently bound to a solid support, usually Sepharose 4B, using conventional coupling techniques (e.g. see Hudson and Hay, 1980) and the immunoadsorbent packed into a small column, for example in a Pasteur pipette or plastic pipette tip. A soluble preparation of the antigen is then passed through the column slowly, perhaps repeatedly, and after extensive washing in detergent-containing buffer (e.g. 0.5% Nonidet or Triton X-100), the bound antigen is eluted with salt, detergent, etc. (e.g. Fraser and Venter, 1980; Stallcup *et al.*, 1981). The affinity of the $M_{ab}$ will determine what conditions are required to

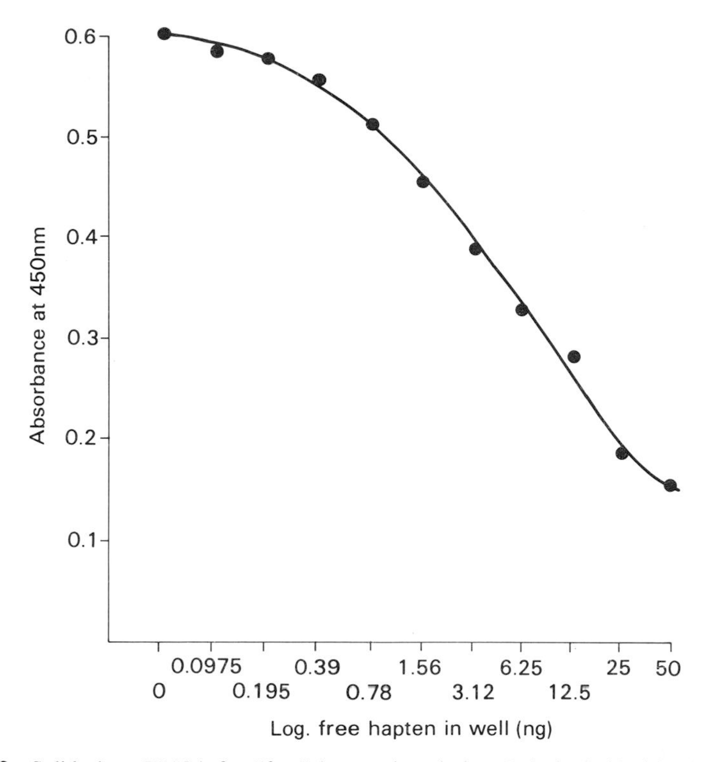


**Fig. 5.10** Solid-phase ELISA for "free" hapten in solution. Polyvinyl chloride plates were coated with 0.5 $\mu$g hapten-conjugated BSA (25 ng hapten). Polyclonal rabbit antibodies to hapten-conjugated chicken $\gamma$-globulin (final dilution 1:20 000) were incubated in each well with free hapten (0–50 ng/well) for 24 h at 4 °C. After washing, plates were incubated with horseradish peroxidase–anti-rabbit Ig for 2 h at room temperature. After washing, colour development was for 10 min with 3,3′,5,5′-tetramethylbenzidine. The reaction was stopped with 2.5 M $H_2SO_4$. Absorbance was measured on a Microplate reader at 450 nm.

elute the antigen; sometimes 2% SDS and 6 M salt are needed. If this method is to be used for immunoaffinity purification of an antigen, $M_{ab}$s with low affinities are preferred to enable efficient high retrieval. If biologically active antigen is required, very careful selection of the $M_{ab}$ is often needed. Examples of useful antibodies for purification include conformation-specific antibodies (Hansen and Beavo, 1982) and antibodies which release their antigens on addition of a ligand (e.g. propranolol releases $\beta$-adrenergic receptors from immunoaffinity columns of a $M_{ab}$, FV-104) (Fraser and Venter, 1980).

There are a few problems with the method, in common with affinity columns made with polyclonal antibodies; for example, the hydrophobic nature of membrane proteins causes non-specific adsorption to the gel, despite detergent solubilization, and even extensive washing in detergent-containing buffers may not successfully remove proteins bound non-specifically. In addition, not all $M_{ab}$s will retain their activity following coupling, and sometimes the antigen cannot be retrieved from the immunoadsorbent.

An interesting example of affinity chromatography is provided by a $M_{ab}$ to a human urinary tetrasaccharide-containing glucose. The purified $M_{ab}$ either covalently bound to cyanogen bromide-activated Sepharose or non-covalently bound via its Fc portion to *S. aureus*. Protein A-Sepharose binds the reduced tetrasaccharide avidly at 4 °C, but releases the oligosaccharide when the temperature is raised to 37 °C. This immunoadsorbent provides a nice example of a temperature-sensitive affinity adsorbent for affinity chromatography (Lundblad *et al.*, 1984).

## 5.2.17. Case study: monoclonal antibodies to human mitochondrial antigens

Some details on the production of $M_{ab}$s to human mitochondrial antigens, including monoamine oxidase, will illustrate some of the problems of obtaining $M_{ab}$s to antigens of interest. The studies show that during attempts to obtain $M_{ab}$s to the antigens of interest, other interesting $M_{ab}$s can be obtained after immunization with partially purified preparations of macromolecules (Billett *et al.*, 1984; Billett and Mayer, 1986).

### 5.2.17.1. Monoclonal antibodies to human liver mitochondrial outer membrane

During an attempt to isolate $M_{ab}$s to human liver mitochondrial outer membrane (MOM), an antibody, 1H6/C12, was isolated which reacts to an epitope of the mitochondrial inner membrane (MIM) of hepatocytes. In addition, an antibody, 3F12/F2, which binds a protein in semi-purified human MOM preparations was isolated, but after further characterization it was clear that it was recognizing an epitope on a contaminating cell type, namely the granulocyte. This epitope is not present in monocytes, lymphocytes and red cells from human blood.

In addition to reacting with their epitopes on human liver, these $M_{ab}$s react with those on rat liver, showing useful and specific binding in tissue sections. 1H6/C12, an $IgG_{2a}$, is a useful marker for MIM and is the first $M_{ab}$ against the human MIM to be isolated. 3F12/F2 is a particularly interesting antibody, since its epitope is inside the granulocyte, unlike the epitopes of most granulocyte $M_{ab}$s. It is hoped that 3F12/F2 may prove useful for diagnostic purposes and in the study of granulocyte biology.

*(a) Purification of MOM, immunization and fusion*

MOM vesicles were prepared from 100 g of a healthy postmortem liver (<24 h old) basically using the method of Martinez and McCauley (1977). Unlike the findings of Martinez and McCauley, however, the MOM from the 1.2 M sucrose cushion was mixed with 2 volumes of cold 1 M NaCl to remove loosely bound proteins, and centrifuged at 100 000$g_{av}$ for 30 min at 4 °C. The final pellet (2.5 mg of protein/ml) was sonicated in phosphate-buffered saline and stored at −70 °C in small portions. The sonicated pellet contained no lactate dehydrogenase activity, but did contain succinate dehydrogenase and aryl esterase activities, indicating contamination with MIM and endoplasmic reticulum. BALB/c mice were injected intraperitoneally with 250 $\mu$l of sterile phosphate-buffered saline containing 200 $\mu$g of human MOM emulsified in an equal volume of Freund's complete adjuvant; 3 weeks later they were injected both intraperitoneally and intravenously with 100 $\mu$g of MOM in sterile saline. Fusion was performed 3 days later. Spleen cells ($10^8$) were fused with mouse myeloma P3NS1-Ag4 cells ($10^7$). The mouse myeloma cell line was grown in RPMI medium supplemented with 10% (v/v) heat-inactivated foetal calf serum and 8-azaguanine (15 $\mu$g/ml). Fusion was achieved with poly(ethylene glycol) 1500, adjusted to approximately pH 7.6 with NaOH, in RPMI medium containing dimethylsulphoxide (5%, v/v, final concentration). The final cell pellet was resuspended directly in HAT medium (15% (v/v) foetal calf serum/0.4 $\mu$M aminopterin, 0.1 mM hypoxanthine and 16 $\mu$M thymidine) and distributed into four 96-well microtitre plates. All media contained streptomycin (60 $\mu$g/ml) and penicillin G (100 $\mu$g/ml). Macrophages obtained from 200 g Wistar rats by peritoneal lavage with 20 ml of RPMI medium containing 10% (v/v) heat-inactivated newborn calf serum were used as feeder cells.

*(b) Screening of culture supernates by trace-binding solid-phase assay*

This was performed in round-bottomed poly(vinyl chloride) wells. The wells were usually precoated with 50 $\mu$l of poly(L-lysine) ($M_r$ > 70 000; Sigma type P1886) in phosphate-buffered saline (3 mg/100 ml) for 30 min before the addition of 1 $\mu$g of human MOM in 50 $\mu$l of phosphate-buffered saline. The MOM was incubated overnight at 4 °C; glutaraldehyde was then added (0.25%, v/v, final concentration) and left for 15 min, and the wells washed twice in phosphate-buffered saline. Non-specific adsorption was then prevented by incubation with 100 mM glycine in 0.1% (w/v) bovine serum albumin for 30 min and with 1% (w/v) bovine serum albumin in phosphate-buffered saline for 4 h. The wells were then washed twice with 1% (w/v) bovine serum albumin in phosphate-buffered saline, filled with the same solution, sealed with Parafilm and stored at −20 °C until needed (maximum time approximately 6 weeks). All the steps for the preparation of wells were at room temperature unless otherwise stated.

The assay was performed as follows. A 50 $\mu$l portion of hybridoma supernatant (or serial dilutions of supernatant in phosphate-buffered saline) or, in controls, 50 $\mu$l of P3NS1 myeloma spent medium, were added to each well and left for 2–3 h at room temperature or overnight at 4 °C. Wells were then washed twice in phosphate-buffered saline and twice in phosphate-buffered saline containing 1% (w/v) bovine serum albumin before addition of 50 $\mu$l of $^{125}$I-labelled rabbit anti-(mouse IgG) antibody (specific radioactivity 5 $\mu$Ci/$\mu$g) in phosphate-buffered saline containing 1% (w/v) bovine serum albumin (approximately 60 000 cpm, 12 ng). After 2–3 h at room temperature the wells were washed as before, cut up and counted for radioactivity in a gamma-radiation counter.

Other antigens were bound to the poly(L-lysine)-coated wells, including rat MOM and rat MIM, but the procedure used was identical. In some experiments the wells were not precoated with poly(L-lysine) and glutaraldehyde fixation was omitted.

The $^{125}$I-labelled rabbit anti-(mouse IgG) was prepared as follows. Rabbits were immunized four times with an IgG fraction from mouse serum at 2-weekly intervals (0.25 mg/treatment). The rabbit antibody was subjected to $(NH_4)_2SO_4$ precipitation followed by dialysis against 10 mM potassium phosphate buffer, pH 8.0, to precipitate lipoproteins. Mouse IgG was coupled to CNBr-activated Sepharose 4B (0.5 mg/ml of Sepharose) and the rabbit anti-(mouse IgG) was affinity-purified by allowing it to bind to this column and eluting with 0.1 M glycine-HCl, pH 2.5. Fractions were neutralized immediately with Tris base. For iodination, 80 $\mu$g of the affinity-purified IgG was labelled with 1 mCi of [$^{125}$I]Na by the Chloramine-T method.

### *(c) Properties of antibodies*

Two $M_{ab}$s were generated by this fusion of mouse myeloma cells with spleen cells from mice immunized with human liver mitochondrial membranes. One antibody, 1H6/C12, an immunoglobulin $G_{2a}$ (Ig$G_{2a}$), binds to the inner membrane of rat hepatocyte mitochondria, and immunoperoxidase staining demonstrated that its epitope has an intracellular particulate distribution within rat and human hepatocytes and human brain neurones. The epitope reactive with 1H6/C12 is partially sensitive to proteinase digestion. The second antibody, 3F12/F2, an Ig$G_1$, binds to a contaminating cell type, namely the granulocyte, but it does not bind to monocytes, lymphocytes and red cells in human blood. This antibody reacts with cells in the portal tract and sinusoids of rat and human liver, as shown by immunoperoxidase staining. The epitope for 3F12/F2 is extremely sensitive to proteinase digestion and is only exposed when granulocytes are fixed in acetone, indicating an internal localization (Billett *et al.*, 1984).

These studies demonstrate several principles in $M_{ab}$ production. Immunization with impure antigen preparations gives "novel" immune responses, in that the $M_{ab}$s produced are often unexpected. In these studies antibodies to MOM were wanted: an antibody to MIM and to a totally unexpected epitope for an intracellular protein antigen from contaminating granulocytes were produced. For biochemists and cell biologists the salutory lesson is that the immune system can pick out immunodominant epitopes from a "contaminating" intracellular (liver) fraction (MIM) to the "wanted" fraction (MOM) as well as to epitopes on organelles from totally unrelated contaminating cell types (granulocytes). Neither of these antibodies reacted in immunoprecipitations with solubilized antigens when examined by polyacrylamide gel electrophoresis in denaturing conditions. Furthermore, no reaction in Western blotting reactions was observed, or in solid-phase immunoaffinity chromatography. The antibodies, like many $M_{ab}$s, may recognize conformational epitopes on the antigens which are disrupted in immunoblotting or immunoaffinity chromatography.

#### 5.2.17.2. Monoclonal antibodies to human liver monoamine oxidase

Monoamine oxidase (monoamine:oxygen oxidoreductase (deaminating) (flavin-containing) EC 1.4.3.4) is an integral protein of the outer mitochondrial membrane and is the major enzyme through which catecholamines and indoleamines are degraded by oxidative deamination. It is found in two different forms, with distinct catalytic properties and tissue distribution. MAO A is inhibited by the irreversible active-site inhibitor clorgyline and oxidizes 5-hydroxytryptamine, while MAO B is inhibited by deprenyl and pargyline, and oxidizes phenylethylamine and benzylamine. Despite its selectivity for MAO B, [$^3$H]pargyline will covalently label the active sites of both MAO A and MAO B under appropriate conditions. Studies with [$^3$H]pargyline-labelled enzyme have shown that the FAD-containing subunits of human MAO B from platelets and MAO A from placenta migrate in SDS/polyacrylamide gels with apparent $M_r$ values of approximately 64 000 and 60 000 respectively. This difference in apparent $M_r$ of the two subunits suggests that the FAD-containing peptides in the two forms of MAO are different. However, attempts to distinguish MAO A and B with conventional antisera have given equivocal results.

For these reasons, and in order to further topological studies on the distribution of the enzyme(s), attempts were made (Billett and Mayer, 1986) to raise $M_{ab}$s to the partially MAO B (70%).

### *(a) Purification of human liver MAO, immunization and fusion*

MAO was purified from mitochondrial outer membranes (MOM) isolated from normal liver at postmortem (<24 h after death, 100 g weight); the membranes were prepared as described previously (Billett *et al.*, 1984). MAO was purified by a modification of the method of Dennick and Mayer (1977).

Triton X-100 extracts of MOM were loaded directly on to a Sepharose 6B column (3.2 cm internal diameter × 90 cm long; approximately 0.6 ml/min) pre-equilibrated in 20 mM potassium phosphate buffer, pH 7.2, containing 0.1 mM mercaptoethanol and 0.1% (w/v) Triton X-100. The fractions with the highest specific activity were pooled and the Triton X-100 removed by means of Bio-Beads SM-2 previously equilibrated in 20 mM potassium phosphate, pH 7.2. The enzyme was then precipitated with solid $(NH_4)_2SO_4$ and the precipitate was dialysed extensively against 20 mM potassium phosphate, pH 7.2, containing 0.1 mM mercaptoethanol. The dialysed preparation was then applied to a column (2.8 cm × 6 cm) of DEAE-cellulose previously equilibrated with 20 mM potassium phosphate, pH 7.2, containing 0.1 mM mercaptoethanol. Bound protein was eluted by means of a linear gradient (200 ml) of Triton X-100 (0–1%, w/v) in 20 mM potassium phosphate containing 0.1 mM mercaptoethanol. Fractions containing MAO activity were pooled and Triton X-100 was removed by means of Bio-Beads SM-2. The enzyme was again precipitated with $(NH_4)_2SO_4$ and the precipitate was dialysed extensively against 20 mM potassium phosphate buffer, pH 7.2. The resultant preparation contained about 2 mg of protein. Polyacrylamide gel electrophoresis (10%) of a sample (40 μg) in the presence of SDS showed a major band of $M_r$ 57 000 and two fainter bands of $M_r$ 63 000 and 54 000. After labelling with 1.5 μM [$^3$H]pargyline, 70% of the radiolabel was recovered in the band of protein of $M_r$ 57 000 and 30% in the band of $M_r$ 63 000.

A BALB/c mouse was injected intraperitoneally with 250 μl of sterile phosphate-buffered saline containing 25 μg of purified human liver MAO emulsified in an equal volume of Freund's complete adjuvant; 21 days later it was injected intravenously with 20 μg of purified MAO in sterile saline; a similar injection was given on days 22 and 23. Fusion was performed on day 24.

Spleen cells ($10^8$) were fused with mouse myeloma P3NS1-Ag4 cells ($10^7$) as described previously (Billett *et al.*, 1984), using 50% (w/v) poly(ethylene glycol) 1500 adjusted to approximately pH 7.6 with NaOH in RPMI medium containing dimethylsulphoxide (5%, v/v, final concentration). The final cell pellet was distributed into four 96-well microtitre plates containing a macrophage feeder cell layer (Billett *et al.*, 1984).

### *(b) Screening and cloning of culture supernatants*

This was performed using an indirect trace binding solid-phase assay in round-bottomed poly(vinyl chloride) wells. The wells were precoated with 50 μl of poly(L-lysine) ($M_r$ > 70 000; Sigma type 1274) in phosphate-buffered saline (3 mg/100 ml) for 30 min (Kennett *et al.*, 1981) before addition of 2 μg of human MOM (or 0.2 μg of purified MAO) in 50 ml of phosphate-buffered saline. The antigen was incubated overnight at 4 °C and then fixed with glutaraldehyde (0.25%, v/v, final concentration) for 15 min. Non-specific adsorption was then prevented by incubation with 100 mM glycine in 0.1% (w/v) bovine serum albumin for 30 min and with 1% (w/v) bovine serum albumin in phosphate-buffered saline for 4 h. All steps for the preparation of wells were at room temperature unless otherwise stated.

Aliquots (50 μl) of conditioned media from wells containing growing clones were incubated in the washed wells for 3 h at room temperature. Bound mouse immunoglobulin was detected after a further 3 h incubation of the washed wells with 50 μl of $^{125}$I-labelled rabbit anti-(mouse IgG) antibody, specific radioactivity 5 μCi/μg (Billett *et al.*, 1984) in phosphate-buffered saline containing 1% (w/v) bovine serum albumin (approximately 60 000 cpm, 12 ng). The wells were then washed out and the bound radioactivity counted in a gamma-radiation counter.

Culture 3F12 was cloned by limiting dilution and was found to be clonal in origin. One subclone, 3F12/G10, was injected into pristane-primed mice in order to generate ascites fluid. The IgG subclass of 3F12/G10 was found to be type 1 by using a solid-phase assay with rabbit anti-(mouse Ig subclass) antibodies.

### *(c) Properties of monoclonal antibodies to monoamine oxidase*

The $M_{ab}$ to human monoamine oxide B (3F12/G10) is an immunoglobulin $G_1$, and reacts with its antigen in cryostat sections of human liver, showing an intracellular particulate distribution as demonstrated by immunoperoxidase staining. The antibody indirectly precipitates [$^3$H]pargyline-labelled human MAO B both from liver and platelet extracts but fails to precipitate MAO A from liver extracts. The antibody does not recognize rat liver MAO B, showing that the determinant is not universally expressed on MAO B. The antibody has no effect on the catalytic activity of MAO B.

The $M_{ab}$ to MAO B recognizes the enzyme when it is embedded in the mitochondrial outer membrane or when it is solubilized in Triton X-100. The $M_{ab}$ may be used to study the topographical distribution of the enzyme in the mitochondrial outer membrane, e.g. by immunogold electron microscopy (see Chapter 7). Since the $M_{ab}$ reacts with [$^{3}$H]pargyline-labelled enzyme, i.e. is rendered catalytically inactive by the suicide inhibitor, it cannot react with the FAD binding site. Furthermore, since the $M_{ab}$ does not prevent binding of the substrate, tyramine, it cannot be directed to the catalytic site. The $M_{ab}$ only recognizes monoamine oxidase B (not A), and can therefore be used to separate the enzyme for further studies on MAO B, e.g. screening of recombinant expression libraries containing the CDNA for the enzyme.

A further $M_{ab}$ was isolated which immunoprecipitates a 54 kD protein from a crude human liver extract and shows a particulate distribution in cryostat sections of liver, i.e. a $M_{ab}$ was produced to some contaminant of the partially purified enzyme.

## 5.3. Uses of Monoclonal Antibodies

The literature on the applications of $M_{ab}$s has been classified in the manner illustrated in Table 5.1. The information is presented in this tabular form to emphasize: immunizing macromolecular antigen(s); species donating spleen cells and myeloma clones used in the fusion, e.g. mouse–mouse (P3-NS1); the $M_{ab}$ assay (where described); the objectives of the study; special $M_{ab}$ properties; other facts of interest; references (authors).

### 5.3.1. Enzymes and proteins

The application of $M_{ab}$s to enzymes and proteins is shown in Table 5.5. Many studies (nos 20, 27) were commenced by using purified proteins as immunogens, whereas in the remaining studies impure tissue preparations (i.e. Table 5.5, nos 11, 17, 19, 26) were used to immunize the spleen donor animal. In $M_{ab}$ studies on enzymes and proteins (Table 5.4) and on other macromolecular antigens of interest (Tables 5.6–5.11) these two strategies ((a) and (b) below) are routinely used. The two strategies derive from: (a) the obvious advantage of using a purified macromolecule as immunogen (as in most polyclonal antibody production strategies (Mayer and Walker, 1978, 1980) for use in biochemical, cell and molecular biological investigations); (b) the inherent theoretical advantage of the $M_{ab}$ approach which decrees that the major effort put into macromolecular (e.g. protein) purification can be reduced and even replaced by the effort required for the hybridoma screening and cloning procedures, e.g. to obtain a $M_{ab}$ to an epitope on a single macromolecular species.

However, the use of impure preparations (strategy (b)) means that the differential immunogenicity of macromolecular antigenic epitopes will decide the donor spleen cell heterogeneity before spleen cell–myeloma fusion, and, therefore, determine the $M_{ab}$ specificity of cloned hybridoma supernatants. In this way, when rat myofibrils are the immunizing preparation, $M_{ab}$s to intermediate filaments are produced (Table 5.5, no. 11), human myoblasts beget anti-fibronectin monoclonal (Table 5.5, no. 17), and crude soluble protein from Alzheimer diseased hippocampi (Table 5.5, no. 26) results in anti-neurofilament $M_{ab}$s. The idea that the cloning principle in $M_{ab}$ production removes the need completely for macromolecular purification is therefore compelled in practice by the immunogenicity of antigenic determinants on macromolecules in the preparations. Consideration of this question is necessary when using whole cells or organisms as immunizing species (Table 5.5 and below).

At this juncture it is probably again worth saying that it is best to obtain the purest preparation possible before immunization if specific immunochemical objectives are to be realized. Immunodominant macromolecular species, other than the antigen(s) of interest, can otherwise complicate $M_{ab}$ production to the detriment of the major effort (e.g. Section 5.2.17.1).

The scientific objectives of the work shown in Table 5.5 vary considerably and indicate the value of the immunochemical approach in the life sciences. Monoclonal antibodies to enzymes and proteins were produced: (a) to analyse recombinant DNA gene products in bacteria (Table 5.5, no. 12) or after transfection into target eukaryotic cells (Table 5.5, no. 1); (b) to study enzyme synthesis and degradation (Table 5.5, nos 3, 5, 16); (c) for immunocytochemistry (Table 5.5, nos 6, 17, 18) where $M_{ab}$s can give much better signal/noise ratios than polyclonal antibodies (e.g. Table 5.5, no. 18); (d) to study protein structure (Table 5.5, nos 2, 8, 13, 14, 17, 18, 20, 22, 23, 25, 27); (e) to give conformation-dependent $M_{ab}$s (e.g. Table 5.5, no. 21) which are of obvious value for immunoaffinity chromatography, e.g. where macromolecular antigen elution can be made dependent on ligand type or concentration, thereby raising the possibility of routine preparation of active enzymes, proteins or receptors in workable quantities; (f) to examine specific epitopes on unknown macromolecular antigens in complex cytopathological inclusions (Table 5.5, no. 26), or to isolate and characterize macromolecular antigen(s) present in minute quantities (Table 5.5, no. 4) or which are very difficult to purify (Table 5.5, no. 18) by the specific epitopic interactions of the monoclonal antibody(ies).

Interestingly, several papers comment on the nature of epitopes for $M_{ab}$s (e.g. conformational determinants, see Chapter 6). Immunoblotting after

**Table 5.5** Monoclonal antibodies to enzymes and proteins

| | Immunizing macromolecular antigen(s) | Fusion | Antibody assay | Objective | Special $M_{ab}$ properties | Other facts | Authors |
|---|---|---|---|---|---|---|---|
| 1. | Ovalbumin | Mouse–mouse (P3-NS1) | Immunobeads with anti-mouse Ig | To examine fidelity of transcription and translation of transfected genes | — | — | Bjercke *et al.* (1981) |
| 2. | Type II collagen | Mouse–mouse (MPC 11 45.6 TG1.7) | Passive haemagglutination | To study extracellular matrix | Epitope in helical region | — | Linsenmayer and Hendrix (1980) |
| 3. | Pyruvate kinase | Mouse–mouse (P3-NS1) | Solid phase | To study enzyme synthesis and degradation | — | — | Hance *et al.* (1980) |
| 4. | Cardiac myosin | Mouse–mouse, rat–mouse (P3-X63-Ag8) | Double antibody immunoprecipitation | To produce $M_{ab}$s to microheterogeneous proteins present in mixtures at bioconcentrations | Three $M_{ab}$s to myosin H-chain epitopes | Unique sequence specificity observed | Clark *et al.* (1980) |
| 5. | Yeast catalase T | Mouse–mouse (P3-X63-Ag8) | Protein A–*S. aureus* with 35 S yeast extract followed by SDS-PAGE[a] | To study enzyme synthesis | $M_{ab}$ recognizes catalase T and apoprotein | — | Adolf *et al.* (1980) |
| 6. | Tyrosine hydroxylase | Mouse–mouse (P3-NS1) | ELISA with alkaline phosphatase[a] | Immunocytochemistry | — | Localizes enzyme in catecholamine-containing neurones | Ross *et al.* (1981) |
| 7. | Glycolipid–*Salmonella minnesota* | Mouse–mouse (P3-NS1) | Solid phase with $^{125}$I-labelled Protein A and image-intensified autoradiography | To obtain anti-carbohydrate | — | — | Young *et al.* (1979) |

| | | | | | | | |
|---|---|---|---|---|---|---|---|
| 8. | Myoglobin | Mouse–mouse (P3-NS1) | Solid-phase ELISA with peroxidase | To characterize $M_{ab}$ binding sites on a protein | — | — | Berzofsky *et al.* (1980) |
| 9. | *Dilochos biflorus* lectin | Mouse–mouse (SP2-0-Ag14) | Double antibody immunoprecipitation with $^{125}$I-labelled lectin | To produce $M_{ab}$ to lectin | $M_{ab}$ reacts with *C*-terminal epitope | — | Borrebaeck and Etzler (1981) |
| 10. | Microtubule-associated protein ($MAP_2$) | Mouse–mouse (SP2-0-Ag14) | Solid phase with $^{125}$I-labelled Protein A and image-intensified autoradiography | To produce $M_{ab}$ which avoids polyclonal analytical problems | — | MAPs cut from SDS-PAGE gels for immunization | Izant and McIntosh (1980) |
| 11. | Rat skeletal myofibrils | Mouse–mouse (P3-NS1) | Indirect immunofluorescence | To produce $M_{ab}$ to proteins present in low concentration and difficult to purify | Two $M_{ab}$s which decorate intermediate filaments | — | Lin (1981) |
| 12. | Human interferon (IFN-$\beta$) | Mouse–mouse (FO) | Indirect antiviral assay and ELISA | To purify IFN-$\beta$ from recombinant DNA in bacteria | Two $M_{ab}$s | — | Secher and Burke (1980) |
| 13. | Phenylalanine hydroxylase | Rat–rat (R.210.RC.TG.1) | Enzyme activity in crude tissue extracts | To study $M_{ab}$–enzyme activity interactions | — | — | Choo *et al.* (1980) |
| 14. | Cytochrome P-450 LM2 | Mouse–mouse | Double antibody immunoprecipitation | To prepare $M_{ab}$s to cytochrome P-450 isozymes | — | — | Park *et al.* (1980) |
| 15. | Dinitrophenyl (DNP) keyhole limpet haemocyanin | Mouse–mouse (P3-NS1) | — | To study esterase activity of $M_{ab}$ | — | — | Kohen *et al.* (1980) |
| 16. | Human fibronectin | Mouse–mouse (P3-X63-Ag8) | Solid phase with $^{125}$I-labelled anti-mouse Ig | To analyse cellular fibronectin expression | Eight $M_{ab}$s which react with plasma and cellular fibronectin | Solid-phase assay performed with methanol-fixed (−20 °C) cells | Yardi *et al.* (1980) |

**Table 5.5** Monoclonal antibodies to enzymes and proteins (*continued*)

| | Immunizing macromolecular antigen(s) | Fusion | Antibody assay | Objective | Special $M_{ab}$ properties | Other facts | Authors |
|---|---|---|---|---|---|---|---|
| 17. | Human myoblasts | Mouse–mouse (P3-X63-Ag8) | Solid phase with foetal muscle myoblasts and $^{125}I$-labelled anti-mouse IgG $[F(ab^1)_2]$ | To define human muscle surface antigens | $M_{ab}$ to human fibronectin | $M_{ab}$ reacts weakly by indirect immunofluorescence with rat muscle cells | Walsh *et al.* (1981) |
| 18. | *Drosophila* RNA polymerase II | Mouse–mouse (SP2-0-Ag14) | Solid phase with RNA polymerase II and $^{125}I$-labelled anti-mouse IgG | To identify enzyme subunits in a complex of unknown stoichiometry | $M_{ab}$ detects the two large enzyme subunits | $M_{ab}$ excellent in indirect immunofluorescence and detects two regions containing RNA polymerase in heat-shock chromosome puff | Kramer *et al.* (1980) |
| 19. | *Drosophila* crude hydroxyapatite fraction of nuclear proteins | Mouse–mouse (P3-NS1) | — | To produce $M_{ab}$s to two nuclear membrane antigens which are very difficult to purify | One $M_{ab}$ detects 80 000 mol. wt antigen | Obtained several clones which produced $M_{ab}$s which in indirect immunofluorescence stained polytene chromosomes | Risau *et al.* (1981) |
| 20. | Platelet monoamine oxidase | Mouse–mouse (P3-X63-Ag8) | — | To distinguish genetically closely related enzyme forms | $M_{ab}$ recognizes monoamine oxidase B | — | Denney *et al.* (1982) |
| 21. | Cyclic nucleotide phosphodiesterase | Mouse–mouse (P3-NS1) | Solid phase with $^{125}I$-labelled Protein A. Precipitating $M_{ab}$s identified with *S. aureus* | To produce conformation-dependent $M_{ab}$s | One $M_{ab}$ reacts with conformationally generated epitopes | $M_{ab}$ reacts with phosphodiesterase $Ca^{2+}$–calmodulin complex. $M_{ab}$ used for immunoaffinity chromatography | Scott-Hausen and Beavo (1982) |

| | | | | | | | |
|---|---|---|---|---|---|---|---|
| 22. | *E. coli* $\beta$-galactosidase | Mouse–mouse (P3-NS1) | Solid phase with anti-mouse Ig-paper (CNBr) + $M_{ab}$ + $\beta$-galactosidase + chromosomic substrate | To study enzyme mutants | Several $M_{ab}$s | $M_{ab}$s can activate, inactivate and protect enzyme | Frackelton and Rotman (1980) |
| 23. | Complement $C_3$-inulin | Rat–rat (Y3-Ag.1.2.3) | Erythrocyte-bound complement intermediates with $^{125}$I-labelled anti-rat Ig | To study individual proteolytically sensitive regions of complement | Three $M_{ab}$s | None of $M_{ab}$s inhibits $C_3$ functions | Lachmann *et al.* (1980) |
| 24. | Ovalbumin | Mouse–mouse (P3-NS1) | — | To study the interaction of $M_{ab}$s generated in oocytes from microinjected $M_{ab}$ mRNA with antigen similarly generated | Two $M_{ab}$s | Immune reaction takes place in endoplasmic reticulum | Valle *et al.* (1982) |
| 25. | Sperm whale myoglobin | Mouse–mouse (P3-NS1) | Polyethylene glycol precipitate of $M_{ab}$–[$^3$H]myoglobin complex | To study $M_{ab}$ binding sites | Six $M_{ab}$s | — | Berzofsky *et al.* (1980) |
| 26. | Triton X-100 insoluble rat brain material or crude soluble protein from Alzheimer hippocampi | — | — | To characterize molecular nature of unknown material, i.e. neurofibrillary tangles | Five $M_{ab}$s to neurofilaments | $M_{ab}$s detect neurofilament determinants in neurofibrillary tangles | Anderton *et al.* (1982) |
| 27. | Hen lysozyme | Mouse–mouse (S194-5XXO BU.1) | Solid phase | To examine molecular basis of idiotypy | Two $M_{ab}$s show that idiotypy is not always directly related to antigen specificity | $M_{ab}$s tested on lysozyme proteolytic fragments | Metzger *et al.* (1980) |
| 28. | Pig heart mitochondrial $F_1$-ATPase | Mouse–mouse (NS1) | Solid phase with $F_1$-ATPase. Wells treated with acetone and air dried. + anti-mouse Ig + [$^{125}$I]Protein A | To study protein assembly and topology | Four $M_{ab}$s; three $M_{ab}$s for $\beta$-subunit; one $M_{ab}$ for $\alpha$-subunit; two $M_{ab}$s conformation-specific | $F_1$-ATPase glutaraldehyde cross-linked before immunization. Cross-linking enhances immune response. | Moradi-Ameli and Godinot (1983) |

**Table 5.5** Monoclonal antibodies to enzymes and proteins (*continued*)

| | Immunizing macromolecular antigen(s) | Fusion | Antibody assay | Objective | Special $M_{ab}$ properties | Other facts | Authors |
|---|---|---|---|---|---|---|---|
| 29. | Rat kidney $Na_1^+K^+$ ATPase | Mouse–mouse (NS1-1.Ag4.1) | Solid-phase ELISA with (a) purified ATPase, (b) primary cultures of rat hepatocytes, (c) inhibition of ATPase | To block the functions of surface ion transporting structures | Nine $M_{ab}$s; two $M_{ab}$s fully inhibit ATPase by binding to conformation-specific epitopes on $\alpha$-subunit | $M_{ab}$s purified with immobilized rabbit anti-mouse (IgG, IgM, IgA) and elution with low pH | Schenk and Leffert (1983) |
| 30. | Purified muscle protein phosphatases C-I and C-II | Mouse–mouse (SP2/0-Ag14) | Solid phase with rabbit anti-mouse Ig on plate, then culture supernatant, then $[^{125}I]F(ab^1)_2$ of sheep anti-mouse Ig *and* with antigens on plate, culture supernatants, rabbit anti-mouse Ig and $[^{125}I]F(ab^1)_2$ of goat anti-rabbit IgG | To probe structure–function relationships of phosphatases | Ten $M_{ab}$s to C-I; eight $M_{ab}$s to C-II | Cross-reactivities of $M_{ab}$s to C-I and C-II indicate that the enzymes may be structurally related | Speth *et al.* (1984) |
| 31. | Porcine eye lens vimentin | Mouse–mouse (PA.I) | Immunofluorescence microscopy with cell lines and tissue sections, solid-phase ELISA with vimentin, Western blots | To study conformational properties of intermediate filaments | Seven $M_{ab}$s differential staining with respect to species, fixing technique | $M_{ab}$s do not bind to desmin and glial fibrillary protein | Osborn *et al.* (1984) |
| 32. | Purified myosin (*Dictyostelium discordium*) | Mouse–mouse (SP2/0-Ag14) | Solid phase with myosin and $[^{125}I]F(ab^1)_2$ of rabbit anti-mouse IgG | To study the structure and function of myosin | Ten $M_{ab}$s; nine $M_{ab}$s to the heavy chain (210 kD); one $M_{ab}$ to the light chain (18 kD) as detected by Westerns | $M_{ab}$s recognize seven sites on the head and tail of myosin. Three $M_{ab}$s bind to myosin filaments | Peltz *et al.* (1985) |

| | | | | | | | |
|---|---|---|---|---|---|---|---|
| 33. | Haptenic nitroxide-spin label | — | — | To determine the distances of tyrosine residues from a $M_{ab}$ combining site by n.m.r. | $M_{ab}$ labelled by growing hybridoma on deuterated amino acids | n.m.r. resonances from unique protons in single amino acids | Anglister *et al.* (1984) |
| 34. | Homogenized hypothalami | Mouse–mouse (P3-X63-Ag8-653) | Immunocytochemical screening with sections fixed in Bouin's fluid | To define brain antigens without chemical isolation | 37 hybridoma supernatants to neuronal elements. 44 hybridomas to non-neuronal elements. | No antibodies to astrocytes | Sternberger *et al.* (1982) |
| 35. | Sheep tracheal mucus | Mouse–mouse (SP2/0) | Immunofluorescence with sections of trachea, intestine and salivary gland (paraformaldehyde fixed) | To obtain markers for serous, mucous, goblet and ciliated cell products | $M_{ab}$s to tracheal goblet cell, serous cell, ciliated cell | $M_{ab}$s can be used to monitor secretions in different physiological conditions | Basbaum *et al.* (1984) |
| 36. | Human Igs | — | — | To study antigen-binding characteristics of $M_{ab}$s | $M_{ab}$s against $\kappa$, $\gamma$ and $\delta$ chains of Igs | Only one $M_{ab}$ reacted with antigen in Western blots after electrophoresis in denaturing conditions | Thorpe *et al.* (1984) |

[a] SDS-PAGE, polyacrylamide gel electrophoresis in the presence of sodium dodecyl sulphate; ELISA, enzyme-linked immunoadsorbent assay.

polyacrylamide gel electrophoresis in the presence of sodium dodecyl sulphate (Table 5.5, no. 29), in the presence and absence of 2-mercaptoethanol (Table 5.5, no. 36), and immunoblotting on nitrocellulose after simple denaturation with sodium dodecyl sulphate alone (Table 5.5, no. 28), serves, at least in part, to define the role of conformation in defining epitopic specificity. It is interesting that at least half of the $M_{ab}$s to membrane antigens react with antigens in immunoblots after antigen treatment with sodium dodecyl sulphate (Table 5.5, no. 28) or sodium dodecyl sulphate and 2-mercaptoethanol (Table 5.5, no. 29), whereas $M_{ab}$s to soluble proteins (immunoglobulins) predominantly (3/4) do not react in the presence of 2-mercaptoethanol (Table 5.5 no. 36). It is possible, as pointed out in Chapter 6, that the reduction in—S—S—bridges in soluble proteins causes a collapse of conformational epitopic features, whereas this is not always the case for membrane proteins, perhaps because they have stabilized conformational structures in the presence of reducing agents which can behave as epitopes more frequently in immunoblotting procedures.

The usefulness of $M_{ab}$s in all of these studies depends on the specific immunochemical properties of each $M_{ab}$ (see Section 5.2). Identification of $M_{ab}$s of interest by the judicious design of initial screening procedures, i.e. designing a screening assay which specifically detects antibodies with the wanted properties, e.g. immunoprecipitation with *S. aureus*, or immunofluorescence, is the only guaranteed way of obtaining $M_{ab}$s with the desired properties (see Section 5.2).

Clearly, the absolute epitopic specificity of $M_{ab}$s for macromolecular protein antigens offers distinct advantages over polyclonal antibodies for most uses, e.g. where increased specificity may be obtained only through extensive adsorption procedures (Mayer and Walker, 1978, 1980), except where multi-antibody macromolecular interactions are essential, e.g. in precipitin formation, or alternatively where common epitopes on different macromolecular antigens may make interpretation difficult.

### 5.3.2 Membrane antigens (including receptors)

The application of $M_{ab}$s to membrane antigens (including receptors) is shown in Table 5.6. Whole cells or isolated plasma membrane preparations offer an attractive mixture of immunizing macromolecular antigens for $M_{ab}$ production, both theoretically, since the cell surface represents the interface between cellular self and non-self, and therefore contains the macromolecular components which mediate and regulate information exchange, i.e. receptors, and practically, since whole cells or even plasma membranes are relatively easy to prepare in good yield. It is also arguable that one might expect macromolecular antigens in the cell surface to be particularly immunogenic, since normally *in vivo* such cell surfaces are the prime target of the immune response when recognizing self and non-self. There is a burgeoning literature on $M_{ab}$s to the cell surface, particularly for cells in the immune system, as well as for markers of normal and abnormal cell differentiation (see cell-surface markers, Section 5.3.3). In the sections on $M_{ab}$s to membrane antigens (including receptors), and membrane antigens (cell-surface markers), information is again selected to draw attention to the principles and objectives of these myriad studies.

The primary objectives of producing $M_{ab}$s to membrane antigens have been to:

(a) study and isolate receptors (often present in very small quantities and very difficult to isolate biochemically) and to study receptor-mediated endocytosis (Table 5.6, nos 3, 11, 16, 18);

(b) study membrane biogenesis and turnover (Table 5.6, nos 2, 9);

(c) isolate and characterize membrane components (Table 5.6 nos 5, 6, 7, 10);

(d) study antigen orientation (Table 5.6, no. 8);

(e) prepare monoclonal immunotoxins (Table 5.6, nos 12, 13). This is an expanding area of particular clinical interest since the "magic-bullet" concept of targeting some cytotoxic substance to specific cell types dates back to the earliest studies on chemotherapy. Tumour imaging with $M_{ab}$s is also an area of great interest, e.g. with tumour metastases. The delivery of toxins as covalent conjugates (Table 5.6, no. 12) or contained in $M_{ab}$-targeted liposomes (Table 5.6, no. 13) provides great potential for the treatment of disease, particularly in tumour biology.

(f) study autoimmune disease (Table 5.6, no. 14). Spleens from animals with autoimmune disease provide a source of spleen cells involved in mediating the disease process. Therefore, generation of hybridomas with these cells provides a means of obtaining monoclonally the antibodies involved in the pathogenesis of the disease. In this way the individual $M_{ab}$s may be evaluated alone, or in groups, in order to understand immune mechanisms in the autoimmune process, e.g. organ and tissue specificity in the autoimmune response.

(g) characterize surface antigens associated with cell proliferation (Table 5.6, nos 1, 14, 17).

Several studies have been carried out on the production and use of $M_{ab}$s to the acetylcholine receptor (Tables 5.6, nos 27–33), which is a much-studied receptor species. Libraries (Table 5.6, no. 30) of $M_{ab}$s have been used to study the topography of the receptor in isolated preparations *in situ*. Again, $M_{ab}$s have been used to study, indeed induce, autoimmune responses (experimen-

**Table 5.6** Monoclonal antibodies to membrane antigens (including receptors)

| | Immunizing macromolecular antigen(s) | Fusion | Antibody assay | Objective | Special $M_{ab}$ properties | Other facts | Authors |
|---|---|---|---|---|---|---|---|
| 1. | Human erythrocytes | Mouse–mouse (P3-NS1) | Erythrocyte with $^{125}$I-labelled anti-mouse IgG $[F(ab^{1})_2]$ | Identification of cell lineages | Two $M_{ab}$s to band 3; three $M_{ab}$s to glycophorin A | One anti-band 3 binds to same epitope on all bone marrow cells | Edwards (1980) |
| 2. | Enterocyte microvillus membrane | Mouse–mouse (P3-NS1) | Immunoadsorbent Sepharose anti-mouse IgG + $M_{ab}$ + detergent microvillus extract (assay for enzyme activity) | To study enzyme biogenesis and postnatal development | Two $M_{ab}$s to intestinal sucrose–isomaltase | Enzyme purified by $M_{ab}$ immunoadsorbent | Hauri *et al.* (1980) |
| 3. | Purified asialoglycoprotein receptor | Mouse–mouse (P3-X63-Ag8) | Antigen + $M_{ab}$ + $^{125}$I-labelled anti-mouse (Fab) or FACS[a] | To identify a receptor independent of ligand binding | $M_{ab}$ inhibits radioligand binding | $M_{ab}$ shows receptor may constitute 1–2% of hepatocyte surface proteins | Schwartz *et al.* (1981) |
| 4. | Purified oestrogen receptor from MCF-7 human breast cancer cells | Rat–mouse (P3-X63-Ag8) and (SP2-0-Ag14) | Double antibody with crude MCF-7 cytosol | Radioimmune assay and immunocytochemistry | Three $M_{ab}$s to oestrophilin | $[^{35}S]M_{ab}$ prepared in cloned hybridomas | Greene *et al.* (1980) |
| 5. | NIH-3T3 cells or plasma membranes | Rat–mouse (P3-X63-Ag8) and (P3-NS1) | Glutaraldehyde-fixed cells + $[^{125}I]$anti-rat IgG | To characterize major protein components of dividing eukaryotic cells | Two $M_{ab}$s (mol. wt 220 000 and 90 000) in surface of many cell lines | 13 $M_{ab}$s characterized | Hughes and August (1981) |
| 6. | Membrane glycoprotein from human neuroblastoma (IMR-5) cells | — | — | To isolate membrane components | — | $M_{ab}$ used in immunoadsorbent (Sepharose-anti-mouse IgM-$M_{ab}$) | Momoi *et al.* (1980) |

| | | | | | | | |
|---|---|---|---|---|---|---|---|
| 7. | Human platelets | Mouse–mouse (SP2-0-Ag14) | Platelet assay with $^{125}$I-labelled anti-mouse IgG | To isolate platelet surface glycoproteins | $M_{ab}$ reacts with platelet glycoprotein IIb–IIIa complex | Antigen complex not on platelets in Glanzmann's thrombasthenia | McEver *et al.* (1980) |
| 8. | Human platelets | Mouse–mouse (SP2-0-Ag14) | Filter paper detergent-solubilized platelet membrane (CNBr) + $M_{ab}$ + $^{125}$I-labelled anti-mouse IgG | To study structure, orientation and function of membrane proteins | — | — | Newman *et al.* (1981) |
| 9. | Rat liver plasma membranes | Mouse–mouse (P3-NS1) | Plasma membrane + $M_{ab}$ + $^{125}$I-labelled anti-mouse IgG | To study disposition and turnover of membrane proteins | — | 20 cell lines cloned twice | Siddle and Soos (1981) |
| 10. | Milk fat globule membrane | Mouse–mouse (SP2-0-Ag14) | Radioimmune assay with $^{125}$I-labelled anti-mouse Igs (IgA, $IgG_1$, $IgG_{2a}$, $IgG_{2b}$ and IgM) | To study secretory cell-surface antigens | Four $M_{ab}$s to xanthine oxidase | Nine $M_{ab}$s produced | Mather *et al.* (1980) |
| 11. | — | — | — | To isolate Fc receptor from J 774 cells | $M_{ab}$ is rat anti-mouse | Immunoadsorbent with $M_{ab}$ (Fab) binds detergent-solubilized antigen | Mellman and Unkeless (1980) |
| 12. | — | — | — | To study monoclonal immunotoxins | Anti-Thy 1,2 $M_{ab}$ (IgM) | $M_{ab}$ coupled to *Diphtheria* toxin A chain | Blythman *et al.* (1981) |
| 13. | — | — | — | To target liposomes in cytotoxic therapy | Anti-human $\beta$-microglobulin | $M_{ab}$ coupled to liposomes with *N*-hydroxy succinimidyl 3(2-pyridylthio)-propionate | Leserman *et al.* (1980) |
| 14. | Rat pancreatic islet cell line (RIN) | Mouse–mouse (P3-X63-Ag8) | RIN cells in microtitre plates + $^{125}$I-labelled anti-mouse IgG $[F(ab^1)_2]$ or $^{125}$I-labelled Protein A | To recognize islet cell differentiation antigens | 12 $M_{ab}$s | $M_{ab}$s assessed in complement-mediated cytotoxicty test | Eisenbarth *et al.* (1981) |

**Table 5.6** Monoclonal antibodies to membrane antigens (including receptors) (*continued*)

| | Immunizing macromolecular antigen(s) | Fusion | Antibody assay | Objective | Special $M_{ab}$ properties | Other facts | Authors |
|---|---|---|---|---|---|---|---|
| 15. | Spleen cells from Reovirus Type 1 infected mice (SJL/J) | Mouse–mouse (P3-X63-Ag8) | Fixed paraffin sections of mouse tissues with indirect immunofluorescence | To study cytoimmune disease | Seven $M_{ab}$s multi-organ specific | $M_{ab}$s may react with some macromolecule or common epitope on different macromolecule | Haspel *et al.* (1983) |
| 16. | Partially purified turkey erythrocyte $\beta_1$-adrenergic receptor or calf lung $\beta_2$-adrenergic receptors | Mouse–mouse (SP2-0-Ag14) | Polyethylene glycol precipitation with $M_{ab}$ + $^3$H-labelled ligand ± propranolol | To study and purify $\beta$-receptors | Four $M_{ab}$s to $\beta_1$-receptor; one $M_{ab}$ to $\beta_2$-receptor | $M_{ab}$ cross-reactivity suggests $\beta_1$ and $\beta_2$ related | Fraser and Venter (1980) |
| 17. | Purified human transferrin receptor | Mouse–mouse (S199-5XX0-BU.1) | Cell receptor block (thymus derived leukaemic cells) + $M_{ab}$ + $^{125}$I-labelled human transferrin | To study cell proliferation associated antigens | One $M_{ab}$ | $M_{ab}$ inhibits growth of a human T-leukaemic cell line *in vitro* | Trowbridge and Lopez (1982) |
| 18. | Partially purified bovine adrenal cortex low-density lipoprotein receptor | Mouse–mouse (SP2-0-Ag14) | — | To study receptor-mediated endocytosis | — | $M_{ab}$ affinity for receptor 74 times lower at 37 °C than at 4 °C | Beisiegel *et al.* (1981) |
| 19. | Partially purified rat liver glucocorticoid receptor | Mouse–mouse (SP2/0-Ag14) | ELISA with glucocorticoid receptor and peroxidase-linked rabbit anti-mouse Ig and second antibody immunoprecipitation with [$^3$H]triamcinolone | To study fine domain structure of the receptor | Ten $M_{ab}$s which recognize receptor in domain which does not bind DNA or ligand | No $M_{ab}$ reacted with human lymphocyte glucocorticoid receptor | Okret *et al.* (1984) |

| | | | | | | | |
|---|---|---|---|---|---|---|---|
| 20. | Calf brain muscarinic acetylcholine receptor | Mouse–mouse | — | To study properties (agonist-like) of $M_{ab}$s | One $M_{ab}$ reacts with digitonin-solubilized receptor; one $M_{ab}$ reacts with denatured (SDS) receptor | Both $M_{ab}$s cause contraction of guinea-pig myometrium | Leiber *et al.* (1984) |
| 21. | CNBr fragment of cross-linked fibrin | Mouse–mouse (P3-NS1-4Ag-4.1) | Solid-phase assay with $^{125}$I-goat anti-mouse $F(ab^1)_2$ and fibrin fragments and fibrinogen | To study fibrin cross-linking | $M_{ab}$ to region of fibronectin involved in cross-linking to human fibrin | Fibronectin covalently attached to $\alpha$ chains of fibrin | Sobel *et al.* (1983) |
| 22. | Tissue extracts from foetal mouse brain | Rat–mouse (P3-X63-Ag-U1) | Morphological changes in cultured brain cells induced by supernatants | To prepare $M_{ab}$ to molecules involved in calcium-dependent brain cell–cell adhesion | One $M_{ab}$ to adhesion molecule(s) | $M_{ab}$ disrupts calcium-dependent cell–cell adhesion in brain | Hatta *et al.* (1985) |
| 23. | Hepatocytes, plasma membrane sheets, surface glycoproteins | Mouse–mouse (P3-X63-Ag8-U1) | Solid-phase with plasma membrane, supernatants, and $^{125}$I-labelled goat anti-mouse $F(ab^1)_2$ | To generate surface domain-specific markers for hepatocytes | 11 $M_{ab}$s to plasma membrane; seven $M_{ab}$s recognize denatured plasma membrane polypeptides on Westerns | One $M_{ab}$ recognizes bile canalicular surface; three $M_{ab}$s recognize lateral and sinisordal surfaces | Hubbard *et al.* (1985) |
| 24. | Rat liver purified cytochrome P450c | Mouse–mouse (P3-X63-Ag8.653) | ELISA | To study cytochrome P450 heterogeneity | Nine $M_{ab}$s which react with six epitopes on antigen | One epitope shared with cytochrome P450d | Thomas *et al.* (1984) |
| 25. | Bleached bovine rod outer segment disc membranes | Mouse–mouse (NS1) | Solid-phase assay with bleached disc membranes, culture supernatants and [$^{125}$I]anti-mouse Ig | To study rod outer segment disc membranes | Two $M_{ab}$s to rhodopsin; one $M_{ab}$ to *C*-terminal segment; one $M_{ab}$ to *N*-terminal segment | Bleached rhodopsin 13 times more sensitive to the *N*-terminal $M_{ab}$ | Molday and MacKenzie (1983) |
| 26. | — | — | — | To isolate detergent-solubilized membrane antigen | Prepare biotinyl-$M_{ab}$ | Streptavidin-agarose used with high affinity. Overcomes variability of Protein A adsorbents | Updyke and Nicholson (1984) |

**Table 5.6** Monoclonal antibodies to membrane antigens (including receptors) (*continued*)

| | Immunizing macromolecular antigen(s) | Fusion | Antibody assay | Objective | Special $M_{ab}$ properties | Other facts | Authors |
|---|---|---|---|---|---|---|---|
| 27. | Purified *Torpedo* acetylcholine receptor | Rat–mouse (P3-NS1) and (S194 5XXO.BU.1) | — | To prepare specific toxin-like probes to different parts of the acetylcholine receptor | 17 $M_{ab}$s obtained | — | Tzartos and Lindstrom (1980) |
| 28. | Purified *Torpedo* acetylcholine receptor | Rat–mouse (S194-5XXU.BU.1) | Immunoprecipitation assay with $^{125}$I-labelled $\alpha$-bungarotoxin + anti-rat Ig | To prepare immunoreagents for the dissection of macromolecular structure | — | Epitope remote from acetylcholine binding site | Lennon *et al.* (1980) |
| 29. | Purified *Torpedo* acetylcholine receptor | Rat–mouse (P3-NS1) | Haemagglutination with receptor-coated erythrocytes | To assess the contribution of a single antibody species to an immune reaction *in vivo* | Six $M_{ab}$s | Two $M_{ab}$s separately produces experimental myasthenia gravis | Richman *et al.* (1980) |

| | | | | | | | |
|---|---|---|---|---|---|---|---|
| 30. | Purified *Electrophorus* acetylcholine receptor | (S194-5XXU.BU.1) and (rat Lou myeloma Y3) | Immunoprecipitation with $M_{ab}$ + $^{125}$I-labelled $\alpha$-bungarotoxin + anti-rat Ig | To develop $M_{ab}$ library direct to epitopes on denatured and intact receptor | 40 $M_{ab}$s | Nine immunogenic regions on receptor. The main immunogenic region is least species-specific and found on the extracellular surface of $\alpha$-subunit | Tzartos *et al.* (1981) |
| 31. | Haptenic trans-3,3′-bis [$\alpha$(trimethyl-ammonia)methyl]-azobenzene bromide (Bis Q) | Mouse–mouse (P3-X63-Ag8.65) | Radioimmune assay with Bis Q-bovine serum albumin or polyclonal anti-Bis Q or acetylcholine receptor + peroxidase-labelled anti-mouse Ig | To produce anti-acetylcholine receptor $M_{ab}$s by anti-idiotypic response | One $M_{ab}$ | $M_{ab}$ is anti-idiotype to anti-Bis Q; therefore binds to acetylcholine receptor | Cleveland *et al.* (1983) |
| 32. | Purified *T. californica* acetylcholine receptor | Rat–mouse (SP2/0-Ag14) | Passive haemagglutination assay | To probe non-cholinergic sites on the acetylcholine receptor | One $M_{ab}$ which binds to ion channel domain | Epitope maintained in SDS | Donnelly *et al.* (1984) |

[a] FACS, fluorescent activated cell sorter.

tal myasthenia gravis) in animals. Single $M_{ab}$s (Table 5.6, no. 29) can induce the disease.

Again, as indicated previously, a large percentage of $M_{ab}$s to membrane antigens appear to react with epitopes after polyacrylamide gel electrophoresis in the presence of sodium dodecyl sulphate and 2-mercaptoethanol followed by immunoblotting (Table 5.6, nos 20, 23, 24, 25, 32).

### 5.3.3. Membrane antigens (cell-surface markers)

The application of $M_{ab}$s to cell-surface marker antigens is shown in Table 5.7. As already discussed (Section 5.3.2.), intensive efforts have been devoted to the production and use of $M_{ab}$s to surface components. $M_{ab}$s have been produced to cell-surface antigens, which may define stages of cell differentiation (cell lineages), distinguish differentiated cell types or define tumour cell surfaces. Classical biochemical studies on such macromolecular surface antigens, which are usually found in very small amounts and are very difficult to purify in the absence of specific probes (cf. receptors, Section 5.3.2.), are extremely difficult. Therefore, in spite of the chance aspect of the approach, much effort has been gambled on the production of $M_{ab}$s to cell surfaces from tumour cells, embryonic cells, cells from the central and peripheral nervous systems and blood cells. Table 5.7 considers but a small part of the large literature in order to again highlight the objectives and principles of this type of work. It might be argued that difficulties which were posed for such studies before the advent of $M_{ab}$s justify the speculative use of such impure immunizing macromolecular mixtures to produce $M_{ab}$s of interest, although judicious use of precise hybridoma supernatant screening assays can enable antibodies of interest to be easily identified. One major problem for studies using whole cells for plasma-membrane preparations is that blood group antigens and other surface glycolipids can often provoke the major immune responses (e.g. Brown *et al.*, 1983). The primary objectives of producing $M_{ab}$s to surface-marking membrane antigens have been to:

(a) characterize tumour cell-surface markers (Table 5.7, nos 1, 2, 3, 16) so that ultimately such $M_{ab}$s can be used to diagnose (e.g. tumour imaging) and treat (e.g. with antibody–toxin conjugates, Table 5.6, nos 12, 13) malignant tumours or metastases clinically *in vivo*;

(b) distinguish and identify cell types in complex cellular mixtures. Possibly $M_{ab}$ technology has been more fully exploited by neuroscientists than by almost any other scientific group in order to unravel the mysteries of cell identification and cell–cell interaction in the central and peripheral nervous systems. In this way $M_{ab}$s have been prepared which can distinguish

**Table 5.7** Monoclonal antibodies to membrane antigens (cell-surface markers)

| | Immunizing macromolecular antigen(s) | Fusion | Antibody assay | Objective | Special $M_{ab}$ properties | Other facts | Authors |
|---|---|---|---|---|---|---|---|
| 1. | Human melanoma cells | Mouse–mouse (P3-NS1) | *S. aureus* binding assay to detect $M_{ab}$s ($IgG_2$) in preliminary screen | To detect cell-surface marker antigens | Seven $M_{ab}$s to melanoma surface | Immunoprecipitation with *S. aureus* of cell $[^{125}I]M_{ab}$ or cell-lysate $[^{125}I]M_{ab}$ conjugates | Brown *et al.* (1980) |
| 2. | Melanoma-associated antigens | — | Double-determinant immunoassay (DDIA) with two $M_{ab}$s to two epitopes on one macromolecular antigen | To analyse tissues for cell-surface antigens | — | For DDIA, one $M_{ab}$ on immunoadsorbent + tissue detergent extract + $^{125}I$-labelled $M_{ab}$ to other epitope | Brown *et al.* (1981a) |
| 3. | Melanoma-associated antigens | — | DDIA | To diagnose and treat melanoma | $M_{ab}$ to mol. wt 97 000 surface protein | — | Brown *et al.* (1981b) |
| 4. | Rat astrocytes | Mouse–mouse (P3-NS1) | Cells + $M_{ab}$ + $^{125}I$-labelled anti-mouse Ig | To identify nerve cell types | Two $M_{ab}$s | Rat-specific antigen is not on neurones | Bartlett *et al.* (1981) |
| 5. | Mouse F9 embryonal teratocarcinoma cells | — | — | To define developmental surface markers | $M_{ab}$ to stage specific embryonic antigen | Antigen is complex glycolipid | Nudelman *et al.* (1980) |
| 6. | Intermediate filaments trophoblastoma cells | Rat–mouse (SP2-0-Ag14) | — | To obtain an embryonic cell type marker | $M_{ab}$ decorates filament network in trophectoderm only | — | Brulet *et al.* (1980) |
| 7. | *Drosophila* imaginal discs | Mouse–mouse (P3-NS1) | Indirect immunofluorescence with imaginal discs | To characterize developmentally regulated cell-surface components | — | Antigen is specific for the diploid epithela of *Drosophila* | Brower *et al.* (1980) |
| 8. | Whole leech nerve cords | — | Immunocytochemistry with nerve cord segments containing several ganglia | To obtain nerve cell type immunoreagents | 20 $M_{ab}$s studied | $M_{ab}$s against identifiable sensory and motor neurones | Zipser and McKay (1981) |

**Table 5.7** Monoclonal antibodies to membrane antigens (cell-surface markers) (*continued*)

| | Immunizing macromolecular antigen(s) | Fusion | Antibody assay | Objective | Special $M_{ab}$ properties | Other facts | Authors |
|---|---|---|---|---|---|---|---|
| 9. | Detergent-solubilized Concanavalin A binding neonatal brain glycoproteins | Rat–mouse (P3-NS1) | Live cerebellar cells from neonatal mice with $^{125}I$-labelled anti-rat IgG | To obtain immunoreagents to brain cell-surface components without purification | One $M_{ab}$ | $M_{ab}$ bound to neuronal cells only | Hirn *et al.* (1981) |
| 10. | E63 myoblasts | Mouse–mouse (PP2-O-Ag14) | Fixed E63 cells with $^{125}I$-labelled anti-mouse Ig | To obtain immunoreagents to developmentally regulated myoblast surface antigens | Several $M_{ab}$s | E63 cells have four epitopes which are absent from myotubes | Lee and Kaufmann (1981) |
| 11. | Human acute monoblastic leukaemia cells | Mouse–mouse (P3-NS1) | — | To examine antigens shared by different differentiated cells | $M_{ab}$ reacts with mol. wt 45 000 antigen on monocytes and neurones | — | Hogg *et al.* (1981) |
| 12. | White matter from bovine corpus callosum | Mouse–mouse (P3-NS1) | Immunofluorescence | To produce cell type specific immunoreagents | Four $M_{ab}$s (IgM) which detect oligodendrocytes | $M_{ab}$s interspecific and will work in complement-mediated cytotoxicity tests | Sommer and Schachner (1981) |
| 13. | Adrenergic neuronal cells | Mouse–mouse (P3-NS1) | Neurones + $^{125}I$-labelled anti-mouse IgG $[F(ab^1)_2]$ | To produce immunoreagents to sympathetic nerve antigens | One $M_{ab}$ | $M_{ab}$ binds to adrenergic and cholinergic neurones | Chun *et al.* (1980) |
| 14. | Synaptosomal plasma membrane from adult rat cerebellum | Mouse–mouse (X63-Ag8.6.5.3) | Dot immunoblotting antigen on nitrocellulose + $M_{ab}$ + anti-mouse IgG peroxidase conjugate | To produce neurone-specific immunoreagents | One $M_{ab}$ | Antigen (mol. wt 23 000) is mitochondrial but neurone-specific | Hawkes *et al.* (1982a,b) |
| 15. | Rat thymocyte membranes | Mouse–mouse (P3-NS1) | Fixed thymocytes + $^{125}I$-labelled anti-mouse Ig $[F(ab^1)_2]$ | To study lymphocyte differentiation antigens | Five $M_{ab}$s | $M_{ab}$s define lymphoid subpopulation | Williams *et al.* (1977) |

| | | | | | | | |
|---|---|---|---|---|---|---|---|
| 16. | Human colorectal carcinoma cells | Mouse–mouse (P3-NS1) | Live target cells in radioimmune assay | To identify colorectal-specific antigens | Two $M_{ab}$s which do not bind to normal colonic mucosa or other malignant human cells | — | Herlyn *et al.* (1979) |
| 17. | Human infant thymocytes | Mouse–mouse (P3-NS1) | — | Fortuitous finding of T cell antigen on Purkinje cells | One $M_{ab}$ | $M_{ab}$ recognizes a functionally discrete neuronal population in vertebrate brain | Garson *et al.* (1982) |
| 18. | Rat cerebellar cells | Mouse–mouse (P3-NS1) | Cultured cerebellar cells, radioimmune assay or immunofluorescence | To produce immunoreagents for CNS neurones | One $M_{ab}$ binds to central and not peripheral neurones | — | Cohen and Selvendran (1981) |
| 19. | Membrane pellet from dorsal root ganglia sensory neurones | Mouse–mouse (X63-Ag8.6.5.3) | Fixed sections of dorsal root ganglia + indirect immunofluorescence | To produce immunoreagents to sensory neurone subpopulations | One $M_{ab}$ which is cytotoxic to mammalian central and peripheral neurones | Antigen also on *Trypanosoma cruzi* (causes Chagas disease) | Wood *et al.* (1982) |
| *Histocompatibility antigens* | | | | | | | |
| 20. | Mouse B10 spleen cells | Rat–mouse (P3-NS1) | — | To isolate H-2 antigens | $M_{ab}$ is anti-$H_2$ | — | Stallcup *et al.* (1981) |
| 21. | — | — | — | To study biosynthesis and assembly of human transplantation antigens | $M_{ab}$s to HLA, A, B and C antigens | — | Algranati *et al.* (1980) |
| 22. | DA stain antigens | Rat–mouse (P3-X63-Ag8) | Complement-mediated lysis of $^{51}$Cr-labelled red cells and lymphocytes of DA stain donors | To study complex surface-antigenic structures | Several $M_{ab}$s | — | Galfre *et al.* (1977) |
| 23. | — | — | — | To study HLA antigens in T cell leukaemia virus (HTLV) infected cells | One $M_{ab}$ | In all HTLV infected cells, altered HLA expression was observed | Mann *et al.* (1983) |

between neuronal and non-neuronal cells, and between peripheral sensory and motor neurones (Table 5.7, nos 4, 8, 9, 11, 12, 13, 14, 18).

(c) identify stage-specific antigens in cell lineages (Table 5.7, nos 5, 6, 7, 10);

(d) identify shared antigens on apparently functionally unrelated differentiated cell types (Table 5.7, nos 11, 17, 19). Fortuitously, several intercellular surface components have been recognized in this way, including shared macromolecular antigens on monocytes and neurones (Tables 5.7, no. 11), T cells and Purkinje cells (Table 5.7, no. 17) and neurones and the parasite *Trypanosoma cruzi* (Table 5.7, no. 19). Clearly, as in the latter case, extremely interesting relationships between parasitic cells and host cells may be revealed by this approach which may also reflect on the genetics of invading parasite and host cells which determine the success of the parasitic relationship (Clarke *et al.*, 1983).

(e) carry out detailed studies on the histocompatibility antigens (Table 5.7, nos 20, 22, 23). Obviously there is a large literature on this subject and references in Table 5.7 are simply to illustrate some of the uses of anti-histocompatibility $M_{ab}$s in defining expression and mode of biosynthesis and assembly of the antigens.

### 5.3.4. Viruses and cell-transforming proteins

The application of $M_{ab}$s to viruses and cell-transforming proteins is shown in Table 5.8. The primary objectives of producing $M_{ab}$s to viruses and cell-transforming antigens have been to:

(a) use monoclonals to examine viral taxonomy (Table 5.8, nos 1, 2, 3, 5, 6, 7, 8, 9);

(b) quantitate viral proteins in infected cells (Table 5.8, no. 4);

(c) characterize transformation antigens in chemically or virally transformed cells (Table 5.8, nos 10, 11). Transformation antigens which characterize chemically or oncogene-transformed cells are being increasingly studied so that the regeneration of the malignant phenotype from normal cell phenotype can be understood. $M_{ab}$s to transformation-related proteins (Table 5.8, nos 10, 11) can be used to probe the transitions in specific gene expression which take place.

### 5.3.5. Chromosomes, genes and modified DNA

The application of $M_{ab}$s to chromosomes, genes and modified DNA is shown in Table 5.9. The primary objectives of producing $M_{ab}$s have been to:

(a) detect specific sequences in DNA (Table 5.9, nos 3, 6, 11) or DNA modification (Table 5.9, nos 8, 12, 13);

(b) study autoimmune diseases (Table 5.9, nos 5, 6, 10, 11). These studies have used donor spleens for $M_{ab}$ production from animals which develop a systemic lupus syndrome. Therefore, spleen-donor cells would be expected to produce autoimmune antibodies commonly found in systemic lupus. Subsequently, therefore, hybridoma clones will produce $M_{ab}$ to epitopes on specific lupus antigens, e.g. DNA, RNA. This objective was realized (Table 5.9, nos 5, 6, 10, 11). The principle of using a spleen from an animal with an autoimmune disease to produce $M_{ab}$s may presumably be applied to any disease which provokes the immune system, thereby giving $M_{ab}$s for specific detailed studies on antigens involved in the pathogenesis of disease processes;

(c) map the chromosomal location of antigens (Table 5.9, no. 7);

(d) study enzyme polymorphisms (Table 5.9, nos 2, 4);

(e) characterize non-histone nuclear proteins (Table 5.9, no. 1).

### 5.3.6. Clinical radioimmune assays

Some applications of $M_{ab}$s to the development of clinical radioimmune assays are shown in Table 5.10. Examples have been chosen to illustrate the principles and objectives of such studies. The primary objective of using $M_{ab}$s has been to produce immunoreagents of precise specificity (Table 5.10, nos 1, 2, 5). The great attraction of $M_{ab}$ in the clinical diagnostic field is the production of immunoreagents which are in unlimited supply and which therefore guarantee the worldwide quality control of immune assays. $M_{ab}$s must be used in combinations (Table 5.10, no. 4) to be effective turbidimetrically (or in precipitation reactions) as predicted by the lattice theory of antigen–antibody interactions.

### 5.3.7. Cloning and screening of recombinant expression systems

The use of antibodies (both monoclonal and polyclonal) for studies on gene cloning can be divided into two parts: (a) the use of antibodies for selection of low-abundance mRNA (for polyclonal antibodies see Chapter 4); (b) the use of antibodies to screen recombinant expression libraries or the expression of transfected genes. Clearly, antibodies will have increasing use in such studies where the immunochemical specificity can be relied upon to examine expressed complete polypeptides or fragments of polypeptides from cloned whole or truncated genes; products of site-directed or deletional mutagenesis should be easily detected in this way. However, as with most biological assay

**Table 5.8** Monoclonal antibodies to viruses and cell-transforming proteins

| | Immunizing macromolecular antigen(s) | Fusion | Antibody assay | Objective | Special $M_{ab}$ properties | Other facts | Authors |
|---|---|---|---|---|---|---|---|
| 1. | Measles virus | Mouse–mouse (SP2-0-Ag14) and (X63-Ag8.6.5.3) | Immunoprecipitation and immunofluorescence | To study antigenic relationships between viruses | Three $M_{ab}$s to haemagglutinin; two $M_{ab}$s to nucleocapsid | — | Birrer *et al.* (1981) |
| 2. | Herpes simplex virus types 1 and 2 | — | — | To examine epitope distribution in viral proteins | $M_{ab}$s to nucleocapsid protein | $M_{ab}$s can immunoprecipitate antigens of different subunit size which share epitope | Zweig *et al.* (1980) |
| 3. | Influenza A virus | Mouse–mouse (P3-X63-Ag8) | Haemagglutination and neuraminidase inhibition | To examine antigenic differences in viral nucleoproteins | Five $M_{ab}$s to nucleoprotein epitopes | Epitope variation in viral nucleoprotein is independent of epitopic changes in surface glycoproteins | van Wyck *et al.* (1980) |
| 4. | Purified influenza virus matrix (M) protein | Mouse–mouse (P3-X63-Ag8) | Solid-phase radioimmune assay with M-protein coated plates | To quantify M protein in infected cells | One $M_{ab}$ which immunoprecipitates M protein from $^{35}S$-labelled virion proteins | — | Hackett *et al.* (1980) |
| 5. | Influenza virus A (Memphis) 102/72 | Mouse–mouse (P3-NS1) and (SP2-0-Ag14) | Solid-phase radioimmune assay with $^{125}I$-labelled anti-mouse Ig or $^{125}I$-labelled Protein A | To determine epitopic topography of viral haemagglutinin | Several $M_{ab}$s to haemagglutinin | — | Breschkin *et al.* (1981) |

| | | | | | | | |
|---|---|---|---|---|---|---|---|
| 6. | Virions of Epstein–Burr virus (B95-8 strain) | Mouse–mouse (P8-X63-Ag8) | Glutaraldehyde-fixed viral-infected cells + $^{125}$I-labelled anti-mouse IgG | To obtain anti-viral $M_{ab}$s | $M_{ab}$ against 250 000 mol. wt glycoprotein | $M_{ab}$ reduces viral infectivity (with two viral strains) | Hoffman *et al.* (1980) |
| 7. | — | — | — | To produce $M_{ab}$s to single components of measles virus | One $M_{ab}$ to viral haemagglutinin; one $M_{ab}$ to viral nucleocapsid protein | $M_{ab}$ to haemagglutinin can induce subacute murine encephalitis | Rammohan *et al.* (1981) |
| 8. | Disrupted mouse mammary tumour virus | Mouse–mouse (P3-NS1) | Virus in solid-phase assay with $^{125}$I-labelled anti-mouse IgG + image intensification | To define mammary tumour virus diversity | One $M_{ab}$ to external glycoprotein; one $M_{ab}$ to major internal polypeptides | $M_{ab}$s demonstrate heterogeneous viral gene expression in tumours | Colcher *et al.* (1981) |
| 9. | Influenza A virus (PR/8/34) | Mouse–mouse (P3-X63-Ag8) | Indirect radioimmune assay with different recombinant viruses | To examine topological sites in viral haemagglutinin | Distinct epitopes on viral haemagglutinin in viral mutants | — | Lubeck and Gerhard (1981) |
| 10. | BALB/c methyl-cholanthrene-induced sarcoma (CMS4) | Mouse–mouse (P3-NS1) | Immunoprecipitation and immunofluorescence with $M_{ab}$ and FITC-conjugated[a] anti-mouse Ig [$F(ab^{1})_2$] | To study transformation protein distribution in mouse and human cells | $M_{ab}$ reacts with 53 000 mol. wt transformation-related protein | Antigen in 13 transformed mouse cell lines | Dippold *et al.* (1981) |
| 11. | Plasma membranes from Kirsten RNA sarcoma virus transformed BALB-3T3 cells | Mouse–mouse (P3-NS1) | Cells + $M_{ab}$ + $^{125}$I-labelled anti-mouse Ig | To characterize transformation-dependent antigens | 20 $M_{ab}$s, eight immunoprecipitating | Five macromolecular antigens detected | Strand (1980) |

[a] FITC, fluorescein isothiocyanate.

**Table 5.9** Monoclonal antibodies for chromosomes, genes and modified DNA

| | Immunizing macromolecular antigen(s) | Fusion | Antibody assay | Objective | Special $M_{ab}$ properties | Other facts | Authors |
|---|---|---|---|---|---|---|---|
| 1. | 60–65 000 mol. wt proteins released by DNase treatment of *Drosophila* embryo nuclei | Mouse–mouse (P3-NS1) | Formaldehyde-fixed polytene chromosomes in salivary gland squashes with immunofluorescence | To obtain $M_{ab}$s to non-histone chromosomal proteins | One $M_{ab}$ which associates with regions of gene activity including heat-shock loci | Antigen 62 000 mol. wt | Howard *et al.* (1981) |
| 2. | Purified human placental alkaline phosphatase | Mouse–mouse (P3-X63-Ag8 and SP2-0-Ag14) | Radioimmune assay with enzyme + $^{125}$I-labelled anti-mouse IgG | To detect enzyme polymorphisms | Six $M_{ab}$s used to screen 295 electrophoretically typed placentas | $M_{ab}$s reveal evidence for allelic differences not revealed electrophoretically | Slaughter *et al.* (1981) |
| 3. | Ribosomal RNA–DNA hybrids with methylated bovine serum albumin | Mouse–mouse (P3-NS1) | Indirect immunofluorescence with *Drosophila* salivary chromosomes + FITC-conjugated anti-mouse Ig | To identify *in situ* RNA–DNA duplexes in human chromosome preparations | One $M_{ab}$ which binds to polytene chromosome 3 | $M_{ab}$ recognizes RNA–DNA hybrids but not other forms of RNA or DNA | Dorsey-Stuart *et al.* (1981) |
| 4. | M and L subunits of human phosphofructokinase | — | — | To distinguish electrophoretically indistinguishable enzymes in somatic cell genetics | — | Gene for human phosphofructokinase is on chromosome 21 | Vora and Francke (1981) |
| 5. | Spleen cells from MRL Mp-1 pr/1 pr mice | Mouse–mouse (SP2-0) | Solid-phase assay with rabbit thymus extract + $^{125}$I-labelled Protein A + autoradiography | To understand mechanisms in autoimmune disease | Three $M_{ab}$s to (a) small nuclear ribonuclear proteins; (b) DNA; (c) rRNA. $M_{ab}$ (a) precipitates U-1, U-2, U-4, U-5 and U-6 RNAs | Mice donors develop a lupus erythematosus syndrome | Lerner *et al.* (1981) |

| | | | | | | | |
|---|---|---|---|---|---|---|---|
| 6. | Spleen cells from NZB/NZW mice | Mouse–mouse (MOPC 315-43) | Solid-phase assay with nucleic acid + $^{125}$I-labelled anti-mouse IgG | To produce sequence-specific DNA antibodies | Six $M_{ab}$s with DNA sequence preferences; no interactions with RNA or single-stranded DNA | Mice donors develop lupus erythematosus syndrome | Lee *et al.* (1981) |
| 7. | — | — | — | To map chromosomal location of an antigen | Two $M_{ab}$s which recognize cortical thymocyte surface antigens | Somatic cell hybrids constructed to assign an antigen to X-chromosome | Goodfellow *et al.* (1980) |
| 8. | Aflatoxin $B_1$-adducted DNA with methylated bovine serum albumin | Mouse–mouse (P3-X63-Ag8) | Solid-phase assay with aflatoxin $B_1$–DNA + anti-mouse Ig + alkaline phosphatase conjugate | To detect DNA modifications | Two $M_{ab}$s | $M_{ab}$ in ELISA detects one aflatoxin $B_1$ in 250 000 nucleotides | Haugen *et al.* (1981) |
| 9. | Crude mixture of HeLa cell microtubule-associated proteins | — | Radioimmune assay and immunofluorescence | To identify regulatory factors in mitotic spindle formation | One $M_{ab}$ | In dividing cells, antigen is associated with mitotic apparatus. At interphase, antigen is in the nucleus | Izant *et al.* (1982) |
| 10. | NZB/NZW $F_1$ splenic lymphocytes | Mouse–mouse (P3.63.Ag.8653) | — | To study a "protein" defect by anti-DNA $M_{ab}$ on cell surface | $M_{ab}$ to DNA reacts with a proteinase K susceptible protein in Raji cells | Hypothesis that surface protein triggers systemic lupus antibody production which can recognize DNA | Jacob *et al.* (1984) |
| 11. | NZB/NZW $F_1$ splenic lymphocytes | Mouse–mouse (SP2/0-Ag14) | Solution phase with [$^{32}$P]ds DNA + supernatant + rabbit anti-mouse Ig | To obtain a $M_{ab}$ to the B-form of DNA | $M_{ab}$ reacts with the B-form of DNA | Helical conformation and nucleotide composition may be involved in $M_{ab}$ recognition | Ballard *et al.* (1984) |

**Table 5.9** Monoclonal antibodies for chromosomes, genes and modified DNA (*continued*)

| | Immunizing macromolecular antigen(s) | Fusion | Antibody assay | Objective | Special $M_{ab}$ properties | Other facts | Authors |
|---|---|---|---|---|---|---|---|
| 12. | Guanine-imidazole ring-opened aflatoxin $B_1$-modified DNA + methylated BSA or KLH | [Mouse–mouse: C57BL/6; BALB/C; BALB/C] (X63-Ag8653) | ELISA with imidazole ring-opened aflatoxin DNA and $\beta$-galactosidase-sheep $(7ab^1)_2$ anti-mouse IgG | To improve methods to make $M_{ab}$s to carcinogen-modified DNA | More hybridomas with maximally modified DNA | KLH (keyhole limpet haemocyanin) best carrier. C57BL/6 best mouse strain | Hertzog *et al.* (1983) |
| 13. | Aflaxotin $B_1$–bovine $\gamma$-globulin covalent conjugate | Mouse–mouse (SP2) | ELISA with $AFB_1$–bovine serum albumin + supernatant + alkaline phosphatase coupled rat anti-mouse $\kappa$ chain | To produce $M_{ab}$s which detect $AFB_1$ in environmental samples | One $M_{ab}$ (IgM) | $M_{ab}$ recognizes $AFB_1$, $AFB_2$, $AFM_1$ | Groopman *et al.* (1984) |

**Table 5.10** Monoclonal antibodies in clinical radioimmune assays

| | Immunizing macromolecular antigen(s) | Fusion | Antibody assay | Objective | Special $M_{ab}$ properties | Other facts | Authors |
|---|---|---|---|---|---|---|---|
| 1. | Human growth hormone | Mouse–mouse (P3-X20) | Polyethylene glycol precipitation with $^{125}$I-labelled growth hormone | To obtain hormone-specific immunoreagents | $M_{ab}$s show minimal reactivity with placental lactogen | — | Bundesen *et al.* (1980) |
| 2. | Digoxin–bovine serum albumin | Mouse–mouse (P3-NS1-1-Ag4) | Polyethylene glycol precipitation with $^{125}$I-labelled digoxin | To assess $M_{ab}$s as serum reagents and to detect xenobiotics | $M_{ab}$s very specific and minimally cross-reactive | Digoxin used since polyclonals difficult to produce and cross-reactivity a problem | Bang *et al.* (1981) |
| 3. | Purified apolipoprotein E | Mouse–mouse (SP2-0) | Solid-phase radioimmune assay with apolipoprotein E + $^{125}$I-labelled anti-mouse Ig.Fab | To study the specificity of lipoprotein interactions with surface receptors | Five $M_{ab}$s | — | Milne *et al.* (1981) |
| 4. | IgG | — | — | To test models of immune-complex formation | — | Two $M_{ab}$s combined to give turbidimetric signal equal to polyclonal antibody | Jefferis *et al.* (1981) |
| 5. | Progesterone–bovine serum albumin | Mouse–mouse (P3-NS1) | Dextran-coated charcoal-binding assay with [$^3$H]progesterone | To study steroid haptens | One $M_{ab}$ | $M_{ab}$ blocks pregnancy in mice | Wright *et al.* (1982) |

systems, care must be taken to know the nature of epitopes or, as we shall see below, certain unexpected, but interesting, surprises may occur!

### 5.3.7.1. Selection of low abundance in RNA

Many cell proteins, including those involved in many types of control functions, are usually present in low amounts. This means that cloning the genes, i.e. starting from enriched mRNA for cDNA cloning, may present formidable problems. Polyclonal antibodies have been used for several years to bind to nascent polypeptide chains on polysomes; the polysomes can then be isolated by several methods, e.g. second antibody precipitation or Protein A-Sepharose (see Chapter 4). These types of studies are described in Table 5.11, nos 3 and 4.

### 5.3.7.2. Screening for expressed gene products

There is no doubt that antibodies used to detect expressed proteins containing antigenic determinants in recombinant DNA containing *E. coli* provide a powerful, very sensitive technique for the detection of recombinant clones before amplification by subcloning procedures. The methods that have been developed include genomic or cDNA cloning in plasmid (pBR 322 derivatives) or phage λ followed by replica plating on to nitrocellulose. Colonies containing expressed proteins are lysed and either detected directly immunologically or after transfer of released peptides on to CNBr-activated filter paper.

*Use of phage λ*

Young and Davis (1983a, b) carefully designed λ expression systems (e.g. λ gt 11) in which the foreign DNA to be expressed is inserted in the β-galactosidase structural gene *lac*2 of phage λ, which promotes the synthesis of hybrid fusion proteins. Special mutants of λ were utilized (i.e. λ gt 11) in which a temperature shift to 42 °C could be used to induce expression of the hybrid proteins, and an amber mutation (S100) renders the phage lysis defective, so that cells containing relatively large amounts (in immunochemical terms) of expressed fusion protein containing the antigen of interest can be expressed. High expression of the foreign gene protein product is dependent on the proper orientation of the insert DNA with respect to the β-galactosidase transcription unit and insertion in the coincident reading frame. Immunochemical detection of the antigenic determinants depends on good antibodies, i.e. preferably purified by immunoaffinity chromatography in the case of polyclonal antibodies. The method also depends on chemical lysis of the phage-containing *E. coli* and extensive washing procedures to obtain a low background and a consequent high signal/noise ratio. The detection system is outlined in Fig. 5.11. In the original methods (Young and Davis, 1983a, b) the use of polyclonal antibodies was recommended to avoid the theoretical possibility of a $M_{ab}$ recognizing expression products unrelated to the protein of interest, since single epitopes may be shared by different unrelated proteins (see later). One great advantage of immunological screening is that only colonies containing cloned coding sequences are detected. Polyclonal antibodies have been used to detect several cloned coding genes, including yeast RNA polymerases (Table 5.11, no. 10). In this study 35 recombinant phages antigenically related to RNA polymerases were isolated with 22 antisera to polymerase subunits; at least 22 genes for RNA polymerase subunits may therefore be cloned. Polyclonal antibodies have also been used to detect phage containing genes for bovine galactosyltransferase (Table 5.11, no. 9). In both cases the polyclonal antibodies were purified by immunoaffinity chromatography before use. In the later example (Table 5.11, no. 9), non-specific background staining was reduced by treating the antiserum with proteins extracted from lysed *E. coli* infected with $\lambda$ $1090r^-$. As little as 30 pg of galactosyltransferase could be detected.

$M_{ab}$s have been used to screen recombinant DNA libraries in phage $\lambda$. To avoid any theoretical non-specific reactions of $M_{ab}$s with shared epitopes in different proteins, Goridis *et al.* (1985) used two $M_{ab}$s to screen for genes for the mouse neural cell adhesion molecule (N-CAM) (Table 5.11, no. 5). In this study, considerably stronger signal/noise ratios were found than with a rabbit polyclonal antibody, presumably due to the greater concentration of the $M_{ab}$ to one epitope compared to the concentrations of antibodies to multiple epitopes with a polyclonal antiserum.

A virus neutralizing $M_{ab}$ to the envelope glycoprotein, gp70, of feline leukaemia virus has been used in a procedure which has considerable general implications for mapping an epitope to a defined protein sequence (Table 5.11, no. 6). The principle is to generate random short fragments of the cloned gene by DNase I action. The random DNA fragments are cloned into the EcoR.1 site of the $\beta$-galactosidase gene of $\lambda$ Charon 16 to obtain expression of random protein fragments as fusion proteins. The $M_{ab}$ is then used in a screening procedure (e.g. Fig. 5.11) to identify clones containing the epitope. The DNA is identified, and clones are subcloned and sequenced to characterize the amino acid sequence of the epitope. Such DNA sequences could be used in the design of synthetic peptides for vaccine production (cf. Chapter 6). The method requires no knowledge of the molecular target of the

**Table 5.11** Antibodies for recombinant DNA cloning and screening expression systems

| | Immunizing macromolecular antigens | Fusion | Antibody assay | Objective | $M_{ab}$ properties | $P_{ab}$ (polyclonal antibody) properties | Authors |
|---|---|---|---|---|---|---|---|
| 1. | — | — | Solid-phase assay with immobilized antigens + $^{125}$I-labelled $M_{ab}$ or $^{125}$I-labelled Protein A | To detect recombinant gene products in *E. coli* | Used in a general procedure for screening clones for antigens | — | Kemp and Cowman (1981) |
| 2. | *α*-amylase and ovalbumin | — | — | To detect cloned genes in λ gt 11 expression system with antibodies | — | $P_{ab}$s detect clones expressing cDNAs | Young and Davis (1983a,b) |
| 3. | — | — | — | To use $M_{ab}$s to isolate rare mRNA by polysome precipitation | One $M_{ab}$ used to isolate mRNA for cDNA production | — | Korman *et al.* (1982) |
| 4. | — | — | — | To isolate polysomes enriched in p53 mRNA | $M_{ab}$ binds to Protein A-Sepharose | — | Benchimol *et al.* (1984) |
| 5. | — | — | — | To identify clones producing N-CAM cDNA. (N-CAM is neural cell adhesion molecule expressed in λ gt 11) | Two $M_{ab}$s to N-CAM detect different clones | $P_{ab}$ binds to each fusion protein as well as one $M_{ab}$ | Goridis *et al.* (1985) |

| | | | | | | | |
|---|---|---|---|---|---|---|---|
| 6. | Feline leukaemia virus ($M_{ab}$ to coat protein gp70) | — | — | To map epitopes recognized by $M_{ab}$s. Random short DNA fragments in λ Charon 16 | $M_{ab}$ detects expressed random protein fragments | — | Nunberg *et al.* (1984) |
| 7. | EL-4/MEL14$^{hi}$ cells | — | — | To identify cDNA for a lymphocyte homing receptor (MEL14) | $M_{ab}$ detects clones expressing ubiquitin! | — | Siegelman *et al.* (1986) |
| 8. | Fibrinogen or prothrombin | — | — | To identify cDNA for fibrinogen and prothrombin in pBR322 vectors | — | $P_{ab}$s to human fibrinogen and prothrombin (denatured proteins) | Plaisancié *et al.* (1984) |
| 9. | Bovine galactosyl transferase | — | — | To identify cDNA clone by expression in λ gt 11 | — | $P_{ab}$ to affinity-purified bovine galactosyltransferase | Shaper *et al.* (1986) |
| 10. | Yeast RNA polymerases | — | — | To identify clones by expression in λ gt 11 | — | $P_{ab}$s purified by affinity chromatography | Riva *et al.* (1986) |
| 11. | Substance P | — | — | To identify transfected gene products by fusion of an oligonucleotide coding for part of substance P to 3′ end of gene | $M_{ab}$ recognizes epitope only after chemical conversion of carboxyl groups to amides | — | Munro and Pelham (1984) |

*E. coli* (BNN 91 or BNN 103)

↓ Infect with λ-containing cDNA of interest, 30 min at 32 °C

Infected *E. coli* ·

↓ Pour on to nitrocellulose filter on growth medium

Growth of infected *E. coli*

↓ 32 °C for 8 h

Make replica of master

↓

Replica on nitrocellulose

↓ 42 °C for 2 h to induce expression

Induced cells

↓ Lyse over $CHCl_3$ for 15 min

Lysed cells

↓ Wash with 0.17 M NaCl, 10 mM Tris-HCl, pH 7.5, containing 0.1 mM phenylmethane sulphonyl fluoride, 0.01% SDS at room temperature

Washed lysed cells

↓ DNase I (2 μg/ml) for 10 min

DNA-free washed lysed cells

↓ Wash with above buffer containing 3% BSA. Now add antibody in 0.1% SDS/0.1% Triton X-100/1 mM EDTA (3 h)
Wash twice with 0.1% SDS/0.1% Triton X-100/1 mM EDTA (10 min)

Washed filter

↓ Add $\simeq 5 \times 10^6$ cpm [$^{125}$I] Protein A

[$^{125}$I] Protein A–antibody–antigen conjugates

↓ Wash for 5 × 15 min with 0.1% SDS/0.1% Triton X-100/1 mM EDTA
Autoradiography overnight

Identified fusion protein containing colonies

**Fig. 5.11** Immunological screening for expressed antigen in λ gt 11 cDNA libraries.

$M_{ab}$, but does require that short sequences should contain enough information to constitute the epitope.

Perhaps the most interesting and certainly fascinating use of a $M_{ab}$ in an immunochemical screen for a cDNA clone concerns the production and use of a $M_{ab}$ called MEL 14. This $M_{ab}$ appears to recognize a lymphocyte-homing receptor which is involved in the localization of lymphocytes in postcapillary high endothelial venules. The $M_{ab}$ was used to screen an expression library in λ gt 11 for cDNA for the receptor by the method of Young and Davis (Fig.

5.11). The $M_{ab}$ identified cDNA clones which on sequencing turned out to be cDNAs, not for the receptor but for the protein ubiquitin, which has been implicated through covalent binding in histone modification and as a marker in the ubiquitin-related proteolytic system in reticulocyte lysates. The sequence analysis demonstrated that cDNAs coding for polyubiquitins, i.e. tandemly repeated ubiquitins, had been cloned. Careful protein analyses of the receptor demonstrated that the lymphocyte-homing receptor is a ubiquitinated branched-chain glycoprotein.

This is a most intriguing and salutary tale, since it points out a pitfall of immunological screening, i.e. detecting unexpected cDNA clones because of the $M_{ab}$, in this case, reacting with a protein which turns out to be covalently conjugated to another protein to form the receptor. On the other hand, it leads into a fascinating story of yet another function for this ubiquitous protein (Table 5.11, no. 7).

$\lambda$ gt 11 clones are not the only expression system in which to use immunological screening. The sensitivity of the immunochemical colony screening techniques is such that expressed peptides can be detected in standard cDNA libraries in pBR322. One such approach has involved the transfer of proteins from bacterial lysates on to CNBr-activated Whatman filter paper (Table 5.11, no. 1), followed by immunochemical screening (Table 5.11, no. 8). The principle involved in the screening procedure is shown in Fig. 5.12. For a good signal/noise ratio the method needs (a) protracted washing steps; (b) millipore filtering of buffers, antibody and working solutions; (c) the use of glycine as a blocking agent; and (d) the addition of transformed *E. coli* total proteins (not expressing antigen) to the detecting reagent ([$^{125}$I]protein A). The method was effectively used to detect clones containing fibrinogen and prothrombin DNA inserts. It could be used to screen existing DNA libraries for gene inserts of interest.

Besides the immunochemical screening of cDNA libraries expressing antigens of interest, antibodies can be used to detect antigens in eukaryotic cells after transfection with genes of interest. One interesting example which illustrates the exacting $M_{ab}$ requirements in terms of epitope structure, as well as a rather novel gene construction, concerns the detection of heat-shock protein gene products by the method of peptide tagging of mutant heat-shock proteins. The strategy consists of fusion of an oligonucleotide coding for part of the neuropeptide, substance P, to the 3′ end of a gene: the protein expressed in transfected cells can therefore be detected by a $M_{ab}$ to substance P. One interesting technical aspect of the use of the $M_{ab}$ to detect the peptide-tagged proteins in transfected cells was that natural substance P has an amide group at its *C*-terminus. The $M_{ab}$ was completely unable to bind to non-amidated tagged protein. Therefore, to detect the tagged proteins by immunofluorescence or protein blots, the protein carboxyl groups of the

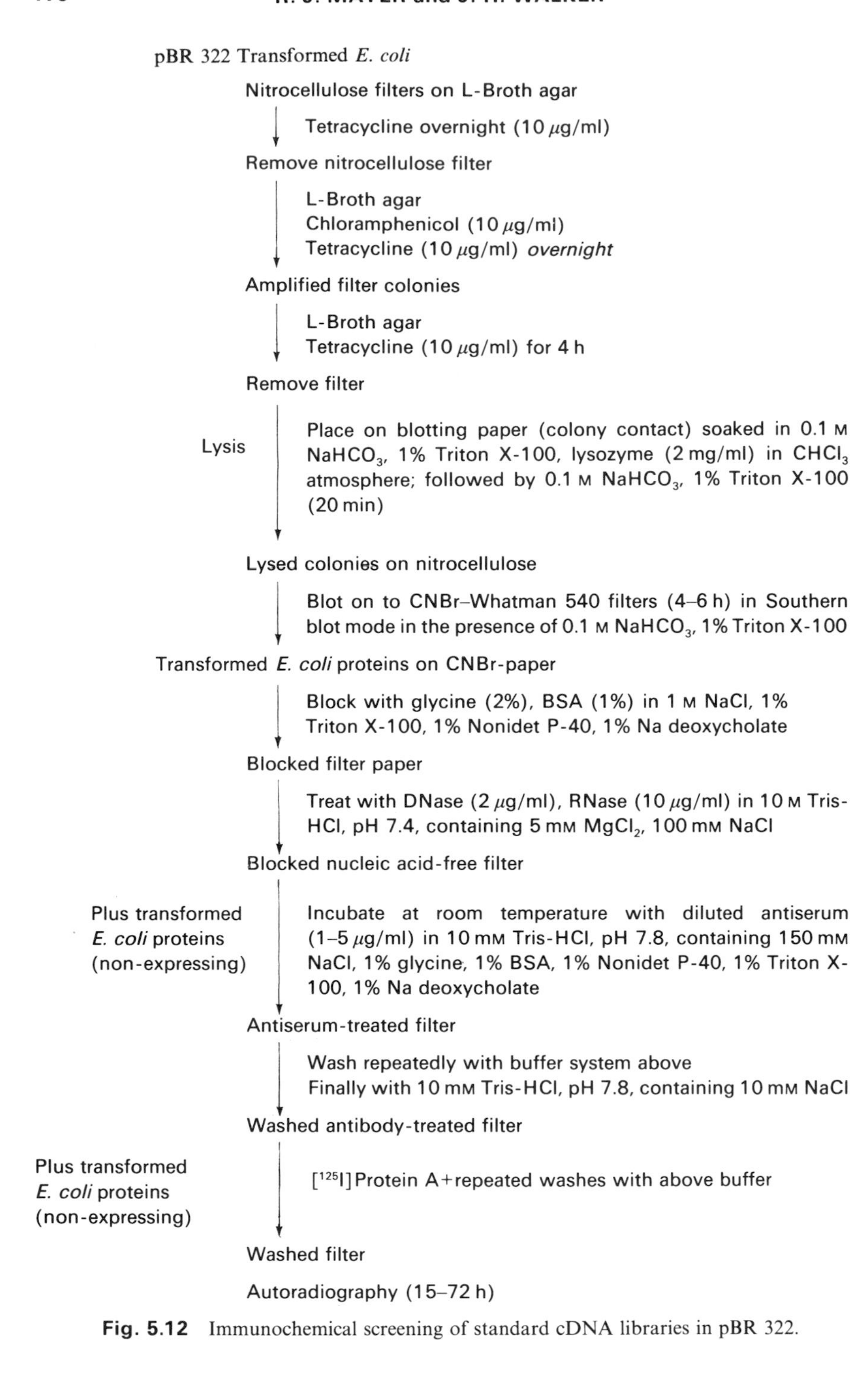


**Fig. 5.12** Immunochemical screening of standard cDNA libraries in pBR 322.

tagged protein were converted to amides. This was simply achieved by incubating fixed cells or blots with a water-soluble carbodiimide and $NH_4Cl$. The technique obviously has wider application in expressed antigen detection after gene transfection in eukaryotic cells (Munro and Pelham, 1984).

### 5.3.8. Microinjection of antibodies to probe cytoskeletal functions

Many of the elaborate interacting protein systems of the cell cytoplasm and nucleoplasm are not understood. These systems obviously function best in intact cells; *in vitro* counterparts are difficult to obtain and manipulate. One approach to studying intracellular functions of protein systems, e.g. the cytoskeleton, is by microinjection of $M_{ab}$s or high-titre affinity-purified polyclonal antibodies (or their fragments) into tissue-culture cells. Several microinjection techniques are available, but the most popular for antibody microinjection is with microneedles (Graessmann and Graessmann, 1976). After microinjection, antibodies interact with the antigens, preventing them from carrying out their functions, and allowing study of the consequences on cell behaviour, e.g. mobility or cytophysiological functions. There are obviously many applications of this approach. The function of non-erythrocyte spectrin in relation to other components of the cytoskeleton provides a good example of the types of study which can be performed by this technique (e.g. Mangeat and Burridge, 1984).

## 5.4. General Conclusions

The specificity of $M_{ab}$s to epitopes on macromolecular antigens, together with the purity of $M_{ab}$s, will guarantee the continued growth in the utilization of these immunoreagents.

Ultimately, the nature and quality of $M_{ab}$s rest with the nature of the immune response in the spleen donor animal, together with the growth and secretory potential of hybridoma clones.

Careful choice of immunizing macromolecular antigen and screening procedures for secreted antibodies can result in $M_{ab}$s with the properties required by the experimenter.

In these circumstances specific monoclonal reagents would be the antibodies of choice for diverse uses in biological sciences, biochemistry, cell biology and molecular biology.

# 6
# Synthetic Peptides: A New Development in Protein Immunochemistry

## 6.1. Introduction

Synthetic peptides are recognized as immunogens (generally conjugated to carriers) of great potential in protein immunochemistry. The concept of using synthetic peptides to mount a monoclonal or polyclonal immune response has evolved with the recent great progress in gene cloning and protein sequencing, which provide the DNA or protein sequences from which appropriately selected sequences can be synthesized for use as immunogens. Naturally, the observations on antibodies to synthetic peptides, particularly the interactions of the antibodies with the sequence-containing proteins, raise several problems concerning interpretation and use of the antibodies. Such problems are hardly surprising when considered against a fundamental aspect of protein structure, namely that protein sequences, although essential for understanding molecular structure, have not led to the laws and rules which govern the conformations adopted by peptides and proteins in solution or when parts of surfaces. The enormous complexities of polypeptide chain folding as it is directed by amino acid side chains, coupled with the existence of sets of "preferred" conformations for peptides and proteins in solution, precludes simple prediction of the properties of antibodies to peptides, particularly in relation to their binding to proteins. Synthetic peptides have been studied as antigens for specific immunological purposes, e.g. vaccine production to intractable pathogens. As is often the case in biological science, the studies may provide as much information on a fundamental conundrum, e.g. the relationship of primary sequence to preferred protein conformations, as they do to vaccine production.

Synthetic peptides as antigens will be considered under the headings of (a) general considerations, (b) antibody production, and (c) properties and uses of antibodies. Finally, a case study will consider the production, screening and use of monoclonal antibodies to synthetic peptides which constitute a sub-sequence of senile plaque core protein which is characteristic and diagnostic of Alzheimer's disease.

## 6.2. General Considerations

Readers determined to probe "synthetic peptides as antigens" in much greater detail would do well to digest from cover to cover the Ciba Foundation Symposium 119 of the same title (Wiley and Sons, Chichester, 1986), since as usual by reading the speaker contributions and ensuing discussions at a Ciba Symposium a concise feeling for the "state of the art" of a subject will be obtained. Indeed, the Symposium inspired one of the authors (RJM) to include a chapter on the subject in this book.

Protein sequences can nowadays be determined directly or inferred from DNA sequences. As such sequences became available, particularly for the proteins or genes of intractable pathogens, many workers saw contained oligopeptide sequences as potential immunogens which could produce antibodies of immense use in understanding the immune response to proteins, and which more practically might lead to successful vaccines, e.g. to foot and mouth disease. Before dealing with specific cases, aspects of primary, secondary, tertiary and quaternary structure must be considered which impinge on the concept that antibodies to synthetic peptides may be at all useful to the study or manipulation of proteins which contain such sequences. Nowadays it is impossible to consider proteins immunochemically without involving monoclonal as well as polyclonal antibodies. Polyclonal antibodies may be conveniently defined as a mixture of monoclonal antibodies to multiple epitopes on a single macromolecular antigen. Consideration of the concept of a monoclonal antibody immediately raises the notion of the single antigenic determinant or epitope to which the monoclonal antibody is produced: the molecular nature of an epitope in a protein and the relationship of the single epitope to overall protein conformation is at the heart of conceptual considerations of synthetic peptides as immunogens and antigens when the sequence is present as part of a protein macromolecule.

An epitope may be defined as the juxtaposition of atoms in space which are contiguous within a polypeptide sequence (i.e. sequential or continuous determinants) or as the juxtaposition of the same number of atoms from amino acid residues which are quite far removed in a polypeptide sequence (i.e. conformational determinants). Given the complexities of folding in a given polypeptide chain, not to mention multiple polypeptide chains, as in quaternary structure, identification of the amino acid side-chain contributors to an epitope may seem a daunting task, certainly in the absence of X-ray crystallographic data, e.g. for logical ordered studies on the effect of sequence alterations on immunogenicity of peptides or antigenicity of sequence-containing protein.

However, the attraction of monoclonal antibodies in investigations on synthetic peptides is that they are "site-specific reagents", which may be used to identify immunogenic regions in the conformation of a protein. Immunological site-specificity is important theoretically and practically when attempting to define regions of high and low immunogenicity in a protein, or study subsequent antibody–macromolecular (protein) antigenic behaviour in solution or solid-phase systems. This attraction has provided the driving force for many of the reported studies. Moreover, synthetic peptides as immunogens genuinely constitute a new approach for solving the problems associated with difficult pathogens, e.g. *Plasmodium* or foot and mouth disease, or for identifying new proteins, inferred from synthetic sequences of cloned genes, e.g. uncharacterized oncogene protein products.

Cursory consideration of peptide and protein conformation may lead to a pessimistic view of synthetic antigens as immunogens. For example, small peptides cannot absorb substantial fractions of antibodies in a polyclonal serum reactive to the native antigen, which implies relatively little antigenic cross-reactivity between small peptides and counterpart sequences in a protein. Competition between monoclonal antibodies to the same macromolecular protein antigen suggests that the immune response is to overlapping epitopes, a cluster of epitopes which define an immunogenic domain. Finally, as is discussed in detail below, conformational determinants, made up of non-contiguous side chains, are the most dominant epitopes, determining the immune response. However, as might be expected on conformational grounds, the immune response is determined by the individual protein macromolecule and presents several interesting and salutary surprises.

It is customary to define terms in scientific subjects: in this case a definition of the major classes of monoclonal antibodies to oligopeptide sequences is needed. Protein conformation makes this no easy matter, as will be seen below, but at least three simple classes can be defined by structural analysis. Class I monoclonal antibodies recognize linear amino acid sequences (sequential, continuous or linear determinants), e.g. $-aa_1-aa_2-aa_3-aa_4$, on the surface of a protein. Class II monoclonal antibodies recognize aspects of secondary structure such as $\alpha$-helix or $\beta$-pleated sheet, whereby only certain amino acids in the above sequence may be brought to a surface of the protein and exposed e.g. $-aa_1---aa_3-$. Class III monoclonal antibodies recognize determinants made up of atoms from spatially adjacent polypeptide chains involving sequentially unrelated amino acid residues. In this class of monoclonal antibodies will be found antibodies which recognize determinants contributed by tertiary and quaternary structure. As usual, no definition is without exceptions, as will be seen below; this is because of the degeneracy of amino acid requirements in epitope structure. Class III monoclonal antibodies would be expected to dominate in any immune response, which reflects the probability of finding residues which are sequentially distant coming together during protein folding. However, antibodies do recognize primary and secondary structural features, e.g. antibodies to synthetic Pro-Gly-Pro react with collagen, i.e. detect the helix.

There are two further major considerations which must be discussed before addressing the production and use of antibodies to synthetic peptides, namely what is the minimum unit of peptide structure which provokes an immune response (when conjugated to a carrier), and most importantly what are the consequences of the conformational alternatives of given sequences as oligopeptides or in proteins on immunogenicity and antigenicity?

The complexities of the immune response in terms of B cell and T cell function in humoral antibody production are not well enough understood to precisely define what is the minumum unit of peptide structure (number of atoms?) which provokes an immune response. However, some observations indicate the sorts of parameter governing the requirement. The minimum requirements for immunogenicity, i.e. that a molecule induces a T cell and B cell response, are: (a) two epitopes; (b) the epitopes may be identical or different; (c) one of the epitopes must be recognized by a T cell. For example, poly(γ-D-glutamic acid) is non-immunogenic in animals but azobenzenearsonate-L-tyrosine-poly(γ-D-glutamic acid) is immunogenic. Interestingly, in this case the azobenzenearsonate seems to behave as a carrier with high polyclonal antibody production to the polyglutamic acid (Alkan *et al.*, 1971). This approach provides a chemical definition of the cellular immune response. However, there are obviously no absolute definitions with peptide sequences, which may contain any combination of approximately 20 amino acids. Studies on synthetic peptides with sequences related to antibody binding sites to foot and mouth disease virus reveal a minimum requirement of three amino acids within the sequence of a peptide of correct identity and position. At least two of the amino acids need to be adjacent to each other (Geysen *et al.*, 1984, 1985).

The final and most important question concerns the role of conformation and conformational alternatives in determining the immunogenicity of a synthetic peptide and the reactivity of an antibody to a synthetic peptide with a protein containing the particular oligopeptide sequence. Generally, the requirement is to produce antibodies to synthetic peptides which recognize and bind to the sequence in a protein. It has already been stated that monoclonal antibodies reacting with conformational determinants (Class III) appear most often. The theoretical problem is that a synthetic peptide can adopt many conformations in solution, some of which will be thermodynamically preferred. Similarly, the oligopeptide sequence can adopt conformations when present in a protein. The sequence in the protein will be subject to conformational constraints imposed by the rest of the protein structure to give thermodynamically preferred conformations. Whether the preferred conformations of a sequence in the peptide and in the protein are the same is conjectural, but of vital importance for the immunochemistry.

Although preferred conformations are thermodynamically predictable, there are an indeterminate number of other conformations which kinetically exist as some fraction of the total conformations which can be adopted by the sequence in an oligopeptide or the protein. Equilibrium considerations mean that sequences in oligopeptides and proteins are continually changing topographically, giving rise to a number of conformational states. Such states have been described as "breathing equilibria", "segmental motility", or a "motif of flexibility". Whatever axiom is used, the concept of an equilibrium mixture of conformational states, including thermodynamically preferred conformations, for a polypeptide sequence both as an oligopeptide and in a protein, is fundamental to understanding the immunogenicity of synthetic peptides and the antigenicity of proteins with antibodies to synthetic peptides.

There is of course a paradox, the disorder–order paradox (Dyson *et al.*, 1986), which is, how does an antibody to a short synthetic peptide (assumed to be disordered or at least with preferred conformations different from the same sequence in the protein) react with the more ordered version of the sequence in the protein? The resolution of the paradox may be that the oligopeptide in solution is more ordered than suspected, and that the antibody target site in the protein is more disordered. Indeed, it might be predicted that preferred conformational (ordered) states for a synthetic peptide in solution are essential to mount an immune response, i.e. B cells and T cells must recognize preferred conformations with lifetimes determined for each peptide by kinetic and thermodynamic considerations. It might also be predicted that regions in proteins with the greatest probability of assuming different conformational (disordered) states might react with antibodies to synthetic peptides which have little conformational similarity to counterpart sequences in proteins. In other words, different preferred conformations will react with the sequences in regions of highest segmental motility, i.e. with sequences having the greatest probability of existing in a large kinetically determined repertoire of conformational states. At least one of the conformational states of the sequence in the oligopeptide must resemble the sequence in the protein for high-affinity interactions of antibody to synthetic peptide with a sequence in the protein to take place.

Certain structural consequences of the above predictions are observed practically. In short, regions in proteins with the best chance of undergoing conformational changes react best with antibodies to synthetic peptides. Such regions in proteins are therefore the best sites for the choice of sequences for synthetic peptide preparation and immunization. The regions include: (a) *N*-termini and *C*-termini which also have the property of generally being at the protein surface (Thornton and Sibanda, 1983); (b) loops and turns in the polypeptide chains, which are accessible at the protein surface. Turns may also be more likely in a small oligopeptide in solution (or as a hapten) than in an $\alpha$-helix or a $\beta$-sheet; and (c) sequences with hydrophilic sections, although in highly mobile loops or turns hydrophobic residues may be exposed for a considerable time and constitute an antigenic site.

One important point to remember is that antibodies to synthetic peptides are generally raised to the conjugated peptide (see later). The synthetic peptide is covalently coupled to a carrier. Therefore it cannot be assumed that the preferred conformations of the peptide attached to the carrier will be the same as in solution: it is possible that conjugation restricts the conformational repertoire of the synthetic peptide. The site and type of chemical coupling could influence the peptide conformation, rendering the subsequent antibody more or less reactive with the sequence in the protein.

Production of high-affinity antibodies is by definition very selective, both naturally and when exploited to produce antisera or vaccines. Given that conjugated synthetic peptides can assume a fixed repertoire of preferred conformations, the immune response progressively identifies the most immunogenically preferred conformations. However, in the process of B cell selection, programming, and B cell–T cell interaction, differentiation and proliferation, it should not be overlooked that low-affinity antibodies may be initially produced which recognize primary and secondary features, including possibly critical folding intermediates of protein structure. However, B cell clones producing antibodies which recognize the major conformational species will finally be selected. As will be seen later, the screening techniques for monoclonal antibodies are obviously selective and will potentiate the natural process, giving cells producing monoclonal antibodies of highest affinity for tertiary and quaternary conformational epitopes.

It must be remembered that some antibodies to synthetic peptides do not react with the sequence in the protein, and that other anti-peptide antibodies may invariably have lower affinities for a protein than anti-protein antibodies. Furthermore, antibodies to short synthetic peptides may preferentially identify epitopes in mobile segments of a protein, because the antibodies are biased to detect mobile regions. It is possible that longer peptides may reveal antigenicity of more structural parts of a protein. It is also worth noting that monoclonal antibodies to proteins usually do not react with peptides, probably because they detect discontinuous conformational determinants not represented in the peptide. The small proportion of monoclonal antibodies to proteins which do react with small peptides in solution may detect sequential (continuous) determinants in the synthetic peptides.

A final thought again concerns the nature of epitopes in peptides and proteins. A fascinating series of experiments by Geysen and colleagues (Geysen *et al.*, 1984, 1985; Ciba Symposium 119, pp. 130–149) challenges the notion of sequence specificity, that specific amino acid residues are targets for the immune response. Monoclonal antibodies were raised against a strain of foot and mouth disease virus and tested for reactivity against viral subunit and the immunologically important coat protein $VP_1$. One monoclonal antibody recognized a serotype-specific neutralizing epitope whose conformational preference was optimal on whole virus. The monoclonal antibody reacted weakly with the viral subunit, and not at all with isolated $VP_1$. The monoclonal antibody failed to react with synthetic overlapping peptides (hexapeptides and nonapeptides) from the published sequence of $VP_1$. The data were interpreted to mean that the monoclonal antibody recognized a discontinuous epitope on the whole virion. Attempts were then made to identify the binding peptide without any assumptions being made about its amino acid sequence. By replacement of amino acid residues one at a time in the antibody-binding sequence, with all 19 other common amino acids, essential residues for antibody binding could be identified. Similarly, non-essential residues could be identified. The results suggested that the peptide-binding site of the monoclonal antibody accommodates Trp-Gln-Met and His-Ser separated by spacer amino acids. It was then shown that the element Trp-Gln-Met was best composed of D-isomers and the element His-Ser of L-isomers. By comparison of the sequences of these peptides with the immunologically important coat protein, the positions of amino acids involved in the discontinuous epitope could be predicted. A set of synthetic peptides was prepared which bound to the monoclonal antibody. It is suggested that the only criterion which needs to be satisfied is that complementarity between the antigen-combining site of the antibody and the molecular surface of the binding peptide is maintained with respect to shape and charge. Therefore, the deduced synthetic antibody-binding peptide is best considered to be a "mimotope" of the epitope which induced the antibody, i.e. a molecule which can bind to the antibody-binding site which is not identical to the epitope. The synthesis of a large number of synthetic peptides (400) under the control of a computer-controlled peptide synthesizer therefore allowed the discovery of the sequence and binding conformation of peptides mimicking an assembled epitope, without knowledge of the protein sequence or X-ray structure. The usefulness of the approach is that it allows definition and manipulation of sequences which make up a discontinuous epitope. It also allows the possibility of "improving" the antigenicity of such a sequence, thus providing a better choice of synthetic peptides as vaccine candidates.

## 6.3 Production of Antibodies to Synthetic Peptides

There are two different strategies which can be employed to produce antibodies to synthetic peptides:

(a) Clone part or all of the gene of interest into a bacterial expression vector, so producing a chimeric product containing the peptide of interest. Subsequent immunization with a preparation of the "xenoprotein" from the recombinant-containing *E. coli* can give rise to polyclonal antibodies or mice ready for monoclonal antibody preparation.

(b) Chemically synthesize a synthetic peptide based on a known protein sequence, or sequence deduced from a cloned gene sequence. The latter possibility means that antibodies can be prepared to peptides and can then be used to identify and study an unknown protein product of a cloned gene, e.g. proto-oncogene protein products.

The first strategy relies on the immune response to the bacterial "xenoprotein" and requires no further comment. The second strategy needs optimization of the immune response to a small oligopeptide, since usually, but not always, small oligopeptides (5–15 amino acids) do not provoke humoral antibody production. The trick is the classical immunological manoeuvre, which is to render the peptide haptenic by chemical coupling to a carrier. Chemical coupling can be relatively non-specific, relying on free amino groups in peptide and carrier, e.g. with glutaraldehyde, or can be through bifunctional cross-linking reagents which can be used in a much more predictable fashion. The unpredictable non-specific methods can also result in insoluble polymer formation. Further, it is hardly worth using uncontrollable cross-linking reactions which would certainly disturb peptide conformation, in view of the obvious importance of conformation in the immune response, or the subsequent use of anti-peptide antibodies.

Controlled covalent coupling reactions are best achieved nowadays with bifunctional reagents which can capitalize on reactions which involve, at least in a part-reaction, thiol groups. A great advantage of this approach with synthetic peptides is that a cysteine residue can be included in the synthetic peptide at any position to suit the required properties of the immunizing peptide, e.g. simply at one end (see Fig. 6.1).

A typical approach would be to react the carrier with *m*-maleimidobenzoyl-*N*-hydroxysuccinimide ester (to derivatize the carrier through free amino groups), and then to react the derivatized carrier with the cysteine-containing peptide. Other approaches are possible, e.g. thiolating a protein with *N*-succinimidyl 3-(2-pyridyldithio)propionate (SPDP), reducing the conjugate and coupling to a peptide derivatized with succinimidyl 4-(*N*-maleimidomethyl)cyclohexane-1-carboxylate (SMCC).

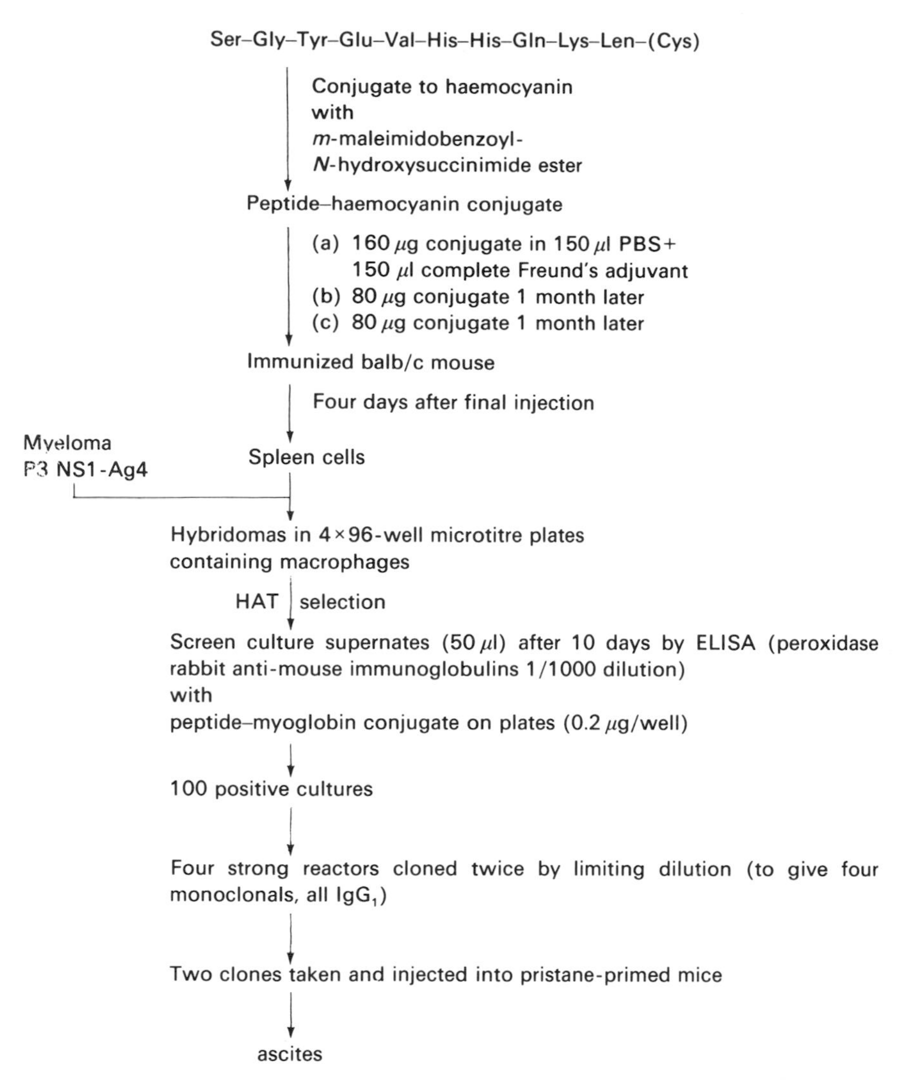


**Fig. 6.1** Production and screening of monoclonal antibodies to synthetic peptides: senile plaque core protein.

Irrespective of chemical coupling procedure, a major decision concerns the choice of the carrier. This problem is much more acute when the development of a vaccine is the ultimate objective than when antibodies are to be produced in animals for subsequent experimental purposes. In human vaccine development great care obviously has to be taken with carrier choice, so that the recipient of the vaccine does not suffer side effects.

Generally, in animal work keyhole limpet haemocyanin and bovine or chicken $\gamma$-globulin have been selected as carriers likely to induce a good immune response for classical haptens or peptides. A synthetic digestible polypeptide may be a good carrier, e.g. polyalanine, since it would be subjected to degradation in the immunized animal and not be likely to raise an immune response, i.e. to produce antibodies which cross-react with host antigens.

Prevention of the suppressor T cell response after the initial injection of immunogen may lead to a good immune response. This may be achieved by mixing colchicine with the immunogen, which may have the effect of inhibiting part of the initial suppressor T cell response. The choice of adjuvant is not a problem in animal studies; the usual Freund's complete and incomplete adjuvant can be used. In vaccination, Freund's adjuvant definitely cannot be used, and water in oil emulsions containing the immunogen have been found to be very efficacious, e.g. squalene–mannide monooleate emulsions.

Standard immunization schedules can give rise to polyclonal antisera or immunized mice (see Fig. 6.1), ready for monoclonal antibody production. Antibody-secreting hybridomas have usually been identified by ELISA screening procedures (see Fig. 6.1).

It is worth noting that the reaction of anti-peptide antibodies with oligopeptides or parent polypeptide need not be straightforward. Problems of antibody–antigen interaction stem from the conformer equilibrium and preferred conformations. These problems are particularly important for solid-phase (e.g. ELISA) and solution-phase reactions; the choice of assay principle and the conditions employed may be crucial for "signal" detection. Polyclonal and monoclonal antibodies to a peptide may react with the parent protein or immunogen (peptide conjugate) in an ELISA but not in solution.

As mentioned previously, the affinities of anti-peptide antibodies for the parent protein can be much reduced compared to anti-protein antibodies; they can be $10^5$–$10^6$ times less strong. Furthermore, the kinetic and thermodynamic parameters affecting the solid-phase reaction of anti-peptide antibodies with antigen may be very different from the solution parameters. At least two points should be considered. First, in a solid-phase ELISA the time period for antibody–antigen interaction may be very important. If the anti-peptide antibody recognizes one or a few related peptide conformations of the immobilized antigen, it may take very extended periods (up to a day) for a large proportion of the peptide sequence to assume the right conformation and therefore bind antibody. On and off reactions will take place continuously until the conformational equilibrium results in the appropriate conformation being assumed by antibody.

A second consideration concerns low-affinity antibodies reacting with antigen in solution and on a solid phase. This consideration affects any low-affinity antibodies, not just anti-peptide antibodies. Competition for antibody between solution- and solid-phase antigen can provide the basis of a competitive immunoassay. The problem is that the solution density of an antigen is uniform, whereas the solid-phase density need not be uniform, because of the deposition of antigen as "islands" or "hot-spots" on plastic surfaces. In this case free competition will not occur with low-affinity antibodies, since the antibody continually dissociating from soluble immune complexes can bind immobilized antigen. Subsequently, the likelihood of antibody dissociated from immobilized antigen binding to other molecules of immobilized antigen is greater (due to the patches of surface antigen) than that of the binding to free antigen and, therefore, eventually all the antibody could bind to the solid surface. Both these considerations should be noted when designing a solid-phase ELISA.

Finally, it should be pointed out that antibody–antigen interaction is the product of antibody concentration and affinity. This may explain why monoclonal anti-peptide antibodies are better than polyclonal anti-peptide antibodies at recognizing parent protein. There is just one monoclonal species in high concentration, which compensates to a degree for low-affinity interactions with the parent protein. In the polyclonal serum there may be large numbers of individual monoclonal antibodies directed at different epitopes, so the concentration of each antibody is low.

## 6.4. Properties and Uses of Antibodies to Peptides

The properties of antibodies to synthetic peptides have been dealt with extensively above. As usual in immunochemistry, the individual properties of a monoclonal antibody are to some extent specific to that antibody molecule. Monoclonal antibodies to synthetic peptides may or may not react with (immunoprecipitate or Western blot) the parent protein. Several reasons for this behaviour have already been mentioned. One point worth making here is that a monoclonal anti-peptide antibody may well immunoprecipitate the parent protein, e.g. in a double-antibody precipitation or with Protein A-bearing *S. aureus*, but may not react in a Western blot if the polyacrylamide gel electrophoresis is carried out in the presence of reducing agent, e.g. along with sodium dodecyl sulphate. In the absence of reducing agent the Western blot may work. The reason is that the disruption of –S–S– bridges will lead to disruption of the protein conformation, thus preventing the conformationally directed monoclonal antibody from reacting with antigen. Monoclonal antibodies can react with their epitopes in the presence of quite high amounts of sodium dodecyl sulphate (e.g. up to 0.5%, Mayer, unpublished), whereas reduction of disulphide bonds may prevent antibody–antigen interaction altogether. It is interesting to speculate that normal disulphide cross-links in proteins may have a major role in driving protein conformational changes during protein biosynthesis.

As mentioned above, monoclonal anti-peptide antibodies have been put to several novel uses. The use of monoclonal anti-peptide antibodies for identifying and characterizing proto-oncogene proteins deserves special mention. Monoclonal anti-peptide antibodies to predicted sequences of cloned genes have been used to detect oncogene-related proteins in human urine (Niman *et al.*, 1985) or characterize the human c-*myc* protein (Evan *et al.*, 1986). These types of study will certainly enable rapid progress in delineating the functions of the protein products of cloned genes.

## 6.5. Case Study: Monoclonal Anti-Peptide Antibodies to Senile Plaque Core Protein

Senile plaques and neurofibrillary tangles are the hallmarks of the brain lesions found in Alzheimer's disease, the commonest cause of senile dementia. The processes leading to the formation of senile plaques and neurofibrillary tangles are poorly understood. In order to understand the relationships between different proteins implicated in plaque and tangle formation, a synthetic peptide (Fig. 6.1) corresponding to part of the *N*-terminal sequence of plaque core protein was prepared, and coupled to haemocyanin, and monoclonal antibodies were prepared (Allsop *et al.*, 1986). An ELISA screening protocol was used and several monoclonal anti-peptide antibodies were eventually prepared.

A principle worth considering in monoclonal antibody production (see Chapter 5) is to subject the antibodies to the final objective test as soon as possible in the cloning procedure, assuming that the final objective is clearly identified. In the case of brain from Alzheimer patients, immunohistochemical staining of senile plaques can be correlated with plaque identification by Congo red birefringence. Therefore, the monoclonal antibodies were rapidly tested in a double-antibody immunoperoxidase system in paraffin and cryostat sections of brain from Alzheimer patients and controls. The immunoperoxidase reaction product showed that the monoclonal antibodies stained plaque and vascular amyloid (Fig. 6.2) but not tangles, suggesting that the polypeptide chain containing the peptide is exposed in the former and is either not present or inaccessible in the latter. For routine immunopathology, antibodies that react in paraffin sections are excellent, leading to rapid routine analyses.

In these studies (Allsop *et al.*, 1986) the peptide chosen for synthesis

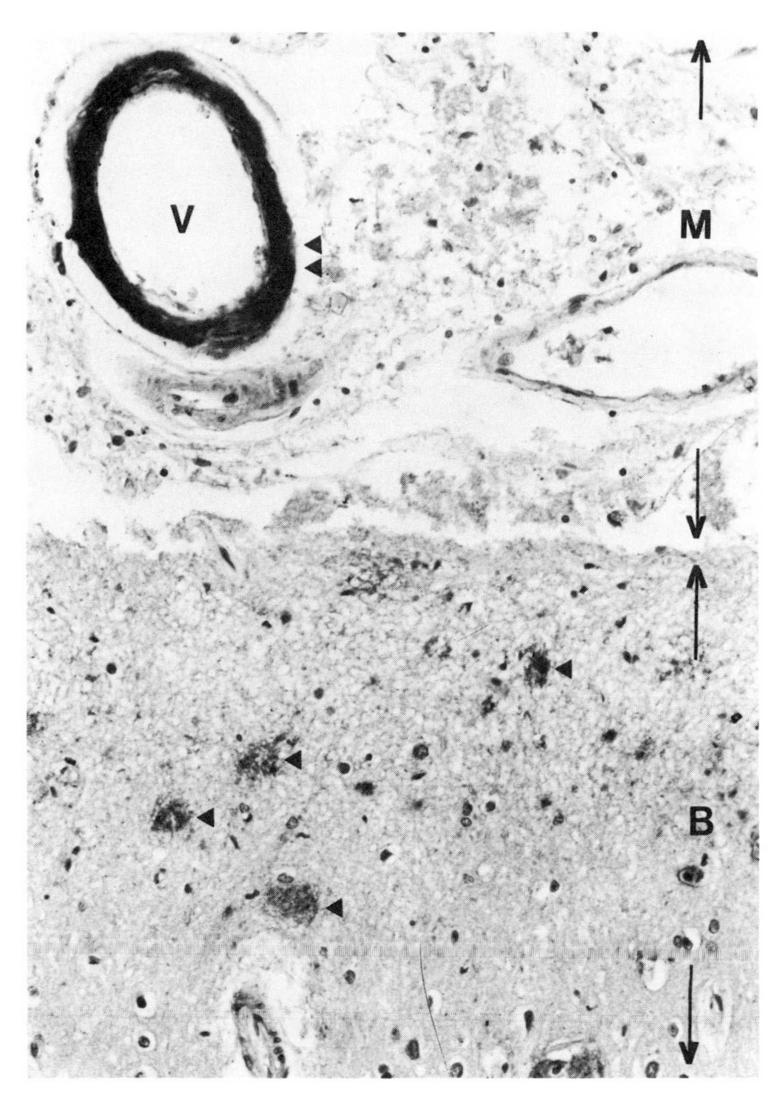


**Fig. 6.2** Immunoperoxidase staining of Alzheimer plaque cores and vessel amyloid with monoclonal anti-peptide antibodies. Paraffin sections were stained by the immunoperoxidase technique and counterstained with haemotoxylin. V = vessel, M = meninges, B = brain. Single arrowheads point to immunoperoxidase-stained plaque cores. Double arrowheads point to vessel amyloid. Small black dots in brain are cell nuclei. (Magnification × 100.)

contained residues 8–17 of the protein. Cysteine was added to the *C*-terminus for the haemocyanin conjugation reaction. The extreme *N*-terminal sequence (residues 1–7) was avoided, since the protein detected at different tissue sites exhibits considerable *N*-terminal heterogeneity.

# 7
# Technical Supplement

This chapter aims to provide a concise collection of specific methods with detailed instructions and practical advice based on the material presented elsewhere in this book. The methods described represent the basic repertoire of immunochemical procedures currently in vogue in cell biological studies.

## 7.1. Gel Electrophoretic Procedures for the Analysis of Antigen Mixtures and the Isolation of Antigens

### 7.1. Introduction

Gel electrophoresis procedures are essential techniques for the isolation of antigens for immunization and also for the characterization of antisera by means of blotting procedures. In this section, detailed protocols are given for one-dimensional and two-dimensional gel electrophoresis procedures.

### 7.1.2. Designs for simple tube and slab gel electrophoresis apparatus

*(a) Tube gel apparatus*

Davis (1964); Maizel (1971)

*(b) Slab gels apparatus*

*(i) Standard size* (gel approx. 12 × 15 cm)

Studier (1973); Adams *et al.* (1969); Reid and Bieleski (1968)

*(ii) Minigel size* (gel approx. 10 × 10 cm)

Amos (1976); Matsuidaira and Burgess (1978)

*(iii) Apparatus for running several gels at once*

5 gels, Kaltschmidt and Wittmann (1970); 8 gels, Dean (1979); 10 gels, Jones *et al.* (1980)

*(c) Power packs*

Davis (1964)

*(e) Gel drier*

Maizel (1971)

*(f) Electrophoretic elution chambers*

Allington *et al.* (1978)

*(g) Manufacturers*

Bio-Rad, Shandon, Raven, Pharmacia, LKB.

## 7.1.3. Sodium dodecyl sulphate—polyacrylamide gel electrophoresis (SDS-PAGE)

REFERENCE: Laemmli (1970).

SDS-PAGE is a high-resolution method used universally for analysing mixtures of proteins according to size. Suitable apparatus is widely available, as are molecular weight standard proteins. To estimate the $M_r$ value of an unknown protein, plot the $\log_{10} M_r$ of the standard proteins against their relative mobilities (distance travelled by the protein/distance travelled by the tracking dye). We present here the solutions required and recipes for preparing 10 ml of separating gel of varying percentage acrylamide and 10 ml of stacking gel. The amounts prepared should be altered according to the apparatus available. For proteins of subunit $M_r$ 15 000 to 200 000 a 10% gel is optimal.

### 7.1.3.1. Solutions

**A.** Acrylamide (warning, toxic, so wear gloves): 30% acrylamide, 0.8% *N,N'*-methylene bis-acrylamide. Make 60 g of acrylamide and 1.6 g of bis up to 200 ml with water. Warm to dissolve. Store in dark bottle.

**B.** Separating gel buffer: 1.5 M Tris-HCl, pH 8.8, 0.4% SDS. Dissolve 45.42 g of Tris base and 1.0 g of SDS in 150 ml of water. Bring the pH to 8.8 (approx. 80 ml of 1 M HCl). Make up to 250 ml.

**C.** Stacking gel buffer: 0.5 M Tris-HCl, pH 6.8, 0.4% SDS. Dissolve 6.0 g of Tris base and 0.4 g of SDS in 80 ml of water. Bring the pH to 6.8 with concentrated HCl and adjust the volume to 100 ml.

**D.** Ammonium persulphate (10%). Dissolve 0.5 g of ammonium persulphate in 5 ml of water. Make up fresh each week.

E. TEMED: *N,N,N′,N′*-tetramethylethylene diamine. Water-saturated isobutanol: Mix with shaking for 2 min 100 ml of water and 150 ml of isobutanol. Allow the phases to separate. Always use the upper phase.

*(a) "Running" buffers*

**1.** 25 mM Tris, 192 mM glycine, 0.1% SDS: 6.05 g Tris base + 28.8 g glycine + 2.0 g SDS made up to 2000 ml.

**2.** 188 mM Tris, 188 mM glycine, 0.1% SDS: 22.75 g Tris base + 14.11 g glycine + 1.0 g SDS made up to 1000 ml.

Use *either* 1 or 2. Buffer 2 often gives sharper bands than the original Laemmli buffer 1.

*(b) Sample buffer*

Mix 2.5 ml of C + 0.2 g SDS, + 2.0 ml glycerol + 0.1 ml 0.1% bromophenol blue, + 1.0 ml β-mercaptoethanol, + 4.2 ml water. Final volume is 10 ml.

*(c) Stain*

Dissolve 2.5 g of Coomassie Brilliant Blue R-250 in 500 ml of methanol (or ethanol). Then add 500 ml of water and 100 ml of glacial acetic acid. Mix for 2 h and filter if necessary.

*(d) Destain*

10% isopropanol/10% acetic acid or 10% methanol/10% acetic acid.

*(e) Storage destain*

7% acetic acid.

### 7.1.3.2. Procedure

**1.** Seal the gel cell. This usually consists of two glass plates separated by plastic spacers 0.5, 1.0, 1.5, or 3.0 mm thick. It is sealed in various ways, one of the most effective being with three spacers—one down each side and one along the bottom as shown in Fig. 7.1. Molten 1% agarose in water is applied to the outer edge of the mould, which is held together with spring clips.

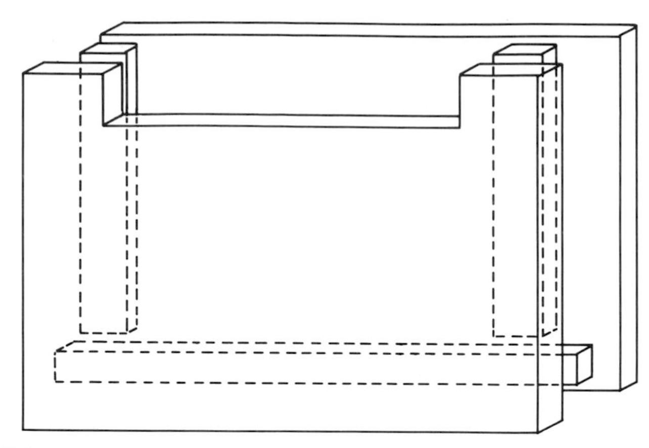

**Fig. 7.1** Assembled glass plates and spacers for preparing polyacrylamide gels.

**2.** Prepare the separating gel. For 10 ml of solution mix as shown in Table 7.1. (Gradient gels may be prepared using a conventional double-chamber gradient maker and a peristaltic pump. Ensure that the flow rate is sufficiently fast that the gel does not polymerize in the gradient making set up (a lower concentration of D is recommended for gradient gels, i.e. 0.01 ml). Wash unpolymerized solution out of the gradient maker and tubing as quickly as possible.)

**Table 7.1**

| Solution | 5% | 7.5% | 10% | 15% | 20% |
|---|---|---|---|---|---|
| A. | 1.67 | 2.5 | 3.33 | 5.0 | 6.67 |
| B. | 2.5 | 2.5 | 2.5 | 2.5 | 2.5 |
| D. | 0.03 | 0.03 | 0.03 | 0.03 | 0.03 |
| E. | 0.005 | 0.005 | 0.005 | 0.005 | 0.005 |
| $H_2O$ | 5.8 | 4.97 | 4.14 | 2.47 | 0.8–10 |

*Note:* all values in ml.

**3.** Quickly transfer the acrylamide solution to the gel cell and overlay with water-saturated isobutanol (upper phase). Allow to set (approx. 15–30 min). An obvious top to the gel can be seen when the gel is set.

**4.** Prepare the stacking gel. For 10 ml mix 2.0 ml A + 2.5 ml C + 5.4 ml water + 0.03 ml D. Do not add E yet.

**5.** Remove the isobutanol by pouring it off into adsorbent paper. Rinse the top of the gel with water, pour off and remove any remaining water with filter paper, taking care not to touch the surface of the gel.

**6.** Add 0.005 ml of E to the stacking gel solution, mix and transfer to the gel cell. Place the casting comb in position and leave to polymerize. Leave a space of between 0.2 and 1.0 cm between the top of the separating gel and the bottom of the sample wells. Mark the position of the wells on the outside of the gel cell.

**7.** Fill the bottom compartment of the gel apparatus with "running" buffer and transfer the gel into the apparatus *after* removing the spacer from the bottom of the cell. Take care to avoid bubbles at the bottom of the gel. If necessary remove bubbles with a jet of "running" buffer squirted out of a hypodermic syringe through a bent needle.

**8.** Fill the upper compartment with "running" buffer, and remove the casting comb and load samples into the wells with a 10 or 50 $\mu$l syringe.

**9.** Connect the power pack such that the anode is at the bottom. For a 1.5 mm thick gel, approx. 12 × 15 cm, run at 40 mA constant current (preferably with cooling) until the dye front reaches the bottom of the gel. For a 0.5 mm thick gel (approx. 10 × 10 cm) run at 20–25 mA.

**10.** Switch off the power pack and disconnect from the gel apparatus. Empty the buffer reservoirs and remove the gel cell from the apparatus. Remove the side spacers and flip open the cell using a spatula with a twisting action.

**11.** If the gel is to be stained for protein, place it into stain solution for 5 min for a small thin gel and 30 min for a 1.5 mm thick gel. Keep the gel moving during staining, then pour the stain solution back into its bottle, rinse the gel in water and destain in 10% isopropanol/10% acetic acid, preferably on a gently moving shaking device. Keep changing the destain until the background is completely colourless and then transfer the gel into 7% acetic acid.

**12.** Gels may be stored in sealed plastic bags without adding destain, or in plastic boxes (as used for staining and destaining) containing 7% acetic acid, or they may be dried down on to Whatman 3MM filter paper.

### 7.1.3.3. Example (see Fig. 7.2)

Gel dimension: Minigel 9 cm wide × 6 cm long × 0.5 mm thick.
% acrylamide: 10.
Run time: 25 min at 25 mA constant current.
Running buffer: 2, see above.
Sample: Sigma high molecular weight standards, 5 $\mu$g protein.

## 7.1.4. SDS-PAGE: an alternative to the Laemmli procedure

REFERENCE: Neville (1971).

This method often gives sharper bands than the Laemmli procedure and does not include glycine in any of the buffers; glycine is especially troublesome if one wants to determine the amino acid composition of proteins isolated by SDS-PAGE.

### 7.1.4.1. Solutions

**A.** Upper reservoir buffer: 0.2 M boric acid, 0.805 M Tris, 0.5% SDS, pH 8.64; 6.18 g boric acid + 48.7 g Tris, + 2.5 g SDS made up to 500 ml with water.

**B.** Upper gel buffer: 0.1335 M $H_2SO_4$, 0.2705 M Tris, 0.5% SDS, pH 6.1. Dissolve 1.64 g Tris + 0.25 g SDS in 30 ml of water. Bring pH to 6.1 with $H_2SO_4$ and adjust the volume to 50 ml.

**C.** Lower gel buffer and lower reservoir buffer: 0.247 M HCl, 0.858 M Tris, 0.5% SDS, pH 8.47 (running pH of 9.05). Dissolve 51.91 g Tris + 2.5 g SDS in 300 ml of water. Bring pH to 8.47 with HCl and adjust the volume to 500 ml.

**D.** Sample buffer: 2 ml of B + 0.2 g SDS + 2.0 ml glycerol + 0.1 ml of 0.1% bromophenol blue + 1 ml of $\beta$-mercaptoethanol + 4.7 ml of water (10 ml final volume).

**E.** Acrylamide for lower gel of final concentration 4.8% acrylamide, 0.2% bis-acrylamide: 9.6 g acrylamide + 0.4 g *N,N′*-bis-acrylamide. Made up to 100 ml.

**F.** Acrylamide for a lower gel of final concentration 10.9% acrylamide, 0.2% bis-acrylamide: 21.80 g acrylamide + 0.2 g *N,N′*-bis-acrylamide. Made up to 100 ml.

**G.** Acrylamide for *upper* gel of final concentration 2.8% acrylamide, 0.4% *N,N′*-bis-acrylamide: 2.8 g acrylamide + 0.4 g *N,N′*-bis-acrylamide. Make up to 50 ml with water.

**H.** Ammonium persulphate: 0.5 g/100 ml, or 50 mg/10 ml.
TEMED: *N,N,N′,N′*-tetramethylethylene diamine.

#### 7.1.4.2. Gel recipes (for 10 ml)

*(a) Lower separating gels*

**1.** *4.8% acrylamide/0.2% bis* (for proteins of $M_r$ 40 000–500 000). Mix 5 ml E + 2 ml C + 1 ml H + 1.99 ml water + 15 μl TEMED.

**2.** *10.9% acrylamide/0.2% bis* (for proteins of $M_r$ 15 000–200 000). Mix 5 ml F + 2 ml C + 1 ml H + 1.99 ml water + 15 μl TEMED.

*(b) Upper stacking gel*

Mix 5 ml G + 2 ml B + 1 ml H + 1.99 ml water + 15 μl TEMED.

#### 7.1.4.3. Procedure

Essentially the same as the Laemmli method except for the buffers used. The lower reservoir of the gel apparatus is filled with Solution C diluted 1:4 with water. The upper reservoir is filled with Solution A diluted 1:4 with water.

### 7.1.5. Gel notes

**1.** *Staining and destaining tips.* For optimal appearance keep the gel moving gently while in the stain and destain solutions, e.g. using a shaking water bath. Regenerate your destain: (a) Wool. Knot a few strands of undyed wool together and add to the destain solution containing the gel; (b) Sponge. Float a small piece of foam rubber in the gel-destaining box; (c) Activated charcoal. After use mix with activated charcoal and then filter.

**2.** *Non SDS-gels.* Many *soluble proteins* give good bands on the discontinuous system of Davis (1964) and Ornstein (1964). This is identical to the Laemmli (1970) system without SDS; urea (1 M to 8 M) may be added to the gel to improve resolution. *Membrane proteins* may also be analysed using Davis (1964) discontinuous gels containing 0.1% Triton X-100 or 0.1% sodium deoxycholate (Dewald *et al.* 1974). Membranes are washed to remove adsorbed proteins (0.25 M sucrose, 0.15 M NaCl, 5 mM Tris-HCl, pH 8.0) and then solubilized in 1% (w/v) detergent in 10% sucrose, 0.1 M Tris-HCl, pH 8.5, at a concentration of 1–2 mg/ml. The samples are centrifuged at $6 \times 10^6 g$ for 7 min and 5 μl of 0.04% bromophenol blue is added per 50–100 μl of sample.

**3.** *Stains for enzyme activity.* Some enzymes, including alkaline phospha-

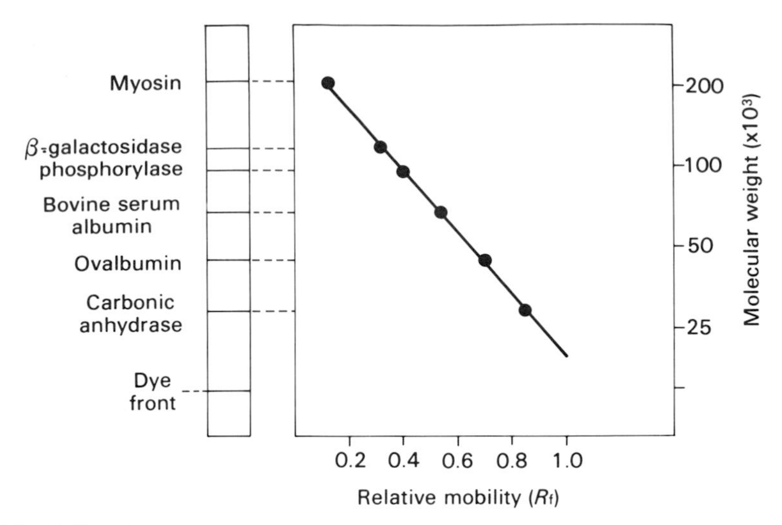


**Fig. 7.2** Calibration curve for SDS-polyacrylamide gel electrophoresis: (a) schematic gel; (b) semi-log plot of molecular weight against the mobility relative to the dye front ($R_f$). The gel concentration was 10%.

tase from human placenta, retain some enzyme activity even after SDS-PAGE. With non-denaturing PAGE, esterases, β-glucuronidase, alkaline phosphatase, α-mannosidase, and various NADH- and ATP-requiring enzymes have been demonstrated by applying histochemical staining procedures (Dewald *et al.*, 1974).

**4.** Other cross-linkers for acrylamide used on a mole for mole basis instead of *N,N'*-bis-acrylamide are: (a) *N,N'*-diallyltartardiamide (Anker, 1970). Gels can be dissolved with 2% periodic acid for 20–30 min at room temperature, (b) *N,N'*-bis-acrylylcystamine (Hansen, 1977) dissolves in reducing agent, 0.1 ml β-mercaptoethanol per g of gel.

**5.** Tracking dye for acid buffer systems: Pyronine G. Tracker dye for DNA sequencing: xylene cyanol FF (Maxam and Gilbert, 1979).

**6.** To run two slab gels at once (e.g. for 2D gel second dimensions) construct two gel chambers by sandwiching two notched glass plates, one non-notched and spacers: seal as normal and run as normal (same voltage, 2 × current).

### 7.1.6. Methods of sample concentration for SDS-PAGE

**1.** *Lyophilization*: Dialyse vs $H_2O$, freeze-dry, and take up in sample buffer. A simple procedure for dialysing several samples at the same time is performed using microcentrifuge tubes. These are held upside down by means of a polystyrene raft. The top of the tube has its central portion removed and is used to hold dialysis membrane in place.

**2.** *Trichloracetic acid (TCA) precipitation*: Maizel (1971). Standard procedure: precipitate for 60 min on ice and wash with ice-cold 10% TCA. Wash with cold acetone or ethanol:ether (1:1). Take pellet up in sample buffer. Adjust the pH to approximately 7.0 if necessary with 1 M NaOH (i.e. add NaOH solution until the bromophenol blue changes from yellow to blue).

**3.** Sephadex G100 or G200 sprinkled on a dialysis bag containing sample will reduce 2 ml to 0.2 ml in a few hours at 4 °C. Alternatively, Carbowaxes or Aquacide (Calbiochem) can be used (Colover, 1960) by placing the sample in a dialysis bag in 40% (w/v) solution of polyvinyl pyrolidone or polyethylene glycol (Carbowax) in water or buffer. The molecular weight of the colloid must be greater than 10 000 (i.e. the pore size of the dialysis membrane). The sample is concentrated in 16 h at 4 °C.

**4.** Pressure filtration devices and centrifugation concentration (Amicon).

**5.** Concentration with hydroxylapatite (HA): Adjust sample to 0.01 M phosphate. Add hydroxylapatite powder. Mix for 15–30 min. Spin, remove supernatant. Reslurry with a minimum volume of 0.5 M phosphate. Spin. The supernatant contains the concentrated protein. For 200 ml of 1 mg/ml albumin add 20 g of hydroxyapatite and subsequently elute in 5 ml of 0.5 M phosphate buffer.

**6.** Acetone precipitation: Mix 1 volume of sample with 5 volumes of cold acetone, mix and keep at −20 °C for 10 min. Spin at 10 000*g* for 5 min and wash if necessary. Dry and take up the pellet in sample buffer.

**7.** Evaporation: Place the sample in a dialysis bag in a stream of warm air from a hair drier. Evaporation keeps the sample cool and in 2–4 h samples of 5–15 ml are concentrated 100-fold.

For methods (2) and (6) the method should be checked to ensure quantitative precipitation. Subsequent addition of urea may help to ensure full solubilization of the precipitated proteins.

### 7.1.7. Protein subunit molecular weights

REFERENCES: Weber and Osborn (1969)
Proteins available from Sigma, Pharmacia, Biorad, Boehringer, BDH, Calbiochem.

**Table 7.2**

| Protein | Source | $M_r$ |
|---|---|---|
| $\alpha_2$-macroglobulin | bovine plasma | [b]340 000 |
| [a]myosin | rabbit muscle | 200 000 |
| $\alpha_2$-macroglobulin | bovine plasma | 170 000 |
| [a]β-galactosidase | *E. coli* | 116 000 |
| [a]phosphorylase B | rabbit muscle | 97 400 |
| transferrin | | 74 000 |
| [a]albumin | bovine serum | 68 000 |
| L-amino acid oxidase | *Crotalus atrox* | 64 000 |
| [a]catalase | bovine liver | 60 000 |
| pyruvate kinase | rabbit muscle | 57 000 |
| glutamate dehydrogenase | bovine liver | 55 400 |
| actin | rabbit muscle | 45 000 |
| [a]albumin | egg white | 43 000 |
| alcohol dehydrogenase | yeast | 37 000 |
| D-amino acid oxidase | pig kidney | 37 000 |
| lactate dehydrogenase | pig heart | 36 500 |
| glyceraldehyde-3-phosphate dehydrogenase | rabbit muscle | 36 000 |
| pepsin | pig stomach | 35 000 |
| [a]carbonic anhydrase | bovine erythrocytes | 30 000 |
| chymotrypsinogen | bovine pancreas | 25 700 |
| [a]trypsinogen diisopropylfluorophosphate | bovine pancreas | 24 000 |
| [a]trypsin inhibitor | soybean | 21 000 |
| [a]β-lactoglobulin | bovine milk | 18 400 |
| myoglobin | horse muscle | 17 200 |
| haemoglobin | bovine erythrocytes | 16 700 |
| [a]lysozyme | egg white | 14 300 |
| α-lactalbumin | bovine milk | 14 200 |
| RNase A | bovine pancreas | 13 700 |
| cytochrome C | horse heart | 12 300 |
| RNase S | bovine pancreas | 11 400 |
| proinsulin | bovine pancreas | 8 800 |
| aprotinin | bovine lung | 6 500 |
| trypsin inhibitor | bovine pancreas | 6 200 |
| glucagon | pig pancreas | 3 480 |
| insulin B chain | bovine pancreas | 3 400 |

[a]Most used in SDS-PAGE; [b]non-reduced
Table in order of decreasing $M_r$

## 7.1.8. Stains for polyacrylamide gels

### 7.1.8.1. "StainsAll"

(1-ethyl-2-(3-(1-ethyl-naphtho(1,2d)thiazolin-2-ylidene)-2-methyl-propenyl) naphtho(1,2d)thiazolium bromide)

**A.** Simultaneous staining of DNA (blue), RNA (bluish purple), protein (red) and acid polysaccharides and mucopolysaccharides (purple). Fix the gel in 25% isopropanol overnight. Then stain for 60 min in 0.2% stain in 0.4 M sodium acetate/0.4 M acetic acid, pH 4.7.

REFERENCES: Kay *et al.*, 1964; Dahlberg *et al.*, 1969.

**B.** For calcium-binding proteins (Campbell *et al.*, 1983): Fix the gel in 25% isopropanol overnight. Then stain in the dark for 48 h in 25% isopropanol, 7.5% formamide, 30 mM Tris, pH 8.8, containing 0.0025% StainsAll. The solution must be violet. Destain in the dark in water for 2–3 h. Change the water until the background is clear. Then dry the gel down, again in the dark. Calcium-binding proteins (e.g. calsequestrin) are blue or purple, and other proteins are red.

### 7.1.8.2 Sensitive fluorescent stains for proteins

*(a)* Dansyl chloride (Talbot and Yaphantis, 1971).

*(b)* Fluorescamine (Stein *et al.*, 1974; Handschin and Ritschard, 1976).

*(c)* MDPF, 2-methoxy-2,4 diphenyl-3(2H)furanone (1 ng sensitivity and non-fading fluorescent label) (Barger *et al.*, 1976; Handschin and Ritschard, 1976).

### 7.1.8.3. Phosphoproteins

*(a)* Staining phosphoproteins on polyacrylamide gels (Cutting and Roth, 1973).

**1.** Run the gel and fix in 10% aqueous sulphosalicylic acid (SSA) for up to 12 h with several changes if the samples contain phosphate.

**2.** Transfer to fresh 10% SSA containing 0.5 M $CaCl_2$ for 60 min. (May store up to 10 days.)

**3.** Rinse in deionized $H_2O$ and place in 0.5 M NaOH at 60 °C for 30 min.

**4.** Rinse twice at 10-min intervals in 1% aqueous ammonium molybdate.

**5.** Place gel in 1% ammonium molybdate in 1 M $HNO_3$ for 30 min (34.6 ml of 65% $HNO_3$ per litre).

**6.** Transfer to 0.5% methyl green in 7% acetic acid for 30 min.

**7.** Destain with 10% SSA.

**8.** Store destained gels in 7% acetic acid.

1 nmole of phosphate is detectable.

*(b)* Determination of the phosphate content of polypeptides isolated from preparative polyacrylamide gels. Ash samples contain 100–400 μg of protein with magnesium nitrate. Analyse phosphate by the method of Bechtel *et al.* (1977).

### 7.1.8.4. RNA

*(a)* Methylene blue. Peacock and Dingman (1967). Fix for 15 min in 1 M acetic acid. Then stain in 0.2% methylene blue in 0.2% sodium acetate/acetic acid (0.4 M, pH 4.7). Destain in 10% acetic acid/10% isopropanol.

*(b)* Acridine orange. Richards *et al.* (1965).

### 7.1.8.5. DNA

*(a)* Ethidium bromide (LePeoq, 1971).

*(b)* Methyl green (Boyd and Mitchell, 1965).

### 7.1.8.6. Lipoproteins

*(a)* Oil red 0 (Sargent and George, 1975).

*(b)* Sudan black (Sargent and George, 1975).

### 7.1.8.7. Glycoproteins

*(a) Periodic acid–schiff (PAS)*

Fix gel in 12.5% TCA for 30 min.
Rinse in $H_2O$.
Place in 1% periodic acid in 3% acetic acid for 50 min.

Wash in several changes of $H_2O$.
Transfer to fuchsin-sulphite solution in the dark for 50 min.

Make a soln of 16.0 g potassium metabisulphate and 21.0 ml of concentrated HCl. Add 8.0 g of basic fuchsin and stir gently for 2 h at room temperature.

Then add acid-washed charcoal and after 15 min filter. (Stored cold, it works for several months.)

Finally, place gel in 0.5% potassium metabisulphate (3 × 10 min changes of 30 ml per gel).
Wash in $H_2O$ until clear background is obtained.
Store in 7% acetic acid.

### *(b) Peroxidase-conjugated lectins*

#### *(i) Concanavalin A* (Olden and Yamada, 1977)

I. SOLUTIONS

**A.** Sigma Type IV ConA 1 mg/ml in PBS.

**B.** Tyrodes solution minus glucose (PBS): For 2 l: 16.0 g NaCl, 0.4 g KCl, 0.54 g $CaCl_2.2H_2O$, 0.42 g $MgCl_2.6H_2O$, 0.1 g $NaH_2PO_4$. When dissolved add 2.0 g $NaHCO_3$ (pH should be approx. 7.4).

**C.** Horseradish peroxidase. Sigma Type I. 500 units per 20 ml PBS.

**D.** Diaminobenzidine-HCl (DAB). 0.5 ml of 1 M Tris-HCl, pH 7.4, and 9.5 ml $H_2O$ + 6 mg DAB + 5 μl of 6% $H_2O_2$. (Warning: DAB may be carcinogenic.)

II. PROCEDURE

1. *For minigels*: *(a)* Stain for protein with Coomassie, destain, photograph and then destain completely in 25% isopropanol/10% acetic acid. Or simply fix in 25% isopropanol/10% acetic acid for 30 min. *(b)* 4 × 15 min in PBS and 10 min in PBS containing 1% bovine serum albumin. *(c)* 60 min in ConA. *(d)* Wash for 10 h in PBS (change solution 3 ×). *(e)* 4 h in horseradish peroxidase. *(f)* Wash 4 × 10 min in PBS. *(g)* Stain with diaminobenzidine for 1–5 min. *(h)* Discard diaminobenzidine and place gel in 5% isopropanol/2.5% acetic acid.

**2.** *For nitrocellulose blots*: *(a)* Transfer protein on to nitrocellulose. *(b)* Block nitrocellulose with 3% bovine serum albumin in PBS for 60 min. *(c)* Wash 10 min in PBS. *(d)* Incubate with ConA for 60 min. *(e)* Wash in PBS 5 × 5 min. *(f)* Incubate in horseradish peroxidase. *(g)* Wash in PBS

5 × 5 min. *(h)* Stain with diaminobenzidene. *(i)* Discard stain solution, rinse the blot in water and dry in the dark.

*(ii) Other lectins conjugated with peroxidase*

Essentially as above without the need for sequential incubation in lectin and peroxidase. Incubate with lectin at a concentration of 10–50 μg/ml.

*(iii) Alcian-blue* (Wardi and Michos, 1972)

Fix the gel for 30 min in 12.5% trichloroacetic acid and for 60 min in 1% perchloric acid in 3% acetic acid. Then treat for 30 min in 5% potassium metabisulphite, and for 4 h in 0.5% Alcian blue in 3% acetic acid. Destain in 10% isopropanol/10% acetic acid.

#### 7.1.8.8. Mucopolysaccharides/Glycosaminoglycans

*(a)* Toluidine blue for mucopolysacharides (Rennert, 1967; Hsu *et al*, 1972). Fix the gel in 25% isopropanol/10% acetic acid, stain in 0.1% Toluidine blue in 10% acetic acid for 30 min and destain in 10% acetic acid.

*(b)* Alcian blue for acid mucopolysaccharides: Stadler and Whittaker (1978). Fix the gel in 25% isopropanol/10% acetic acid for 60 min. Stain for 30 min in 0.5% Alcian blue in 3% acetic acid. Destain in 10% isopropanol/10% acetic acid.

### 7.1.9. Silver stain

REFERENCE: Oakley *et al.* (1980).

The method described below is at least 10 times more sensitive than Coomassie staining. Use deionized water throughout.

**1.** After running the gel treat overnight with 12.5% trichloracetic acid in 65% methanol (i.e. 12.5 g TCA + 65 ml methanol made up to 100 ml with water).

**2.** Wash with water briefly.

**3.** Fix for 60 min with 10% glutaraldehyde (40 ml 25% glutaraldehyde + 60 ml water).

**4.** Wash for 2 h in several changes of water.

**5.** Impregnate for 15 min with silver solution. Mix 80 ml of water with 3.75 ml of 1 M NaOH and *stir vigorously*. Add 2.4 ml of $NH_4OH$ (25% stock solution). Then add dropwise silver nitrate solution (1.2 g dissolved in 20 ml of water). As the silver nitrate solution is added, a precipitate forms, changes colour and then dissolves. When all of the silver nitrate has been added, stop stirring and add the solution to the gel.

**6.** Transfer the gel into developing solution (0.0125 g citric acid + 131.5 $\mu$l 37% formaldehyde + 250 ml water). It is useful at this stage to transfer the gel into a clean plastic container (e.g. a disposable plastic weighing boat). Agitate constantly and replace the developing solution as soon as it becomes cloudy.

**7.** When sufficiently stained, wash extensively in tap water.

NOTE

1. If the gel is overstained it may be possible to reduce the staining.

*Reducer solution*

**A.** 37 g copper(II) sulphate + 37 g sodium chloride in 850 ml water. Add ammonia solution dropwise with stirring until the precipitate formed dissolves to produce a deep blue solution. Make up to 1 l.

**B.** 458 g of sodium thiosulphate in 1 l of $H_2O$.

Mix equal volumes of A and B and dilute 1:3, reducer:$H_2O$. Stop reduction by washing in water.

NOTE

2. Only touch the gel with clean gloves, and try to handle the gel very gently, only on its edges, otherwise there is a danger of artefactual marks over the gel.

NOTE

3. Incubation times may be reduced for minigels.

### 7.1.10. Isoelectric focusing in polyacrylamide tube gels under denaturing conditions

REFERENCES: O'Farrell (1975); Anderson and Anderson (1978); Ames and Nikaido (1976).

The method of O'Farrell involves the separation of proteins, first by isoelectric focusing in low percentage polyacrylamide gels containing 5 M urea and 2% non-ionic detergent. Samples are dissolved in buffer containing sodium dodecyl sulphate, which does not interfere with the isoelectric focusing, but ensures that all the protein in the sample is solubilized. After isoelectric focusing, the first-dimension gel is placed on a second-dimension gel to separate the polypeptides according to molecular mass.

### 7.1.10.1. Solutions for isoelectric focusing

Acrylamide/bis-acrylamide 29.2 g acrylamide + 0.8 g bis-acrylamide made up to 100 ml with water.

10% (v/v) Triton X-100 (or Nonidet P-40).

10% (w/v) sodium dodecyl sulphate.

Gel solution: 48.6 g urea + 28.8 ml water + 11.8 ml acrylamide/bis-acrylamide + 20.3 ml 10% Triton X-100 + 4.5 ml pH 5–7 ampholytes + 0.5 ml pH 3–10 ampholytes. Store in 5.5 ml aliquots at −80 °C.

Overlay buffer: 3.0 g urea + 2.0 ml 10% Triton X-100 + 0.45 ml pH 5–7 ampholytes + 0.05 ml pH 3–10 ampholytes + water up to 10 ml. Store at 4 °C in capped tubes.

Sample concentrate: 0.1 ml 10% SDS + 0.02 ml pH 3–10 ampholytes + 0.18 ml pH 5–7 ampholytes + 0.1 ml 2-mercaptoethanol + 0.2 ml Triton X-100. Mix well and store at 4 °C. Mix each time before use.

Electrolytes: Catholyte, 0.1 M NaOH: 4 g NaOH + water made up to 1 l. Degas before use. Anolyte, 0.06% phosphoric acid, 1.7 ml of 85% made up to 2.5 l.

SDS reducing buffer: 12.5 ml 0.5 M Tris, pH 6.8 + 5.5 ml 2-mercaptoethanol + 30 ml 10% SDS + $H_2O$ to 100 ml. Add bromophenol blue (0.01%).

### 7.1.10.2 Preparation of the gels

Thaw out the gel solution, degas for 1 min and add 5.5 μl of TEMED and 7.25 μl of fresh 10% ammonium persulphate. This will provide sufficient solution for 24 gels 10 cm long and 1.5 mm internal diameter. Use a syringe with thin tubing to introduce gel solution into parafilm-sealed tubes, starting at the bottom and avoiding the introduction of bubbles. Overlay with 20 μl of overlay buffer. Leave 60 min to polymerize.

### 7.1.10.3. Preparation of samples

Add glycerol to make the sample 50% (v/v) in glycerol. Then add 0.5 mg of urea and 0.2 μl of sample concentrate per μl of sample. Place sample on to the gel and overlay with 10 μl of overlay buffer. Fill the tube to the top with 0.1 M NaOH.

#### 7.1.10.4. Electrofocusing

Fill the lower compartment with acid solution. Place the upper compartment in place with gels in place and then fill the upper compartment with alkali solution. Run the gels for 12–15 h at 400 V constant voltage and then for 2 h at 800 V.

#### 7.1.10.5. Second dimension

Remove the gels by gentle air pressure using a rubber teat. Collect the gel in SDS reducing buffer and leave for 10 min. Then freeze or place on to the second dimension using 1% agarose in SDS sample buffer to glue the gel in place. Run until the dye front has passed out of the gel. Then place the gel in 25% (v/v) isopropanol 10% acetic acid for 16 h to fix proteins and remove ampholytes. Then stain for protein and destain.

### 7.1.11. Non-equilibrium pH gradient gel electrophoresis (NEPHGE)

REFERENCE: O'Farrell *et al.* (1977).

This method enables basic proteins to be focused as sharp spots, rather than the streaks usually obtained with the original O'Farrell method. NEPHGE is also faster.

#### 7.1.11.1. Protocol

Prepare gels exactly as described in section 7.1.10. using pH 3–10 or pH 7–9 ampholytes at a final concentration of 2%. Maintain gel lengths uniform at 12 cm (2.5 mm diameter). Leave for1–2 h to allow the gels to set. Place the gels in the electrophoresis tank after filling the lower reservoir with 20 mM NaOH. Do not pre-run the gels. Load samples of protein dissolved in 8 M urea, 0.8% pH 5–7 and 0.2% pH 3–10 ampholines. Overlay with 20 $\mu$l of 6 M urea also containing ampholines, and fill the tube with 10 mM phosphoric acid. Connect the cathode to the bottom and the anode to the top reservoir compartments. Run the electrophoresis for 4–5 h at 400 V. The time required for optimal resolution of specific proteins should be checked. For the second dimension the gels are treated as described in Section 7.1.10.

### 7.1.12. Non-denaturing isoelectric focusing in agarose flatbed gels

REFERENCES: BDH information sheet; Pharmacia instruction booklet; Elkon (1984).

NOTE: Use only agarose specially prepared for isoelectric focusing (e.g. Sigma, Pharmacia, BDH).

Take a glass plate 125 mm × 125 mm and create a mould with four thicknesses of masking tape 7 mm wide all around and heat in the oven to 70–80 °C. Prepare 25 ml of a 1% (w/v) agarose gel as follows. Place 0.25 g of agarose, 2.5 g of sorbitol and 22.2 ml of deionized water in a 50 ml conical flask. For 10 ml (glass plate 10 × 8 cm) use 0.25 g sorbitol + 0.1 g agarose + 9 ml water.

Swirl the mixture and heat to boiling. Maintain at 90–100 °C for 5–10 min. Cool the cloudy solution to 70 °C and add 1.9 ml of ampholytes for 25 ml of mixture or 0.76 ml of ampholytes for 10 ml of mixture. Add the ampholytes with constant swirling. Place the hot glass on a level surface and pour the agarose into the mould. Leave to set for 30 min.

Place the gel on to the cooling plate of the apparatus and place electrode strips soaked in 0.5 M acetic acid (anode) and 0.5 M sodium hydroxide (cathode) in place.

Inject 5 μl samples into the gel, place the electrofocusing lid in place and run on 15 W constant power for 2 h (limiting current 200 mA, limiting voltage 1000 V). The use of coloured p*I* standards helps demonstrate completion of the run. If a power supply is not available which produces constant power use 200 V × 15 min; 300 V × 15 min; 400 V × 15 min; 500 V × 3 hours on constant voltage.

Fix the gel for 15 min in 10% w/v trichloroacetic acid. Then wash with 35% ethanol, 10% acetic acid 2 × 15 min. Then press the gel by placing a piece of wet filter paper on the gel, then 10 sheets of dry filter paper on glass plate and a 2 kg weight for 15 min. Carefully remove the filter paper and dry the gel completely.

Stain for 15 min in 0.5% w/v Coomassie in 35% ethanol/10% acetic acid. Destain in 35% ethanol/10% acetic acid. Gently rub the surface of the gel to remove precipitated stain.

Dry the gel on to the glass plate with a hair drier.

NOTES:

**1.** 8 M urea may also be included in the agarose gel.

**2.** Agarose gels may be carefully transferred on to nitrocellulose sheets and subjected to electrophoretic transfer of proteins, essentially as described for polyacrylamide gels (Section 7.5.2).

**3.** Thin polyacrylamide gels may also be used for flat bed isoelectric focusing—precast gels are available commercially (LKB, Pharmacia) or can be cast in simple moulds constructed from two glass plates, one of which is coated with silicone (e.g. Silane d174, Pharmacia). Silicone rubber tubing of diameter equal to the thickness of the gel required is used to construct the mould and is held in place with clamps. For more details see Hudson and Hay (1976) and Mukasa *et al.* (1982).

**4.** Horizontal isoelectric focusing may also be performed preparatively by using Sephadex as a support medium (Croy *et al.*, 1980).

*Miscellaneous isoelectric focusing references*

Phosphorylated isoelectric variants of polypeptides may be demonstrated by treating phosphoproteins with alkaline phosphatase and then subjecting them to two-dimensional electrophoresis (Julien and Mushynski, 1982).

Home-made ampholytes: Just (1980).

Home-made flat-bed apparatus: Hudson and Hay (1976).

## 7.1.13. Standards for isoelectric focusing

REFERENCES: Radola (1973).

Proteins available from Sigma, Serva, Calbiochem, BDH, Biorad. Coloured markers are indicated with an asterisk (*).

**Table 7.3**

| p*I* marker | p*I* at 24 °C | $M_r$ |
|---|---|---|
| pepsinogen | 2.80 | |
| amyloglucosidase | 3.50 | |
| *Methyl red | 3.75 | |
| glucose oxidase | 4.15 | |
| soybean trypsin inhibitor | 4.55 | 21 000 |
| *phycocyanin | 4.65 | 232 000 |
| β-lactoglobulin B | 5.10 | 18 400 |
| *azurin | 5.65 | |
| human carbonic anhydrase B | 5.85 | 29 000 |
| bovine carbonic anhydrase B | 6.55 | 29 000 |
| *horse myoglobin (2 bands) | 6.85 and 7.35 | 17 500 |
| *whale myoglobin | 8.05 | 17 500 |
| lentil lectin (3 bands) | 8.15, 8.45 and 8.65 | |
| α-chymotrypsin | 8.80 | 25 000 |
| trypsinogen | 9.30 | 24 000 |
| *cytochrome c | 10.60 | 12 200 |

NOTE: Cytochrome c and acetylated cytochrome c kits are available from Calbiochem-Behring and produce p*I* bands at 10.6, 9.7, 8.3, 6.4, 4.9, 4.1. A kit of coloured p*I* markers is also available from BDH.

## 7.1.14. Peptide mapping

### 7.1.14.1. Peptide mapping *Staphylococcus aureus* protease

REFERENCE: Cleveland *et al.* (1977).

*(a) For pure proteins*

Protein sample is 0.5 mg/ml of pure protein in 0.125 M Tris-HCl, pH 6.8, 0.5% SDS, 10% glycerol, 0.0001% bromophenol blue.
Sample is heated 100 °C × 2 min, then cooled and *either*

*(a)* Proteolysis at 37 °C for 0, 0.5, 1.5, 5, 10, 20, 30, 45, 60, 90, 120, 150 min with 0.33 mg/ml protein + 25 μg/ml of *Staphylocuccus aureus* protease; *or*

*(b)* proteolysis at 37 °C × 150 min, 0.33 mg/ml protein with 0, 0.04, 0.16, 0.6, 1.2, 5, 10, 20, 40, 80, 200 μg *Staphylococcus* protease/ml.

Run on 15% gels.

*(b) For gel slices* (e.g. with tubulin $\alpha+\beta$ subunits)

Coomassie stained bands are placed in sample wells of second gel and overlayed with protease.

1.5 mm thick gels are used with long stacking gels up to 5 cm long and with 5.4 mm wide sample wells.

*(i) Protocol*

First stain and destain gel. Then cut out specific bands with a razor blade. Trim to 5 mm wide and soak for 30 min with occasional swirling in 10 ml of 0.125 M Tris-HCl, pH 6.8, 0.1% SDS, 1 mM EDTA. Fill well with this same buffer and push the gel slice to the bottom of the well. Fill spaces around the slice by overlaying with 10 μl of this buffer containing 20% glycerol, then overlay with 10 μl of this buffer containing 10% glycerol and varying amounts of protease (0.025 μg up to 1 μg depending on the source of the protease and the sensitivity of the protein). Run into the stacking gel and switch off for 30 min when bromophenol blue dye reaches the bottom of the stacking gel. Then analyse. Detect peptides by silver staining or autoradiography.

*(ii) Variant* (Tijssen and Kurstak, 1983)

Place an SDS-PAGE gel strip containing several protein bands on a second 15% gel. Overlay with protease, run into the stacking gel, leave for 30 min and then analyse. Breakdown products are visualized by a polychromatic silver staining method.

#### 7.1.14.2. Peptide mapping by cyanogen bromide cleavage of proteins in gel slices

REFERENCE: Pepinsky and Sinclair (1986)

Run a polyacrylamide gel and then isolate gel slices containing unlabelled protein or, in the case of protein kinase substrates, protein which has been phosphorylated *in vitro*. Incubate with CNBr (7–21 mg/ml) in a fume hood for 1 h at room temp, in 0.1 M HCl + 0.1% 2-mercaptoethanol. The authors normally use 10 mg/ml. Wash gel slices 2 × 5 min in $H_2O$, 1 × 5 min in 0.25 M Tris-HCl, pH 6.8 and 1 × 10 min at 37 °C in sample buffer. Then analyse the fraction by SDS-PAGE on 15% acrylamide, 0.4% methylene bis-acrylamide Laemmli gels. Then silver stain and/or do autoradiography and/or immunoblot.

NOTE: **Use great care**: cyanogen bromide liberates cyanide gas under acid conditions.

### 7.1.15. Elution of proteins from SDS–polyacrylamide gels

#### 7.1.15.1 Diffusion

REFERENCE: Higgins and Dahmus (1979).

Locate the protein of interest by placing the gel in 4 M sodium acetate for 40–50 min for a gel 1 mm thick; 60–80 min for a 3 mm thick gel. Alternatively, use (1) 0.1 M KCl until white bands become visible (Nelles and Bamburg, 1976); (2) chilling of the gel (Wallace *et al*; 1974); or (3) protein phosphorescence (Mardian and Isenberg, 1978). View the gel against a black background and with lighting from the side. Bands appear black against the white gel. As little as 0.1 $\mu$g protein/mm$^3$ can be seen in 1 mm thick gels. Alternatively, locate the protein by scanning at $OD_{280}$. Cut out the band of protein, mash it up and leave it in 50 mM $NH_4HCO_3$, 0.1% SDS for several hours. Filter and re-extract the gel. Then dialyse against water and freeze-dry.

### 7.1.15.2. Electroelution

*(a) Methods for locating the band of interest*

**1.** Locate the band as described by Higgins and Dahmus (1979). Cut out the protein band, soak in water for 10 min to reduce the salt concentration and electroelute as described by Nelson *et al.* (1973) or by Hunkapillar (1983)—see below.

**2.** Stain the gel for protein, destain and cut out the band of interest. Then mince the gel to give 1 mm cubes—do not mash. Rinse with water and soak in 0.1% SDS in 50 mM $NH_4HCO_3$ for 15 min. Remove liquid by suction and blotting. Cover the gel in 2% SDS in 0.4 M $NH_4HCO_3$ and add 10% dithiothreitol to give a final concentration of 0.1%. Elute using the apparatus described by Hunkapillar (1983). Soak for 3–5 h and elute for 12–16 h at 50 V d.c. using elution buffer containing 0.1% SDS in 50 mM $NH_4HCO_3$. The buffer in the electrode chambers is mixed gently with a two-channel peristaltic pump (3 ml/min). After 12–16 h replace the elution buffer (0.1% SDS, 50 mM $NH_4HCO_3$) with dialysis buffer (0.02%f SDS, 10 mM $NH_4HCO_3$) and continue electrophoresis for 20–24 h at 80 V.

**3.** Stain the gel, destain and dry it down. Then cut out the band of interest, and re-swell in running buffer. Electrophoretically elute the protein, dialyse against water and freeze-dry. Remove Coomassie by treating with 0.1 M HCl in 90% acetone at 0 °C (Chua and Blomberg, 1979).

*(b) Two simple electroelution devices*

**1.** Anderson *et al.* (1973). The gel slice is placed between cellulose plugs (cotton wool) in a disposable 10 ml plastic pipette capped at the anode end with a dialysis bag. Elution takes place in a tube gel apparatus using Laemmli running buffer or 50 mM $NH_4HCO_3$, 0.1% SDS. Protein is eluted for 6–12 h at 100 V.

**2.** McDonnell *et al.* (1977). The gel strip is placed in a dialysis bag and placed in a horizontal gel apparatus of the type used for DNA electrophoresis. The long axis of the gel strip is placed at right-angles to the field and the gel strip is placed on the cathodal side of the dialysis bag. When elution is complete the gel strip is removed from the dialysis bag, which is resealed and dialysed against water. The contents may then be freeze-dried.

NOTE: Polypeptides eluted from polyacrylamide gels electrophoretically may also be concentrated by precipitation overnight at −20 °C after addition of 1.5 volumes of methanol:acetic acid (100:1, v/v).

*(c) One protocol for the preparation of antigens for immunization by preparative gel electrophoresis*

Antigens may be obtained as described in the preceding section to obtain 50–100 μg of protein for each of several injections.

One particular method is described below which may be used to obtain monospecific antisera to protein subunits starting with relatively crude material (Blomberg *et al.* 1977; Chua and Blomberg, 1979). The method includes:

**1.** Partially purifying the antigen of interest or at least preparing a specific subcellular fraction.

**2.** Using polyacrylamide gel electrophoresis (with or without sodium dodecyl sulphate, or with cholate or urea, depending on the subcellular fraction being examined); or isoelectric focusing in polyacrylamide gels (in the presence or absence of urea) to give optimal purification of the subunit of the antigen of interest.

**3.** Staining the gel for protein, then drying the gel down, cutting out the band of interest and electrophoretically eluting the stained protein, e.g. with the ISCO electrophoretic sample concentrator (Allington *et al.*, 1978) after rehydration of the gel in electrophoresis buffer containing 0.1% (v/v) sodium dodecyl sulphate. With this apparatus the sample is eluted and concentrated.

**4.** Removing the Coomassie stain from the protein sample by treating with 0.1 M HCl in 90% acetone at 0 °C. In this way, 50–100 μg has been prepared for each of several injections.

## 7.2. Immunization Details

### 7.2.1. Antigen amount and immunization sites

To prepare a monospecific antiserum in a rabbit, the animal should be immunized with a total of 0.2–1.0 mg of the antigen injected on four or five occasions. On each occasion the antigen should be injected at two or more sites (Fig. 7.3). Intradermal injection directly over the shoulder blades may be advantageous. To prepare a polyspecific antiserum a total of 1–10 mg of antigen should be used for the immunization schedule.

Very small amounts of protein may be used to raise antisera in rabbits by immunization into the lymph nodes (Huang *et al.*, 1986; Sigel *et al.*, 1983).

Similar amounts of antigens can be used in both sheep and rabbits, although the variations in the immune response to different antigens is so great that no hard and fast rules can be given with respect to the amount of

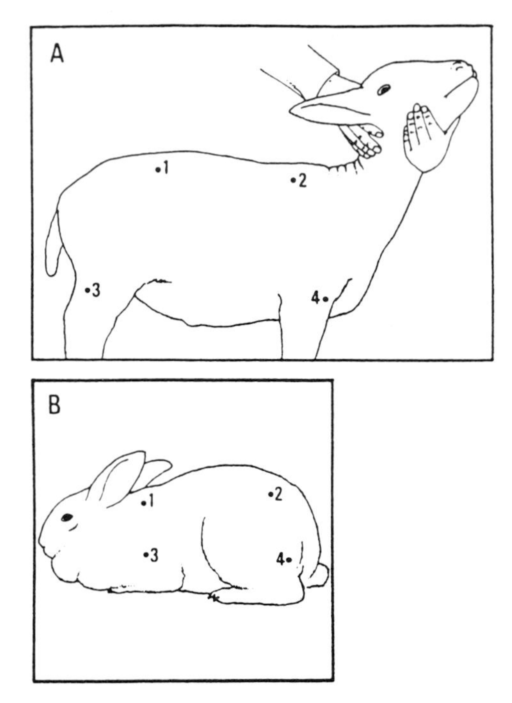


**Fig. 7.3** Recommended sites of injections for multisite injections. (a) Sheep or goat. (b) Rabbit. Injections are given subcutaneously (1,2) and intramuscularly (3,4) on both sides of the animal, i.e. eight injections in all.

antigen required to produce an antiserum. Indeed, in some cases it may be necessary to enhance the immunogenicity of the antigen if an antiserum is to be obtained (see Chapter 2).

One example of an immunization schedule in a sheep is as follows: eight injections of 0.5 ml of antigen preparation emulsified with 0.5 ml of complete Freund's adjuvant at four intramuscular and four subcutaneous sites (Fig. 7.3A), on four occasions, at 2-week intervals, gave excellent precipitating titres to 6–phosphogluconate dehydrogenase, acetyl-CoA carboxylase and fatty acid synthetase (Walker *et al.*, 1976). Blood samples can be processed and the immune response estimated conveniently by immunoprecipitation analyses. Schedules over 8–12 weeks with antigen injection at 2-week intervals can give good precipitating titres. It should be pointed out that high precipitating titres are not always required, e.g. for the preparation of immunoadsorbents of low affinity. In this case large blood samples could be obtained regularly (e.g. every 2 weeks) during an immunization schedule.

Blood samples which are collected after shorter immunization schedules often contain far fewer antibodies to contaminating antigens in the antigen preparation of interest.

### 7.2.2. Preparation of antigen–adjuvant mixtures

Exactly equal volumes of Freund's adjuvant and antigen solution should be mixed vigorously together until a stable emulsion develops (one which will persist as a drop when placed into water). The most convenient mixing device is a double Luer-lock fitting, which enables two syringes to be connected. Very effective mixing of adjuvant and antigen solution can thus be obtained without any loss of material. Alternatively, antigens may be precipitated on to aluminium hydroxide (Hudson and Hay, 1976).

## 7.3 Bleeding Procedures

### 7.3.1. Bleeding rabbits from the marginal ear vein

Bleeding rabbits from the marginal ear vein is very effective and 5–10 ml of blood may be obtained easily. The rabbit should be wrapped up tightly in a blanket (approx. 150 cm × 100 cm) such that only the head and ears are free. An infrared lamp may be used to warm the rabbit and thus stimulate blood flow. While one person holds the animal, a second should shave the margin of the upper side of the ear with a scalpel blade and then apply a thin layer of petroleum jelly. The vein may then be sectioned carefully with the scalpel blade or preferably punctured with a needle. After 1 or 2 min, blood will start to flow through the vein and gentle pressure on the vein between the section and the animal's head should ensure that blood will continue to flow. If the blood flow shows signs of stopping, the site of incision may be rubbed with cotton wool; this should start the flow again. If difficulties are experienced in obtaining blood, then the application of an irritant (e.g. xylene) to the ear in a position distal to the incision will usually induce rapid blood flow. (If xylene is used in this way it should be washed off afterwards with ethanol.) When sufficient blood has been obtained, the blood flow may be stopped by gently holding a piece of cotton wool over the wound. In cases where the blood flow is reluctant to stop, the cotton wool may be held in place with a paperclip. Blood can rapidly be obtained from sheep or goats by venepuncture of the jugular vein with a syringe needle. This method is useful for obtaining a blood volume (10 ml) which is enough for antiserum testing.

### 7.3.2. Bleeding rabbits from the central ear artery

Bleeding rabbits from the central artery of the ear using 18-gauge needles with the hubs removed is recommended by Gordon (1981). The dorsal surface of the ear is moistened with 70% alcohol and then shaved over the central artery. The hub is then broken off an 18-gauge needle and placed to one side while the last cm of the ear tip is moistened with xylene on its upper and lower surfaces. After about 30 s, when the artery is distended, insert the needle into the artery at a shallow angle, while an assistant holds a centrifuge tube behind the needle. A sample of 30 ml of blood may be obtained rapidly with this procedure.

### 7.3.3. Preparative bleeding of rabbits

Preparative bleeding of rabbits is performed as described above and approximately 40 ml of blood can easily be removed by bleeding from the ear or from the heart. If the animal must be sacrificed, most of its blood can be obtained by bleeding from the heart using a cannula (connected to a light vacuum if desired). The animals are first anaesthetized with pentobarbitone sodium, which is usually supplied as Sagatal (May & Baker Ltd), which contains 60 mg pentobarbitone sodium B.Vet.C. per ml. The dose rate in all species is 1 ml for every 2.25 kg body weight and is injected into the ear vein using a fine needle connected *via* fine bore tubing to a syringe. Inject intravenously 50% saline (0.15 M), 50% Sagatal. Alternatively, use a 1:1 mixture of Sagatal and heparin (1000 U/ml; Paines & Byrne Ltd) to obtain plasma. Touch the eye to test for the blinking reflex to ensure that the animal is completely unconscious before continuing.

Blood is removed from the heart by feeling for a pulse between the ribs and then inserting a wide-gauge, teflon-sheathed needle of the sort used for blood transfusion. The needle is connected to 50 cm of plastic tubing 3–5 mm in internal diameter. It is pushed into the heart and held in position as soon as blood begins to flow. The end of the tubing is held close to ground level and blood is collected in a suitable container.

It is also possible to obtain 100–150 ml of blood by cutting the throat of the anaesthetized animal.

### 7.3.4. Preparative bleeding of goats and sheep

Preparative bleeding of goats or sheep is easily performed by the Seldinger Wire technique (Fig. 7.4). A jugular vein is pierced under local anaesthesia

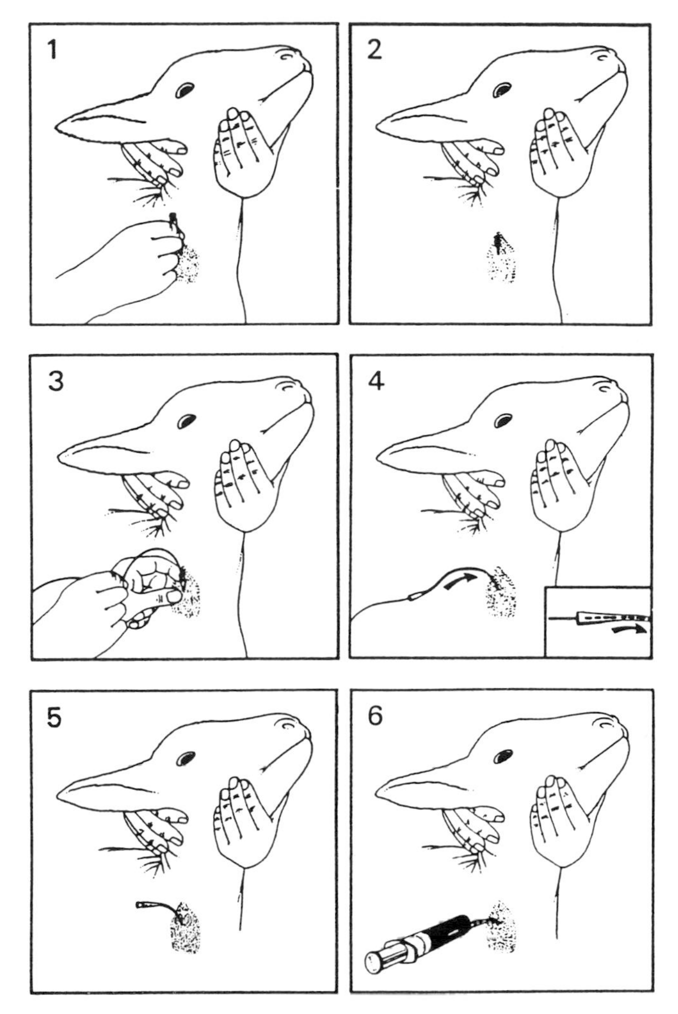


**Fig. 7.4** Bleeding sheep (or goats) from the jugular vein. 1. Piercing of jugular vein with a 14-gauge needle. 2. 14-gauge needle in position. 3. Insertion of guide wire. 4. Insertion of catheter. 5. Catheter (stoppered) in position. 6. Withdrawal of blood (into a 50 ml syringe).

(e.g. with 0.5 ml 2% (v/v) lignocaine given subcutaneously). After venepuncture with a 14-gauge needle, a guide wire is inserted through the needle which allows the subsequent rapid insertion of a catheter over the wire into the jugular vein. A hypodermic syringe (50 ml) may be used to take up to 1 l of blood which should subsequently be replaced with an equal volume of iso-osmotic saline.

Animals may be re-bled at intervals if required. The choice of intervals varies but at least 2–3 weeks should be left between successive bleedings.

Extensive details on immunizing and bleeding various species of animals are given elsewhere (Herbert, 1978).

## 7.4. Antiserum Processing

### 7.4.1. Preparation of IgG

Blood is obtained from immunized animals as described above. It is allowed to clot overnight at room temperature and the serum is carefully decanted from the clot. Ammonium sulphate (Fig. 7.3) is added to the serum at 0 °C to give a final concentration of 291 g/l (50% of saturation). The suspension is centrifuged at 10 000 *g* for 10 min and the immunoglobulin precipitate is washed at 0 °C with a solution of 1.75 M ammonium sulphate until it is white. This procedure will remove albumin, transferrin, and *α*-proteins, including haptoglobin and haemoglobin (Harboe and Ingild, 1973). The final white pellet is dissolved in 10 mM $NaH_2PO_4$, pH 7.0, and dialysed overnight against water. The precipitated lipoproteins may be removed by centrifugation at 10 000*g* for 10 min. The supernatant is dialysed overnight against 10 mM $NaH_2PO_4$ buffer, pH 8.0, and then loaded on to a column (e.g. 4 ml of DEAE-cellulose per ml of serum; Harboe and Ingild, 1973) of DEAE-cellulose previously equilibrated in the same buffer. The column is washed with two column volumes of the buffer and the eluted IgG is concentrated by ammonium sulphate fractionation as described above. Finally, the IgG can be dissolved in a buffer of choice for immunochemical work (e.g. 20 mM $NaH_2PO_4$, pH 7.0, containing 0.15 M NaCl).

It is advisable to check the proportion of IgG which is eluted from the DEAE-cellulose column by this procedure, since it may vary with sera from different species. A suitable way to do this is to elute the column with a linear gradient (10–50 mM) of $NaH_2PO_4$ buffer, pH 8.0. This procedure should elute all the immunoglobulins of interest and the fractions containing each immunoglobulin class can be identified by the appropriate commercially available anti-immunoglobulin serum. Non-immune (control) serum should be similarly fractionated for use in comparative experiments with the IgG to the antigen of interest.

### 7.4.2. Preparation of IgG fragments

The monovalent Fab fragment of IgG is recommended for the localization of intracellular antigens by immunohistochemical techniques (Poole, 1974). The

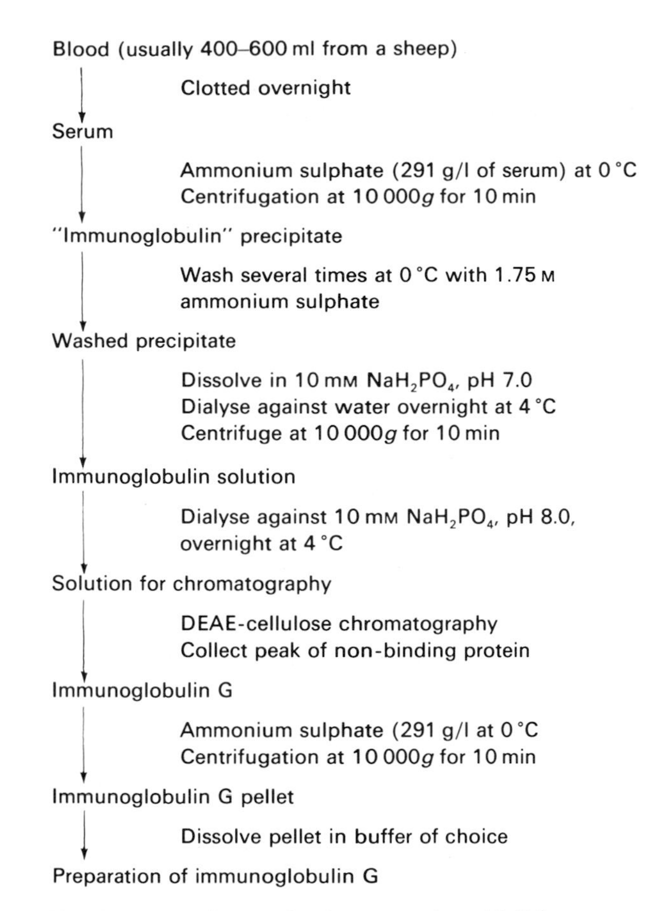


**Fig. 7.5** Flow diagram for the preparation of IgG from serum.

Fab fragment is prepared as follows. First IgG (25 mg/ml, final concentration) is digested with pepsin (0.5 mg/ml, final concentration) for 24 h at 40 °C in 0.2 M sodium acetate buffer, pH 4.5, containing 0.1% (w/v) sodium azide. Then the pH is adjusted to 8.6 with sodium hydroxide to inactivate the pepsin. After centrifugation the pepsin digest is subjected to chromatography on Sephadex G–200 equilibrated in phosphate-buffered saline. Three peaks should be obtained: first, a small IgG peak, second, $(Fab)_2$ and third, Fc. $(Fab)_2$ may then be reduced to Fab with 10 mM cysteine for 30 min.

### 7.4.3. Labelling of IgG and IgG fragments

For many studies involving the use of antisera it is necessary to use labelled antibody molecules. Antibodies may be labelled by covalently coupling them to other proteins, by making them fluorescent, by labelling them with radioactive reagents or by biotinylating them. They may also be labelled by adsorbing them on to colloidal gold particles.

### 7.4.3.1. Labelling with other proteins

In Table 7.4 four examples are given of proteins which may be covalently attached to IgG or IgG fragments to enable enzyme immune assay or immunohistochemistry to be performed.

**Table 7.4** Labelling IgG and Fab with other proteins.

| Protein | Method of Linkage | Use | Reference |
|---|---|---|---|
| Alkaline phosphatase (AP) | 5 mg of AP + 2 mg of IgG in 2 ml of phosphate-buffered saline, 0.06% (v/v) glutaraldehyde for 4 h at room temp. Then dialyse against appropriate buffer before use. | enzyme immune assay + immune blot + immunohistochemistry | Engvall and Perlmann (1971, 1972) |
| Horseradish peroxidase (HRP) | 12 mg of HRP + 5 mg of IgG in 1 ml of 0.1 M sodium phosphate, pH 6.8. Add 50 μl of 1% (w/v) glutaraldehyde. Incubate for several hours and then dialyse against appropriate buffer before use. | immunohistochemistry + ELISA + immune blots | Avrameas (1969) Kraehenbuhl *et al.* (1971) |
| Cytochrome c | 8 mg of cytochrome c + 5 mg of Fab in 1 ml of 0.1 M sodium phosphate, pH 6.8. Add 50 μl of 1% glutaraldehyde. Incubate for several hours and then dialyse against appropriate buffer before use. | immunohistochemistry | Avrameas and Ternynck (1971) Kraehenbuhl *et al.* (1971) |
| Ferritin | 20 mg ferritin + 10 mg affinity-purified antibody made 0.02% with glutaraldehyde 60 min × 23 °C. Then 0.1 lysine and dialyse. | immunohistochemistry | Hackenbrock and Hammon (1975) |

### 7.4.3.2. Fluorescent antibody

(a) Fluorescein isothiocyanate (Poole, 1974; The and Feltkamp, 1970). To 5 ml of IgG solution (20 mg/ml in 0.15 M sodium phosphate buffer, pH 9.5) is added 1 mg of fluorescein isothiocyanate isomer 1. After 1 h the reaction products are separated by means of a small Sephadex G-25 column.
(b) Rhodamine (McKinney and Spillane, 1975).

### 7.4.3.3. Radiolabelled antibody (and antigen)

Labelling can be conveniently carried out in the following ways.

*(a)* Reductive methylation (Rice and Means, 1971). To 0.1 mg of protein in 0.1 ml of 0.2 M sodium borate buffer, pH 9.0, add 10 μl of 40 mM formaldehyde. Then after 30 s, add four 2 ml aliquots of sodium borohydride (5 mg/ml), followed after 1 min by a further 10 μl aliquot of sodium borohydride. Either the formaldehyde ($^{14}C$) or the sodium borohydrate ($^{3}H$) may be radioactive reagents.

*(b)* Iodination (Landon *et al.*, 1967; Greenwood *et al.*, 1963). To 2 mg of protein in 0.5 ml of phosphate-buffered saline in a glass tube add the following: 500 μCi of $Na^{125}I$, 30 μl of Chloramine T (1 mg/ml) and mix. After 2 min at room temperature add 30 μl of sodium metabisulphate (2 mg/ml). Separate the reaction products by gel filtration on Sephadex G-25 or by extensive dialysis. Immobilized Chloramine T may also be used as a gentler labelling procedure (Markwell, 1982, available from Pierce as Iodo beads).

*(c)* Other iodination procedures include the use of lactoperoxidase (Marchalonis, 1969; Hubbard and Cohn, 1975) and the method of Bolton and Hunter (1973a,b). Lactoperoxidase methods provide gentle labelling conditions, especially when immobilized enzyme is used (e.g. Biorad's enzymobeads).

### 7.4.3.4. Other methods

Antibodies may also be labelled by adsorbing them onto colloidal gold particles of varying diameters (Roth, 1983). Biotinylation of antibodies (Billingsley *et al.*, 1985) also provides an excellent means of creating sensitive staining reagents which can be detected with avidin and streptavidin-linked detection systems.

## 7.5. Methods for Testing and Using Antibodies

### 7.5.1. Immunodiffusion and immunoelectrophoresis

#### 7.5.1.1. Apparatus

*(a) Immunodiffusion.* Immunodiffusion is often performed in agar gels cast in Petri dishes, but it is more convenient to use glass plates (10 cm × 10 cm). Smaller plates may be used, e.g. microscope slides. To prevent the gels from drying out they must be kept in a moist environment. Plastic lunch boxes are convenient for this purpose. The immunodiffusion plates are laid on moistened paper tissues or foam rubber in the base of the box. Microscope slides may be placed on a wet paper tissue in a Petri dish.

*(b) Immunoelectrophoresis.* Most standard apparatus designed for horizontal electrophoresis may be converted for immunoelectrophoresis (e.g. the Shandon Model U77). Alternatively, electrode chambers may be constructed from perspex (approximate dimensions 25 cm × 10 cm × 6 cm) and fitted with platinum electrodes. A cooling plate is required to cool the gel during electrophoresis. Cooling plates may be obtained commercially or constructed from perspex or from copper tubing fixed to a sheet of copper (e.g. 10 cm × 20 cm). Clearly, with copper cooling plates, electrical insulation is essential: thus the cooling plate may be placed in a plastic bag. A slow steady flow of water through the cooling plate should ensure adequate cooling. Alternatively, electrophoresis may be performed in a cold room.

*(c) Choice of power pack.* The power supply should be capable of supplying up to 500 V on constant voltage.

*(d)* A horizontal surface is essential for casting gels. A perspex table may be purchased or constructed which can be adjusted to provide a horizontal surface. In some techniques it is necessary to remove strips of gel after cutting through the agar or agarose gel. A perspex bridge is an advantage here to ensure that a straight line is cut through the gel (see below: "crossed immunoelectrophoresis"). Well punches may be constructed from stainless steel or rigid plastic tubing of different diameters and connected to a water suction pump. Whatman 3MM paper is required for electrophoresis wicks and also for pressing gels. Adsorbent paper is also required for pressing gels (e.g. Kleenex dressing towel). Plastic trays (approximate dimensions 21 cm × 11 cm × 2 cm) are required for washing, staining and destaining gels. A boiling-water bath and a water bath set at 50 °C are also required. All of the apparatus required is commercially available, e.g. from Behring, Pharmacia, LKB and Shandon.

### 7.5.1.2. Solutions

*(a)* 2% (w/v) agarose in water (100 ml).

*(b)* Immunodiffusion buffer (2 × concentrated).
0.3 M NaCl containing 40 mM sodium phosphate, pH 7.0; or 0.28 M NaCl containing 103 mM Tris-HCl, pH 7.6; or 0.16 M barbitone buffer, pH 8.2. The optimal conditions for immunoprecipitation vary between antibody–antigen systems but it is usual to include 0.15 M NaCl in a gel buffered between pH 6 and 8. For membrane proteins, include Triton X-100, Lubrol PX or a similar detergent in the buffers (0.2–1.0%, v/v). Also include sodium azide (0.2%, w/v) to prevent bacterial growth.

*(c)* Immunoelectrophoresis buffers.

*(i)* Running buffers (5 × concentrated).
*Svendsen* buffer (Axelsen *et al.*, 1973). Buffer 1: barbitone-Na 13 g + barbitol 2.07 g. Boil the water and then add the solids. Make up to 1 l. Buffer 2: Tris 45.2 g + glycine 56.2 g. Make up to 1 l. Then mix buffers 1 and 2 to provide a 5 × concentrated stock solution from which the diluted buffer may be made when required.

*Chua and Blomberg* (1979) buffer. Tris 48.4 g + sodium acetate 27.22 g + EDTA sodium salt 1.86 g made up to 1 l after bringing the pH to 8.6 with glacial acetic acid.

*(ii)* Gel buffers (2 × concentrated).
From the running buffer stock solution take 100 ml and add 150 ml of water + detergent (e.g. Triton X-100; 0.2% v/v, final concentration). The presence of non-ionic detergent in the gel is essential for membrane proteins, has no effect upon the resolution of soluble proteins, and may be advantageous in later processing of the gel to remove unreacted material.

*(d)* Other solutions.
0.15 M NaCl to wash out untreated material. Stain (2 l): 5 g Coomassie Brilliant Blue (Sigma) + 900 ml methanol + 200 ml acetic acid + 900 ml water. Destain (2 l): 200 ml acetic acid + 900 ml methanol + 900 ml water.

### 7.5.1.3. Casting agarose gels

(The following is for a 10 cm × 10 cm glass plate with a gel 1.5 mm thick.) Place the bottle containing 2% (w/v) agarose gel into a boiling water bath. Meanwhile pipette a 7.5 ml aliquot of the ( × 2) gel buffer into a boiling tube and place this in a water bath at 50 °C. When the agarose is completely molten, transfer 7.5 ml to the boiling tube and mix the two solutions completely. The glass plate must now be "precoated": dry a thin layer of agarose on to the plate to fix the gel firmly to the glass. Fill about $\frac{1}{3}$ of the length of a Pasteur pipette with the 2% (w/v) agarose solution and transfer it quickly on to a glass plate. Working as quickly as you can, use another glass plate to spread the agar over the surface. The agar starts to set quite rapidly so do not try spreading for too long. Now dry this thin coat of gel on to the glass plate with a hairdrier (or in an oven). Alternatively, simply moisten a paper tissue with the hot agarose and swab the surface of the plate with this. The agarose should dry very rapidly. The gel must now be cast. If antiserum is required in the gel (e.g. for rocket immunoelectrophoresis and radial immunodiffusion), it may be added to the agarose in the tube in a water bath at 50 °C; for example, for 1% (v/v) antiserum add 0.15 ml of antiserum to the 15 ml of gel and mix with a glass rod. Lay a glass plate on a horizontal area close to the water bath. Quickly remove the boiling tube and pour the liquid on to the glass plate. The liquid agarose must be poured as a continuous stream into the centre of the glass plate. Start pouring it on to the plate and then increase the rate of flow. Use a glass rod to move the agarose solution to all the edges of the plate, and a Pasteur pipette to remove bubbles from the gel surface. Great care must be taken to ensure that the liquid does not flow over the sides of the plate. Leave the plate for 5–10 min for the gel to set completely. Wells may now be cut in the gel ready for immunodiffusion or immunoelectrophoresis.

When gels are cast on microscope slides, a warm pipette may be used to place 3 ml of agarose directly on to the slide.

### 7.5.1.4. Procedure for immunodiffusion

After casting an agarose gel on a 10 cm × 10 cm glass plate, punch wells according to the pattern required (Fig. 7.6). Addition of 4% (w/v) polyethylene glycol 4000 may enhance immunoprecipitation (Kostner and Holasek, 1972).

With single radial immunodiffusion, ring diameters may be measured after diffusion overnight (e.g. 16 h) or after 1–2 days, by which time all of the antigen should have reacted. For the other techniques, several days of diffusion may be necessary to visualize all of the antigen–antibody systems present. In all cases it is an advantage to wash and stain the gels for protein as described later.

### 7.5.1.5. Procedure for rocket immunoelectrophoresis

After casting a 10 cm × 10 cm plate with agarose gel (agarose is preferred to

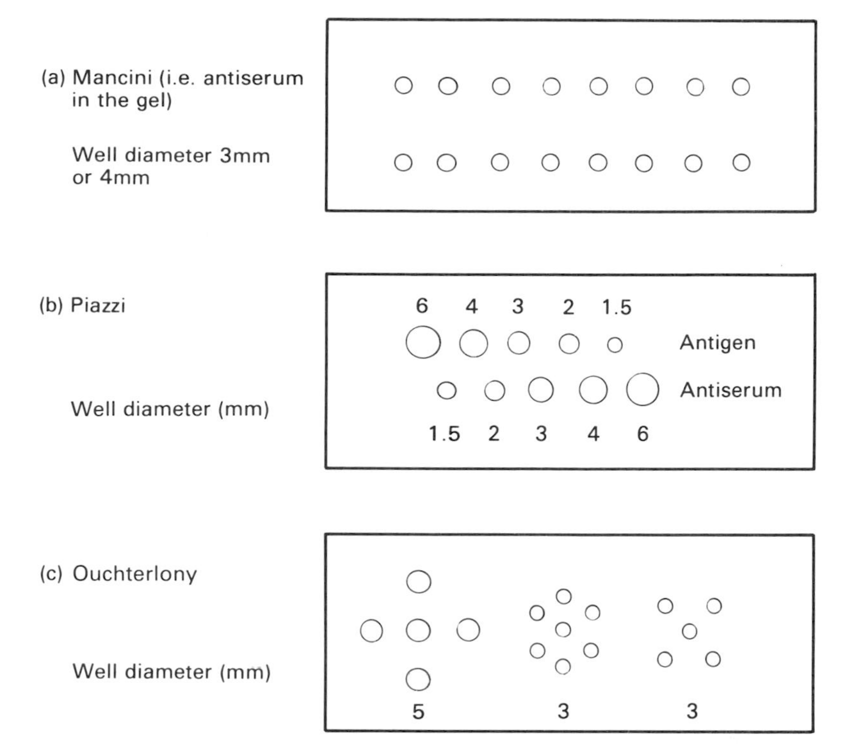


**Fig. 7.6** Well patterns for immunodiffusion. For radial immunodiffusion (a) antiserum is incorporated into the agarose gel and antigen diffuses from the well to produce a ring of immunoprecipitate. The diameter of the ring is proportional to the amount of antigen. For Piazzi immunodiffusion (b) well diameters vary such that a very large range of antigen/antibody proportions are tested. In Ouchterlony immunodiffusion (c) one sample of antiserum in the central well may be tested against several antigen samples or *vice versa*.

agar due to its much lower electro-endo-osmosis) containing antiserum, cut out a row of up to 10 wells (4 mm diameter), 0.5 cm apart and 1 cm from the cathode end of the gel. Place the gel on to the cooling plate and attach the Whatman 3MM wicks (3–5 sheets at each end). The wicks should not overlap on to the wells. With the current switched on (constant voltage, 100 V), place the samples for electrophoresis into the wells. Then place a glass plate on top in such a way as to hold the wicks in contact with the gel, and to prevent evaporation from the gel. Allow electrophoresis to proceed overnight and then process the gel as described below.

An alternative method involves first casting a gel which does not contain antiserum and then removing a section of gel and replacing it with gel containing antiserum (Fig. 7.7). This method has several advantages: a smaller quantity of antiserum is required; and there is no danger of the gel

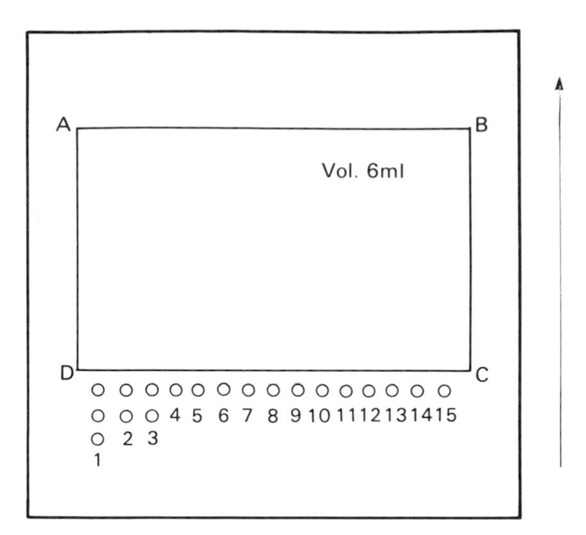


**Fig. 7.7** Rocket immunoelectrophoresis. After casting 15 ml of gel solution on to a 10 cm × 10 cm glass plate, the section labelled ABCD (5 cm × 8 cm) is removed from the plate and replaced with 6.0 ml of agarose which contains antiserum (i.e. 3 ml of buffer ( × 2) and 3 ml of agarose ( × 2)). Wells (2.4 mm; volume 5 $\mu$l) are then punched as shown. Electrophoresis is performed overnight at 100 V (constant voltage at $\leqslant$15 °C).

containing antiserum flowing off the plate. The template in Fig. 7.7 is one pattern that may be used for the analysis of 15 samples. Clearly, if desired two different antisera could be used in two sections each with a volume of 3.0 ml and each sufficient for the analysis of 7 samples. As illustrated in Fig. 7.7, larger samples may be applied by increasing the number of wells (e.g. samples 1, 2 and 3) and also by increasing the size of the wells.

Fused rocket immunoelectrophoresis is performed essentially as described for rocket immunoelectrophoresis, using the template illustrated in Fig. 7.8. A second series of wells is included in a staggered array and samples from, for example, a gel filtration column, are placed in the wells, allowed to diffuse for 30–60 min and then subjected to electrophoresis.

### 7.5.1.6. Procedure for crossed immunoelectrophoresis

The technique of crossed immunoelectrophoresis (Weeke, 1973b) involves the electrophoretic separation of a complex protein mixture in a buffered agarose gel, and then another electrophoresis at right-angles to the first into a gel containing antibodies which recognize the proteins separated in the first dimension. Thus, for example, an antigen-containing preparation may be

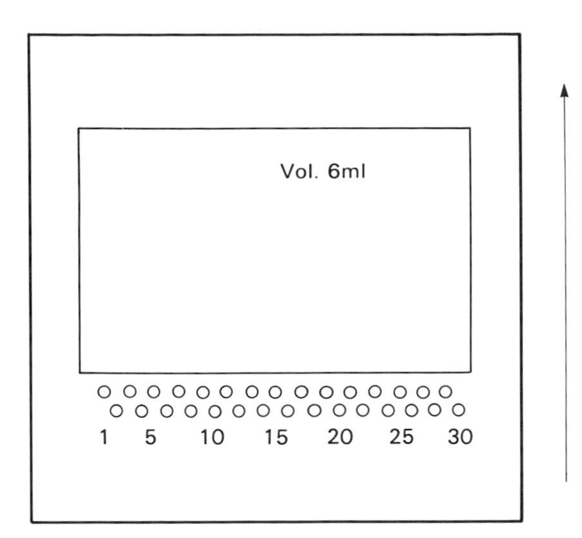


**Fig. 7.8** Fused-rocket immunoelectrophoresis. The procedure for this technique is identical to that adopted for rocket immunoelectrophoresis except that a second row of sample wells are punched. The wells (2.4 mm) are then filled with samples which are allowed to diffuse for 30–60 min prior to electrophoresis overnight (constant voltage, 100 V at ⩽15 °C).

separated by electrophoresis (Fig. 7.9a). The area above the dotted line is then covered with gel containing antibodies which recognize the antigen. The filter paper wicks connecting the agarose gel to the electrode chambers are then placed as shown in Fig. 7.9. Application of a potential difference now results in the proteins being forced into the gel containing antibodies.

The antibodies interact with the antigen to produce an insoluble immunoprecipitate which takes the form of a rocket. The area of the rocket is proportional to the amount of the antigen present. The electrophoresis is carried out at pH 8.6, since at this pH the antibody molecules have a net zero charge and therefore do not move out of the gel. At this pH most proteins are negatively charged and move towards the anode. For electrophoresis in the first dimension, cast a 10 cm × 10 cm gel without antiserum. Then punch a well in one corner of the gel 1.5 cm from each edge. Now place the gel on the cooling plate of the electrophoresis apparatus. Take 4–5 (10 cm × 10 cm) pieces of filter paper and moisten by dipping into the electrode tanks. Use the moistened wicks to connect the gel to the electrode chambers. It is essential that the gel is oriented with the sample well nearest to the cathode chamber. Now place the sample to be analysed into the sample well. Place a glass plate over the wicks in such a way as to force the wicks into contact with the gel and thus establish good electrical contact. Two plates may be run at the same time if desired (Fig. 7.9c). When both samples have been applied place the

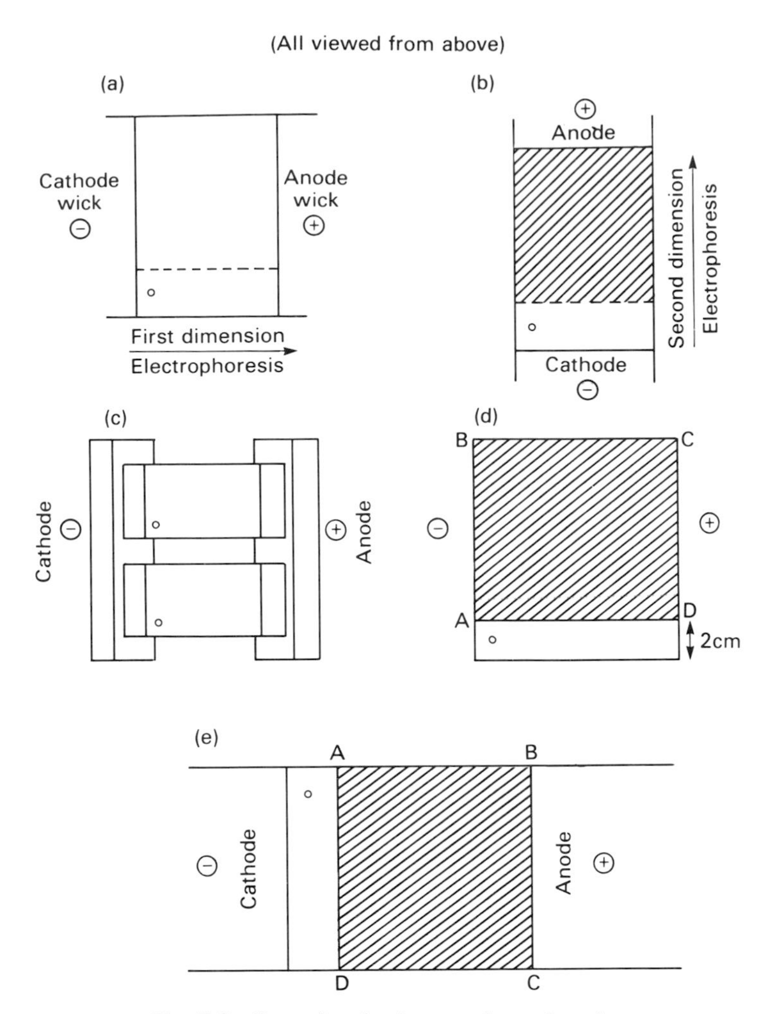


**Fig. 7.9** Crossed-rocket immunoelectrophoresis.

cover over the apparatus and switch on the power pack. Adjust the voltage to 300 V (constant voltage) and leave for 1.5 h. When the first electrophoresis step is completed, switch off the power and remove the plate from the electrophoresis apparatus. Leave the wicks in the apparatus. Now remove most of the gel from the plate as shown in Fig. 7.9d. The shaded area must be removed. This is achieved as follows. First cut along line A–D using a perspex bridge to guide the scalpel blade. Then remove the bridge and use the scalpel to lift the gel away from the plate at corner C (or D). Carefully peel the gel from the plate, taking care not to disturb the gel below the line A–B.

Discard the gel you have removed and put the plate on the horizontal surface. Now go to the water bath and pipette the amount of antiserum required into the 12 ml of melted gel (6 ml gel buffer (×2)+6 ml 2% agarose). Mix well by stirring with a glass rod. Remove the boiling tube from the water bath and quickly pour the agarose on to the glass plate as before. Pour the agarose solution on to the plate in such a way as to prevent gel flowing on to the gel already present. Leave about 10 min for the gel to set. Section ABCD (Fig. 7.9d) should now be covered with gel containing antibodies. The plate must now be put back into the electrophoresis apparatus oriented as shown in Fig. 7.9c. Place the wicks in position and the glass plate on top. Then replace the cover over the electrophoresis apparatus and switch the power pack on. Adjust the voltage to read 100 V, and leave overnight at 15 °C.

Alternatively, as with rocket immunoelectrophoresis, it is usually an advantage to use a template (Figs 7.10 and 7.11) to produce antiserum-containing gel sections. It is also a good idea to mix bromophenol blue

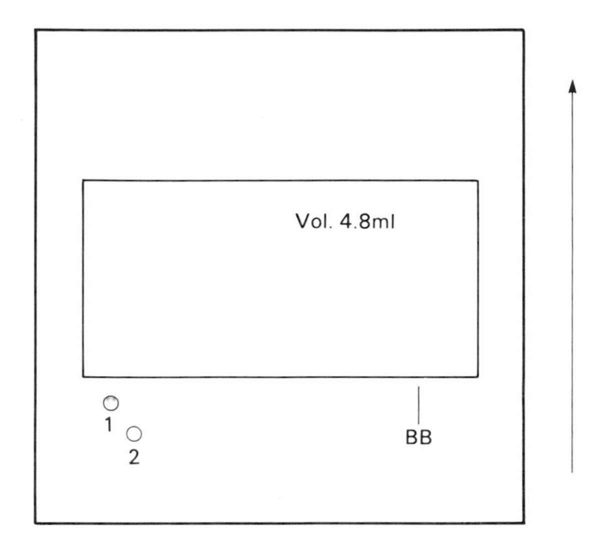


**Fig. 7.10** Crossed-rocket immunoelectrophoresis (alternative method). Procedure: (i) Cast gel over the whole plate, i.e. 7.5 ml of agarose (×2)+7.5 ml of gel buffer (×2). (ii) Punch well (1) and analyse the antigen sample by electrophoresis (300 V, constant V) until the bromophenol blue marker dye reaches the point indicated. (iii) Cut and remove the section outlined on the template (8 cm × 4 cm) and replace with 4.8 ml of gel solution (2.4 ml of buffer (×2) and 2.4 ml of agarose (×2)) containing serum. (iv) Carry out electrophoresis overnight at 100 V, constant voltage, into the antiserum-containing gel. (v) Well 2 is used for tandem crossed-rocket electrophoresis (i.e. with a second antigen sample) as a means of testing immunochemical identity. BB = bromophenol blue.

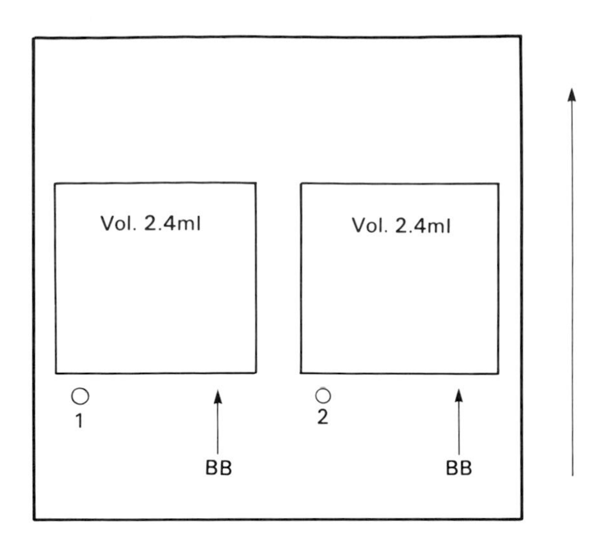


**Fig. 7.11** Crossed-rocket immunoelectrophoresis (for two samples). The procedure is identical to that described in Fig. 7.10, with the exception that two samples are analysed. Electrophoresis in the first dimension is completed in approximately 45 min and then two sections of agarose gel (4 cm × 4 cm) containing antiserum (1.2 ml of buffer ( × 2) and 1.2 ml of agarose ( × 2)) are cast before electrophoresis overnight in the second dimension. BB = bromophenol blue.

tracking dye with the sample. The first-dimension separation may then be terminated when the dye reaches the point marked (BB) in Figs 7.10 and 7.11.

Intermediate gels containing antigen or antiserum may also be used as in Fig. 7.12. Here, after a first-dimension separation by electrophoresis, an intermediate gel is cast between the separated antigen and the antiserum-containing gel. If the intermediate gel contains pure antigen or an antiserum of defined specificity, the antigens recognized by the second antiserum (in gel section B) may be identified.

### 7.5.1.7. Staining gels for protein

Immunoprecipitates are often visible as white lines in the gel when viewed against a dark background. However, for increased sensitivity it is usual to stain the gel for protein; some immunoprecipitates are not seen without staining. Another advantage of processing the gel is that it yields a permanent and easily stored record.

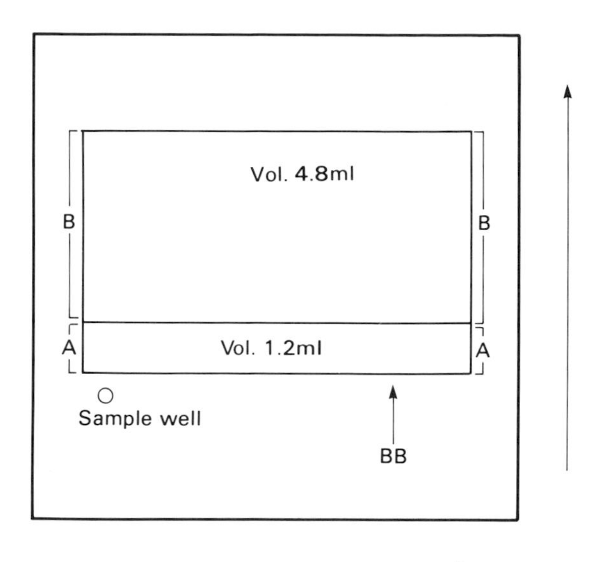


**Fig. 7.12** Crossed immunoelectrophoresis with intermediate gel. (i) Cast 15 ml of gel solution without antiserum and analyse the antigen sample by electrophoresis in the first dimension. (ii) Remove that section of gel labelled A (1 cm × 8 cm) and replace with buffered agarose containing, e.g., pure antigen or monospecific antiserum (0.6 ml of buffer (× 2) and 0.6 ml of agarose (× 2)). (iii) When section A has set, cut around and remove section B (8 cm × 4 cm). Replace this with 4.8 ml of gel containing a second antiserum (2.4 ml of buffer (× 2) and 2.4 ml of agarose (× 2)). BB = bromophenol blue.

After electrophoresis the gel contains a large amount of protein other than that in the immunoprecipitate (e.g. unreacted antibody molecules and non-immune IgG). This protein is removed by pressing and washing the gel. Take the gel and fill the sample well with a drop of saline. Lay the gel on a bench and place a 10 cm × 10 cm square of Whatman 3MM paper over it, taking care not to trap any air. Place a pad of absorbent paper (approx. 50 cm × 50 cm) folded four times to give a pad (approx. 10 cm × 10 cm) over this, then a glass plate and finally a 1 kg weight (reagent bottle, etc.). Leave for 10 min, then remove the weight, the absorbent paper and the Whatman filter paper. Great care is required in removing the Whatman paper square to prevent the gel being removed from the surface of the glass plate. Now immerse the pressed gel in the saline and leave for 20 min. Then pour off the saline carefully. The gel may float free of the glass plate. Pour more saline into the tray and leave the gel for a further 20 min. Again carefully pour off the saline and replace it with distilled water; leave for 20 min. Carefully remove the plate and gel from the tray and press it as described above. After removing the Whatman paper square, dry the gel on to the glass plate with a hairdrier or in an oven. Place the plate into the empty tray and cover the plate with Coomassie blue stain. Leave for 20 min and then pour the stain away. Rinse the plate in distilled water and then add destaining solution. Agitate and replace destaining until the plate shows blue rockets against a reasonably colourless background. Then pour off the destain and dry the plate.

### 7.5.1.8. Techniques for the identification of antigens in immunoprecipitation lines

A wide variety of techniques may be used to decide which of several immunoprecipitation lines obtained with a multispecific antiserum corresponds to an antigen of interest. Some useful techniques are illustrated below. They serve to show the approaches which have been used to identify immunoprecipitation lines corresponding to specific antigens. Several immunodiffusion and immunoelectrophoretic techniques have been described elsewhere in this book by which reactions of identity of a purified antigen (or monospecific antiserum) with an unidentified antigen may be carried out. These techniques will not be considered here.

Staining of immunoprecipitation lines has been extensively used to identify antigens. For many enzymes, simple histochemical staining methods exist. A list of the more common techniques is shown in Table 7.5.

Many other histochemical stains have been used and are described by Owen and Smith (1977). Enzyme activity may be measured in the test tube with cut-out immunoprecipitation lines. If this technique is used it is advisable to carry out immunodiffusion or immunoelectrophoresis in duplicate, so that one plate can be stained for protein and used as a guide for the localization of the otherwise invisible immunoprecipitation lines. Immunoprecipitation lines can be identified by binding of radiolabelled ligands. Examples of these procedures are shown in Table 7.6.

Several techniques have been developed by which antigens radiolabelled *in vivo* or *in vitro* can be identified. Antigens which are radiolabelled may be subjected to immunodiffusion or immunoelectrophoresis. The individual immunoprecipitates may be analysed by polyacrylamide gel electrophoresis in the presence of sodium dodecyl sulphate or by isoelectric focusing in 9 M urea. If the subunit molecular weight or isoelectric point of antigen are known, the immunoprecipitate can be identified. This approach may be used in reverse where protein analysis is carried out first and a known protein subunit is then subjected to immunochemical analysis.

**Table 7.5** A selection of enzyme stains

| Enzyme type | Reagent | Reference | Notes |
|---|---|---|---|
| Dehydrogenase | Substrate + NAD(P) + *p*-Nitro Blue Tetrazolium | Walker *et al.* (1976) | — |
| Esterase | naphthyl acetate + Fast Red TR | Brogren and Bøg-Hansen (1975) | e.g. acetyl choline-esterase |
| Nucleotide phosphatase | Substrate + $Pb(NO_3)_2$ + $(NH_4)_2S$ | Blomberg and Perlmann (1971b) | e.g. AMP, ADP, ATP phosphatases |
| Oxidase | *p*-phenylenediamine | Uriel (1964) | — |
| Phosphatase (acid or alkaline) | α-naphthyl phosphate + Fast Blue B | Blomberg and Perlmann (1971a) | — |
| Peroxidase | (a) diaminobenzide + $H_2O_2$<br>(b) α-naphthol + phenylenediamine (NAD1) | Uriel (1964) | See the immune blotting section<br>e.g. cytochrome oxidase |

**Table 7.6** Identification of antigen of interest by binding of radiolabelled ligands

| Ligand | Antigen(s) | Reference |
|---|---|---|
| [$^{125}$I]Bungarotoxin | Acetylcholine receptor | Teichberg *et al.* (1977) |
| [$^{14}$C]Epinephrine | Plasma membrane proteins | Blomberg and Berzins (1975) |
| [$^{32}$P]phosphate | Casein | Al-Sarraj *et al.* (1978) |
| | Acetylcholine receptor | Gordon *et al.* (1977a,b) |
| | Acetylcholine receptor | Teichberg *et al.* (1977) |

## 7.5.2. Immunoblot procedures

### 7.5.2.1. Apparatus for blotting

Electroblot cells: Towbin *et al.* (1979); Stott *et al.* (1985), standard apparatus.

Kyhse-Andersen (1984), a flat-bed design which minimizes mess and buffer requirements and uses isotachophoretic transfer of proteins.

Washing tray for many strips of blot simultaneously: Tsang *et al.* (1983).

### 7.5.2.2. Diffusion blotting

REFERENCE: Southern (1975).

*(a) Protocol 1*

This method produces two sheets of nitrocellulose, each imprinted with the protein pattern of the gel.

**1.** Run a polyacrylamide gel.

**2.** Place gel in distilled water for 5 min for normal gel. For minigel, simply rinse with $H_2O$.

**3.** Sandwich: 1 kg weight, glass plate, sheets of towel folded up, sheet of blotting paper (Whatman 3MM), sheet of nitrocellulose, polyacrylamide gel, nitrocellulose, filter paper (Whatman 3MM), glass plate. (All made moist with buffer containing 8.4 g/l Tris + 42.9 g/l glycine.)

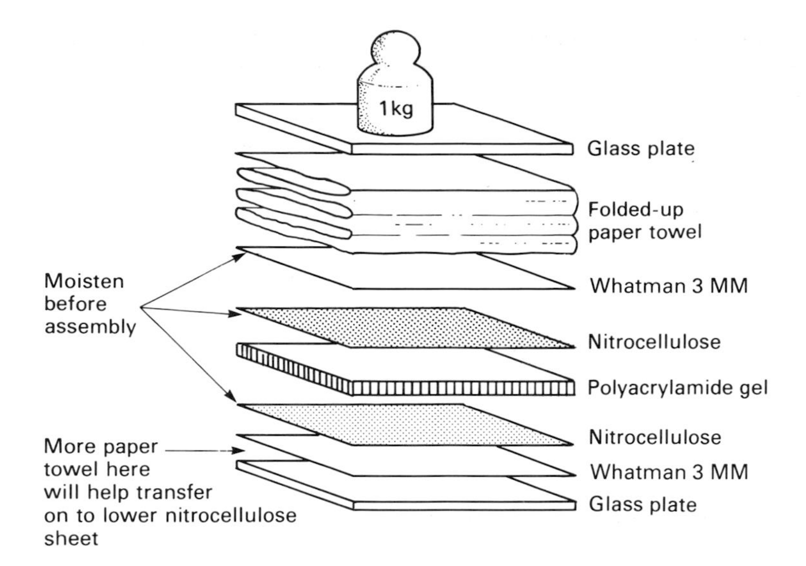


**Fig. 7.13** Diffusion blotting.

**4.** Leave overnight.

NOTE 1: The lower sheet of nitrocellulose takes up less proteins than the upper sheet.

NOTE 2: *Always* moisten the nitrocellulose with buffer before placing it in contact with the gel.

*(b) Protocol 2*

**1.** Run a polyacrylamide gel.

**2.** As above.

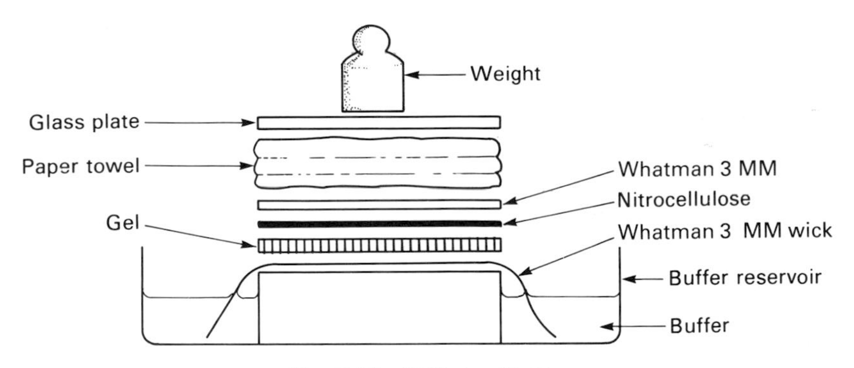


**Fig. 7.14** Diffusion blotting.

**3.** Essentially as above except that buffer is sucked through the gel and one sheet of nitrocellulose by capillary action overnight.

Great care is taken to prevent direct contact of the adsorbent paper or the nitrocellulose with the continuous filter paper sheet, since the buffer must flow *through* the polyacrylamide gel. Sheets of parafilm may be used to ensure this.

*(c) Protocol 3*

Petersen *et al.* (1982) have used a gel drier without heating to transfer protein on to the nitrocellulose by suction.

### 7.5.2.3. Electrophoretic transfer on to nitrocellulose

REFERENCE: Towbin *et al.* (1979); Burnette (1981); Tsang *et al.* (1983).

**1.** Run gel.

**2.** Make sandwich.

**3.** Place into the tank and run for 3 h overnight on 200 mA.

*Transfer buffer*

25 mM Tris, 192 mM glycine, 20% methanol for 4 h: 12.1 g Tris + 56.64 g glycine + 800 ml methanol + water to 4 l.

This buffer can be modified by replacing methanol with ethanol, or eliminating alcohol altogether. The alcohol is supposed to prevent the gel from swelling.

*Alternative transfer buffers*

**1.** 25 mM (or 50 mM) sodium phosphate, pH 6.5. 3 h at 40 V (0.8 A).

**2.** 0.7% acetic acid with nitrocellulose on the cathode side.

**3.** 2 mM sodium acetate 5 mM morphilinopropane sulphonic acid (MOPS), pH 7.5, in 20% ethanol; nitrocellulose on anode side.

**4.** 8 M urea in water with nitrocellulose on the anode side.

*Power pack requirements*

Battery chargers giving 12 V output work well but conventional power packs giving 200 mA or 400 mA may also be used. Many authors recommend 0.8 A for the final 30 min–1 h of transfer, to move the more recalcitrant proteins from the gel.

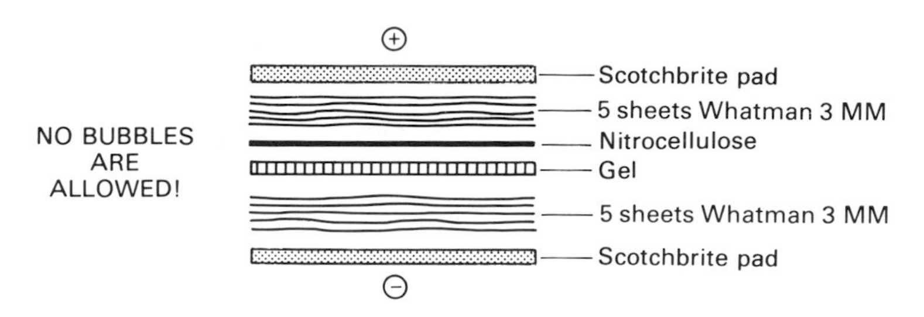


This sandwich is placed between perforated perspex sheets and installed in the electrophoresis tank:

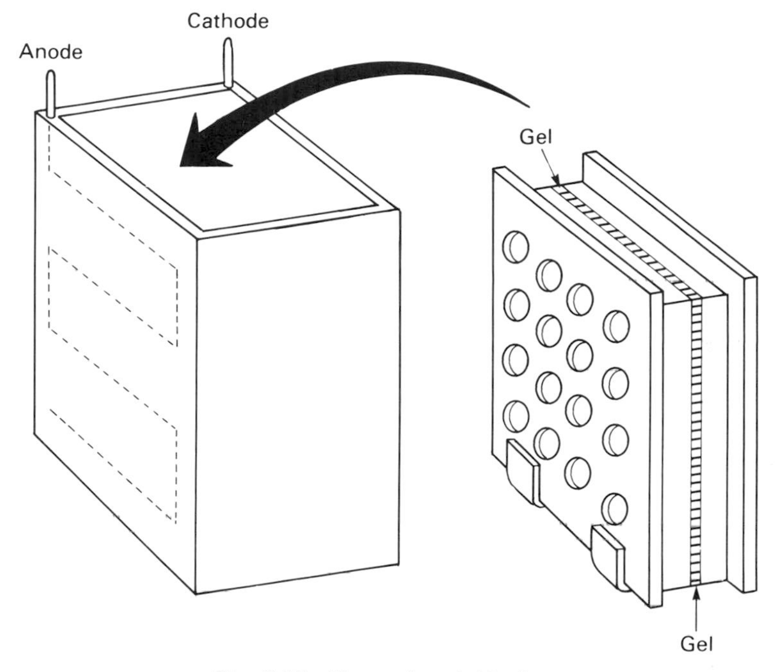


**Fig. 7.15** Electrophoretic blotting.

*Electroblot apparatus*

This may be constructed from a perspex tank fitted with platinum (or stainless steel gauze) electrodes 5–10 cm apart.

### 7.5.2.4. Stains for protein on blots

**1.** *Ponceau S.* Not very sensitive but easily destained completely from protein by buffers of neutral pH. Stain the nitrocellulose blot for 5–10 min in either 0.2% Ponceau S in 3% trichloroacetic acid *or* 0.2% Ponceau S in 50 mM sodium acetate, pH 5.0. Destain the background with water. Photograph if necessary but keep the blot moist. Band positions can be marked with pinpricks. Destain completely in 10 mM Tris-HCl, pH 7.5, 0.15 M NaCl, and then immune stain.

**2.** *Amido Black.* The protocol described for a sensitive protein assay may be used (see protein assays section). The eluant solution may be used to remove the stain from the protein bands and the blot may then be used for antigen localization.

**3.** *India ink* (Hancock and Tsang 1983; Moeremans *et al.*, 1985). Proteins transferred on to nitrocellulose by electrophoresis are visualized as follows:

*(a)* Wash 4 × 10 min in PBS-TW (0.15 M NaCl, 10 mM $Na_2HPO_4/NaH_2PO_4$, pH 7.0, containing 0.3% Tween 20) at 37 °C with shaking. Rinse in deionized water after each wash.

*(b)* Add 100 $\mu$l to 250 $\mu$l of India ink paper per 100 ml PBS-TW. Leave at least 2 h up to 12–16 h.

*(c)* Rinse 5 min in deionized water and dry.

The method is as sensitive as silver staining or staining with colloidal gold. The India ink recommended is Pelikan Fount.

**4.** *Gold staining* (Moeremans *et al.*, 1985) and silver enhancement (Hsu, 1984).

**5.** *Fast Green* (Eichner *et al.*, 1984). Stain the sheet of nitrocellulose for 5–10 min in 0.1% Fast Green in 50% methanol/10% acetic acid, destain in water or 50% methanol/10% acetic acid and subsequently react with the standard immunodetection protocol using diaminobenzidine to produce a brown reaction pocket. The blue Fast Green stained bands disappear almost completely during the immune staining protocol and can be eliminated completely from the photographic record by using Kodak CPP high-contrast film and a blue filter (Tiffin 47B).

**6.** *Ferridye* (Janssen Pharmaceuticals) may be used to stain proteins transferred on to nylon.

### 7.5.2.5. Immune localization of antigens on blots

*(a) Peroxidase- or $^{125}$I-labelled antibody, or Protein A*

REFERENCES: Burnette (1981); Tsang *et al.* (1983).

NOTE: 0.001% thimerosal added to TBS prevent bacterial action. Azide inhibits peroxidase.

**1.** Prepare a nitrocellulose blot of the samples of interest.

**2.** Block sticky sites on the nitrocellulose with TBSA: Tris-buffered saline (TBS, 10 mM Tris-HCl, pH 7.5, containing 0.15 M NaCl) containing 3% bovine serum albumin for 3 h with shaking. Other sticky site blockers are listed in section 7.5.2.6. "Blotto" works superbly well.

**3.** Incubate with the first antibody diluted between 1:50 and 1:500 in TBSA for between 1 h and overnight. With affinity-purified antibodies or monoclonals, try 1–50 μg/ml IgG.

**4.** Wash the blot 5 min × 5 min in TBS (large volumes).

For detection with peroxidase-conjugated reagents:

**5.** Reincubate in 1:200 to 1:1000 diluted second antibody (peroxidase conjugate) in TBSA.

**6.** Wash 5 min × 5 min in TBS (large volumes).

**7.** Stain for peroxidase activity with 0.01% (v/v) $H_2O_2$ and 0.05% diaminobenzidene hydrochloride in 50 mM Tris, pH 7.6. Dissolve 6 mg of 3,3′-diaminobenzidine-HCl in 10 ml of 0.05 M Tris-HCl, pH 7.6 (0.5 ml 1 M Tris-HCl, pH 7.6+9.5 ml $H_2O$). Add 0.1 ml of 3% (v/v) $H_2O_2$. Filter if necessary. Staining should only require 2–15 min, but the mixture may be left up to 30 min in the dark.

**8.** Wash the blot several times in tap water and store until dry between several sheets of filter paper in the dark.

For detection with radiolabelled second antibody:

**9.** Incubate with 100 000 cpm/ml of radiolabelled reagent. Then wash extensively and dry the blot. Then process for autoradiography.

*Variants*

**1.** Peroxidase- or $^{125}$I-conjugated Protein A may be used to detect most immunoglobulin G subclasses of antibodies from mice, rats, rabbits and guinea-pigs (see Table 7.7). Use peroxidase-conjugated Protein A instead of second antibody at 10 μg/ml. Protein G ($^{125}$I-labelled, available from Amersham) reacts with a wider range of immunoglobulins than Protein A, in particular with immunoglobulins from goats and sheep (Ackerstrom *et al.*, 1985; Bjorck and Kronvall, 1984).

**2.** Peroxidase–Anti-Peroxidase (PAP): Steps (1) and (2) as above. First antibody, e.g. Rb anti-albumin for 1 h at 1:40 to 1:1000 dilution. Wash as (4) above. Linker antibody, e.g. sheep anti-Rb IgG for 60 min at 1:50. Wash. PAP complex diluted 1:50 for 30 min. Wash and stain for peroxidase activity. All antibodies should be diluted in TBSA or some similar sticky-site blocking solution.

*Alkaline phosphatase-labelled second antibody*

REFERENCE: Turner (1983).

Transfer proteins on to nitrocellulose. Treat the blot to block unoccupied protein-binding sites. Monoclonal antibody is used neat as tissue-culture supernatants or diluted 1:1 with blocking buffer for 1 h at 4 °C. Wash. Incubate with 1:100 diluted alkaline phosphatase-labelled second antibody for 1 h at 4 °C. Wash.

*Stain*

Solution A, 10 mg Fast Blue B (0-dianisidine) in 10 ml of distilled water; Solution B, 4 mg naphthol AS TR phosphate and 24 mg $MgSO_4$ in 10 ml of 50 mM boric acid/50 mM KCl, pH 9.2 with NaOH. The two solutions are mixed just prior to use. Terminate reaction with 70% ethanol, rinse in water and air-dry.

*Alternative stain*

Solution A, 4 mg/ml of 5-bromo-4-chloro-3-indoxylphosphate (Sigma B-8503) in 2:1 methanol:acetone. Solution B, 1 mg/ml nitroblue tetrazolum (Sigma N-6876). Solution C, 1 M $MgCl_2$. Solution D, 0.1 M ethanolamine-HCl buffer, pH 9.6. Mix 0.75 ml of A with 5 ml of B and 0.2 ml of C. Make up to 50 ml with D. Store A at −20 °C and make B up fresh. Stain for 5–10 min, wash with $H_2O$ and dry between sheets of filter paper.

*Alkaline phosphatase–anti-alkaline phosphatase*

REFERENCE: Blake *et al.*, 1984.

Increased sensitivity is obtained by using soluble immune complexes in a manner analogous to the peroxidase–anti-peroxidase method.

### 7.5.2.6. Blot buffer variants

**1.** Tween 20 is as effective as BSA for blocking sticky sites (Batteiger *et al.*, 1982).

**2.** 4% Bovine serum albumin (BSA), 0.15 M NaCl, 10 mM NaPi, 1 mM EDTA, 0.2% TX100, 1 mM $NaN_3$, pH 7.1, is an efficient sticky-site blocker. The wash solution can be made without BSA (Baines and Bennett, 1985).

**3.** Cadbury's Marvel, 3% in TBS. Instant fat milk powder (Johnson *et al.*, 1984). "Blotto", 0.5% (w/v) non-fat dry milk (Carnation), 0.001% Antifoam A (Sigma), 0.001% thimerosal (Merthiolate) in PBS. According to some authors, 1–5 min exposure to "Blotto" by immersion is sufficient to block all non-specific binding.

*Other peroxidase stain recipes*

**1.** *Diaminobenzidine* (Warning: diaminobenzidine may be a carcinogen): 0.1% (w/v) imidazole, 0.01% (w/v), 3,3′-diaminobenzidine in PBS+0.1% (v/v) Triton X-100. Add 50 μl of 30% $H_2O_2$ per 100 ml of substrate solution. Imidazole intensifies the staining. Incubate for 2–10 min. The reaction product is brown.

**2.** *3-amino-9-ethylcarbazole*: 0.02% aminoethylcarbazole in 50 mM sodium acetate, pH 5.0, 0.2% $H_2O_2$. Incubate 5–15 min at room temperature. The reaction product is red.

**3.** *4-chloro-1-naphthol*: 30–40 mg of 4-chlor-1-naphthol in 0.2–0.5 ml of absolute ethanol. Add with stirring 100 ml of 50 mM Tris-HCl, pH 7.6, containing 50–100 μl of 30% $H_2O_2$. Filter off the white precipitate. Incubate for 5–15 min, at room temperature. The reaction product is blue.

**4.** *Hanker-Yates*: Dissolve 75 mg of Hanker Yates reagent (1 part *p*-phenylenediamine: 2 parts pyrocatechol) in 100 ml of 0.1 M Tris, pH 7.6, buffered saline. Add 100 μl of 30% $H_2O_2$. Incubate for up to 30 min. The colour of the end product is blue-black.

### 7.5.2.7. Biotin/(strept)avidin methods

REFERENCE: Amersham protocols booklet; Hsu *et al.* (1981).

Biotin-labelled antibodies provide versatile reagents which can be coupled to a variety of detection systems based on the egg-white protein avidin or a similar bacterial protein streptavidin. Streptavidin is to be preferred because it shows less non-specific binding. Chemically modified forms of avidin are also available (e.g. from Dako immunoglobulins) which also have low non-specific binding. Streptavidin can be obtained (Sigma, Amersham) labelled with fluorescein, rhodamine, various enzymes, including peroxidase, and radiolabelled with $^{125}I$. Alternatively, streptavidin, being multivalent, may be used to provide a bridge between biotin-labelled antibody and biotin-conjugated enzymes.

Finally, a third possibility is the use of complexes formed by mixing avidin or streptavidin with biotin-labelled enzymes in proportions which ensure an excess of biotin-binding sites in the formed complex. Such preformed complexes may be obtained commercially (Amersham) or prepared from commercially available kits (Vectastain, Vector Laboratories), and give much stronger staining. The basic procedure is identical to those described above (Section 7.5.2.5).

*Recommended dilutions*

*(a)* Biotin-labelled second antibody. 1–5 μg/ml of affinity-purified antibody for 60 min at room temperature.

*Then either:*

*(b)* Preformed streptavidin–biotin enzyme complexes (Amersham). Use 1:100 for β-galactosidase and 1:400 for peroxidase for 60 min. The complexes can be made by incubating 10 μg/ml (strept)avidin with 2.5 μg/ml biotin-labelled peroxidase for 30 min in 50 mM Tris-HCl buffer, pH 7.6, for 30 min at room temperature prior to use. This is the most sensitive approach;

*or*

*(c)* Streptavidin peroxidase. 1:300 (Amersham) for 30 min;

*or*

*(d)* Streptavidin 2.5–5.0 μg/ml for 15 min at room temperature and then biotin-labelled peroxidase diluted 1:500 (Amersham), i.e. 5.0 μg/ml for 30 min.

### 7.5.2.8. Immunogold staining of blots

REFERENCES: Surek and Latzko (1984); Moeremans *et al.* (1985).

Transfer proteins on to nitrocellulose as described above.
Block sticky sites as above.
Incubate in first antibody (1–2 μg specific antibody/ml) for 120 min.
Wash extensively with buffered saline.

Incubate overnight with gold-labelled second antibody (Janssen Life Sciences Products) diluted 1:100 in buffered saline containing 0.1% bovine serum albumin, 20 mM sodium azide and 0.4% (w/v) gelatin.
Wash several times in buffered saline.
Labelled antigens appear as red bands. These may be intensified by precipitation of silver on to the gold. Use double-distilled, deionized water.

*Silver-enhancement solutions*

*Solution 1.* 2 M sodium citrate buffer pH 3.85, 23.5 g trisodium citrate $(2H_2O) + 25.5$ g citric acid $(1H_2O)$ in 100 ml of water.

*Solution 2.* Dilute 50 ml of solution 1 to 500 ml.

*Solution 3.* "Developer". Prepare immediately before use. 77 mM hydroquinine, 5.5 mM sodium lactate in 200 mM citrate buffer, pH 3.85. Make the solutions up as follows in containers entirely covered in aluminium foil. Mix 10 ml of solution 1 with 60 ml of water (Solution A). Dissolve 0.85 g of hydroquinone in 15 ml of water (Solution B). Dissolve 0.11 g of silver lactate in 15 ml of water (Solution C). Add B to A. Add C and mix well.

*Solution 4.* Standard photography fixer.

*Silver-enhancement procedure*

*Step 1.* Wash nitrocellulose filter 2 × 10 min in water.

*Step 2.* 2 min in Solution 2.

*Step 3.* 5–15 min in Solution 3. Developer in a Petri dish entirely covered with aluminium foil.

*Step 4.* 5 min in fixative (standard photography fixer).

*Step 5.* Wash in water 3 × 5 min and air-dry.

### 7.5.2.9. The "Dot Blot" or "Spot Test"

REFERENCE: Towbin *et al.* (1979).

This method enables rapid detection of antigen in various samples and can be used as a semi-quantitative method when $^{125}$I-labelled Protein A is used. The method is especially useful for testing column eluants. Samples containing antigen are spotted on to nitrocellulose and allowed to dry. The nitrocellulose is then treated to block protein-binding sites, and then reacted with antiserum and finally with peroxidase or $^{125}$I-labelled Protein A or second antibody.

*Procedure*

**1.** Spot samples (1–2 μl) containing 2.5–100 ng of antigenic protein on to nitrocellulose and allow to dry for 60 min.

**2.** Treat the nitrocellulose with 5% (w/v) bovine serum albumin in PBS for 30 min (or alternatively, with 2% non-fat dried milk powder in PBS or TBS).

**3.** Wash the nitrocellulose blot with PBS 2 × 5 min.

**4.** Incubate with 3–5 μg/ml of affinity-purified antibody or with a 1:1000 dilution of antiserum in PBS containing bovine serum albumin for 2 h at room temperature.

**5.** Wash the blot 10 × 2 min.

**6.** Incubate with $^{125}$I-labelled Protein A (2 μCi/ml in PBS-BSA) for 60 min, then wash 10 × 2 min, dry the blot and use autoradiography to assess cross-reactivity. Positive spots can then be cut out of the blot and counted in a gamma-counter.

Alternatively, incubate the blot with peroxidase-conjugated second antibody diluted 1:1000 in PBS containing BSA for 60 min. Then wash and react with diaminobenzidine (0.05%) and $H_2O_2$ (0.01%) in 50 mM Tris-HCl, pH 7.6. The peroxidase reaction product may be quantitated approximately by using a densitometric scanner to scan the blot and then measuring the area under the peaks.

### 7.5.2.10. Affinity isolation of specific antibodies from blots

REFERENCES: Olmstead (1981); Talian *et al.* (1983).

Yields of antibody may be very low and therefore it is recommended that you run plenty of antigen on your gels. It is probably also best to use antisera obtained early in an immunization schedule to optimize recovery from the blot.

**1.** Run a slab gel with one sample along the entire length of the gel. Transfer the separated proteins on to nitrocellulose electrophoretically and localize the proteins with Ponceau S as described above (Sections 7.5.2.4).

**2.** Use side strips to localize the antigens of interest.

**3.** Cut out the specific bands of interest from the major part of the blot.

**4.** Block sticky sites for 90 min in TBSA containing 0.2% Triton X-100 (TBSA TX100).

**5.** Incubate in 1:20–1:200 antibody in TBSA overnight at 4 °C.

**6.** Wash 4 × 15 min in TBSA-TX100.

**7.** Elute antibody by placing the strip in 1.5 ml of 0.2 M glycine-HCl, pH 2.8, containing 0.2% gelatin at room temperature for 2 min.

**8.** Place the strip into a further 1.5 ml of eluting buffer.

**9.** Neutralize the eluted antibody by adding 75 μl of 1 M unbuffered Tris per 1.5 ml of eluant. Check that the pH is now 7–8.

Use the antibody neat or concentrate if necessary. The strips of antigen may be re-used if stored at −20 °C between sheets of parafilm. Restart the protocol at step 4.

NOTE: In the protocol of Talian *et al.* (1983) antibody may be labelled with fluorescein before eluting the antibody from the antigen on the nitrocellulose blot.

#### 7.5.2.11. Other ideas

**1.** Blots can be stored for prolonged periods of time at −20 °C. It is therefore often convenient to have a stock of blots from gels where one sample has been run along the entire width of the gel. This can be done for various samples—crude fractions and purified proteins. Then when you want to test a new antiserum you simply cut strips from the appropriate blots and test your antibodies.

**2.** Nylon membranes may also be used for protein blotting and can be stained for protein with Ferridye, a colloidal iron stain from Janssen Pharmaceuticals.

**3.** Covalently bonding transfer membranes are also available based on cellulose (Biorad) or nylon (PALL Biodyne immunoaffinity membranes). These are especially useful if you want to affinity-isolate specific antibodies.

**4.** Nitrocellulose membranes may be cleared with immersion oil or with EUKITT mounting medium.

### 7.5.3. Immunoprecipitation procedures in solution using Protein A immunoadsorbents

Protocols are given below for isolating immune complexes with Protein A.

The specific and strong interaction of Protein A with serum immunoglobulin (Ig) has been known since Jensen (1959) isolated Protein A from the cell walls of *Staphylococcus aureus*. Subsequently, the reactive ligand was identified as the $F_c$ portion of IgG molecules (Forsgren and Sjöquist, 1966; Kronvall and Frommel, 1970). The proportion of serum globulins which is bound by Protein A has been reported to vary from more than 90% in rabbits to a minor proportion of immunoglobulin molecules in sheep and goats (Lind *et al.*, 1970). However, rabbit, sheep, goat and mouse polyspecific antisera to mouse and human immunoglobulins (Ig) were all similarly effective in binding to *Staphylococcus aureus* cells, when used to give immune complexes with Ig in detergent-solubilized mouse or human lymphocytes (Kessler, 1976). This study indicates that the *Staphylococcus* adsorbent performs well with antibody complexes, involving antisera from mammalian species whose serum immunoglobulins do not react with Protein A as quantitatively as do those of rabbits (Kessler, 1976). The binding phenomenon means that Protein A or Protein A-containing *Staphylococcus aureus* was initially used in radioimmunoassay to quantitate $\alpha$-foetoprotein in normal adult serum (Jonsson and Kronvall, 1974).

The increasing need for rapid specific immunoisolation methods to isolate soluble and membrane antigens in many fields of cell biology has led to detailed studies of the use of Cowan I strain of *Staphylococcus aureus* as an immunoadsorbent for antibodies complexed with radiolabelled antigens from mammalian cell lysates. The studies culminated in the development of a technique where Protein A-bearing strains of the bacterium could be substituted for a second antibody in indirect double-antibody immune precipitation reactions (Kessler, 1975). Immune complexes are tightly bound within seconds to the Protein A-containing bacteria, whereas maximum binding of free IgG is a much slower process. Tight binding of immune complexes means that bacterial-washing procedures may be conveniently carried out before antigen elution. The elution of antigen can be achieved with a number of denaturing reagents. The final analytical step of most immunoisolation procedures is polyacrylamide gel electrophoresis in the presence of SDS. Antigens are therefore eluted from the bacteria by methods giving preparations ready for electrophoresis, e.g. by treatment with sample volumes of SDS-urea and reducing agent (Kessler, 1976). The bacteria are simply pelleted by low-speed centrifugation and the antigen-containing sample is then immediately ready for electrophoresis.

With two provisos this technique is to be generally recommended for soluble or membrane antigen isolation. The first requirement is that antibodies in a specific antiserum should have high affinities for Protein A-bearing *Staphylococcus aureus* (e.g. with antisera from sheep and goats). The second requirement is that non-specific adsorption should be minimal.

An alternative to Protein A-bearing *Staphylococcus aureus* is Protein A-Sepharose. Protein A-Sepharose has been prepared and allowed to react with antibodies to a subunit of cytochrome oxidase. After covalent binding of the antibodies to the Protein A-Sepharose, the immunoadsorbent has been elegantly used to isolate cytochrome oxidase precursors (Werner and Machleidt, 1978).

For immunoglobulins which do not react with Protein A (Table 7.7) immunoadsorbents may be used which have specific antibodies bound to Sepharose.

**Table 7.7** Immunoglobulins binding protein A (from Lindmark *et al.*, 1983)

| Species | Ig class and subclass | Reactivity with Protein A |
|---|---|---|
| Mouse | $IgG_1$ | (+) |
| | $IgG_2$ | + + + |
| | $IgG_3$ | + + + |
| | $IgG_4$ | + + + |
| | IgM | (+) |
| | IgE | — |
| | IgA | — |
| Rat | $IgG_1$ | (+) |
| | $IgG_{2a}$ | — |
| | $IgG_{2b}$ | (+) |
| | $IgG_{2c}$ | + + + |
| | IgM | — |
| | IgA | ? |
| Rabbit | IgG | +(+) |
| | IgM | — |
| Guinea-pig | $IgG_1$ | + + |
| | $IgG_2$ | + + |
| | IgM | — |
| Sheep (goat) | $IgG_1$ | −(+) |
| | $IgG_2$ | (+)+ |
| | IgM | — — |
| | IgA | — — |

#### 7.5.3.1. Immunoisolation of externally oriented membrane antigens with Protein A-bearing *Staphylococcus aureus*

A convenient scheme for the immunoisolation of externally oriented membrane antigens involving the use of *Staphylococcus aureus* is shown below. This scheme has been adapted from the method described by Kaplan *et al.* (1979).

*(a) Cells in monolayer culture*

Incubate with [$^{35}$S]methionine to label the proteins for 15 min at 37 °C.

Cells in monolayer culture

↓ Incubate with [$^{35}$S]methionine to label proteins for 15 min at 37 °C

[$^{35}$S]-labelled cells

↓ Incubate with 10 mM trinitrobenzene sulphonic acid (TNBS) for 30 min at 40 °C in phosphate-buffered saline, pH 7.8. Wash cells in phosphate-buffered saline, pH 7.8

TBNS-derivatized [$^{35}$S]-labelled cells

↓ Incubate at 4 °C for 20 min in phosphate-buffered saline, pH 7.8, containing rabbit anti-DNP (dinitrophenyl) IgG (0.5 mg/ml) and 1 mg/ml of bovine serum albumin

Anti-DNP-Trinitrophenyl (TNP)-[$^{35}$S]-labelled cells

↓ Cells washed with phosphate-buffered saline, scraped from plate, sedimented and lysed with phosphate-buffered saline (200 μl), containing 0.5% Nonidet P-40 and protease inhibitors.

Cell lysate

↓ Centrifuge at low speed to obtain plasma membrane protein containing supernatant. Incubate supernatant with 100 l of 10% (w/v) formaldehyde-fixed heat-killed *S. aureus* for 15 min at 4 °C in 12.5 mM potassium phosphate buffer, pH 7.4, containing 0.2 M NaCl, 0.02 % $NaN_3$ and 1 mg/ml of bovine serum albumin

*S. aureus*-Anti-DNP-TNP-[$^{35}$S]-labelled externally oriented membrane proteins

↓
1. Wash in sequence with incubation buffer, 0.1% SDS, 0.05% Nonidet P-40 and finally with incubation buffer
2. Wash bacteria with electrophoresis sample buffer containing 2% SDS and 40 mM dithiothreitol±6 M urea. Sediment bacteria.

TNP-[$^{35}$S]-labelled externally oriented membrane proteins for polyacrylamide gel electrophoresis

**Fig. 7.16** Immunoisolation of externally oriented membrane antigens with Protein A-bearing *Staphylococcus aureus*.

### *(b) [$^{35}$S]-"labelled" cells*

Incubate with 10 mM trinitrobenzene sulphonic acid (TNBS) for 30 min at 40 °C in phosphate-buffered saline, pH 7.8. Wash the cells in phosphate-buffered saline, pH 7.8.

### *(c) TBNS-derivatized [$^{35}$S]-"labelled" cells*

Incubate at 4 °C for 20 min in phosphate-buffered saline, pH 7.8, containing rabbit anti-DNP (dinitrophenyl) IgG (0.5 mg/ml) and 1 mg/ml of bovine serum albumin.

*(d) Anti-DNP-Trinitrophenyl (TNP)-[$^{35}$S]-"labelled" cells*

Wash the cells with phosphate-buffered saline, scrape from the plate, sediment, and lyse with phosphate-buffering saline (200 $\mu$l) containing 0.5% Nonidet P-40 and protease inhibitors.

*(e) Cell lysate*

Centrifuge at low speed to obtain plasma-membrane protein containing supernatant. Incubate the supernatant with 100 $\mu$l of 10% (w/v) formaldehyde-fixed heat-killed *S. aureus* for 15 min at 4 °C in 12.5 mM potassium phosphate buffer, pH 7.5, containing 0.2 M NaCl, 0.02% $NaN_3$ and 1 mg/ml of bovine serum albumin.

*(f)* S. aureus-*Anti-DNP-TNP-[$^{35}$S]-labelled externally oriented membrane proteins*

**1.** Wash in sequence with incubation buffer, 0.1% SDS, 0.05% Nonidet P-40 and finally with incubation buffer.

**2.** Wash bacteria with electrophoresis sample buffer containing 2% SDS and 40 mM dithiothreitol ± 6 M urea. Sediment bacteria.

*(g) TNP-[$^{35}$S]-labelled externally oriented membrane proteins for polyacrylamide gel electrophoresis*

### 7.5.3.2. Immunoprecipitation protocol for radiolabelled tissue extracts using Protein A-Sepharose

REFERENCE: (Kessler, 1975)

Make tissue extracts with 40 mM Tris-HCl, pH 7.5, 4 mM EGTA, 2 mM EDTA, 20 mM NaCl, 1% aprotinin and 75 mg/ml PMSF. For membrane antigens, include Triton X-100 (1%, final concentration). Centrifuge the extract at 100 000*g* for 20 min (for small samples use a Beckman airfuge). Add 25 $\mu$g of specific antibody and incubate for 2 h at 4 °C. In parallel, incubate an identical sample of extract with non-immune IgG. Isolate the antigen–antibody complex by adding 35 $\mu$l of Protein A-Sepharose and incubating for a further 60 min. Wash the Protein A-Sepharose by centrifugation five times in buffer A (15 mM Tris-HCl, pH 7.2, 140 mM NaCl, 5 mM $MgCl_2$, 1 mM EGTA, 0.5% Triton X-100, 0.5 mg/ml bovine serum albumin). Then wash the Protein A-Sepharose by centrifugation through a cushion of

1 M sucrose in the same buffer and finally three further wash steps in buffer A containing 300 mM NaCl. Treat the final pellets with 50 μl of SDS sample buffer prior to gel electrophoresis.

#### 7.5.3.3. Immunoprecipitation of antigens from cells in culture using fixed *Staphylococcus aureus*

Cells ($10^7$/ml), 1 ml + 0.5 ml of 0.1 M NaCl in RIPA buffer (20 mM Tris, 1% Triton X-100, 0.1% SDS, 0.1 mM PMSF, 0.5% sodium deoxycholate).

Leave for 60 min on ice.

Spin for 5 min in a microcentrifuge.

Add antiserum, 20–50 μl of monoclonal supernatant.

Add 10 μl of rabbit anti-mouse IgG (to ensure binding of the immune complex to Protein A).

Mix for 60 min at 4 °C.

Add 50 μl of 10% *S. aureus* (i.e. formaldehyde-fixed, washed in 0.1 M NaCl in RIPA).

Mix for 60 min on a blood mixer wheel at 4 °C.

Wash ppt 2 × 0.5 M NaCl in RIPA, 3 × 0.1 M NaCl in RIPA.

Microfuge, 2 min spins.

Resuspend by sonication in a water bath sonicator and vortex.

Resuspend in 15 μl of electrophoresis sample buffer.

Boil for 2 min, spin to remove Protein A. Load on to SDS-PAGE.

### 7.5.4. Enzyme-linked immunoassays

#### 7.5.4.1. Basic procedure (Voller *et al.*, 1979)

The following method may be used to determine whether samples of serum contain antibodies to a particular antigen. In this case the wells are coated with antigen (e.g. viral protein) and serial dilutions of antisera are placed in the wells and thus a titre is obtained, i.e. the highest serum dilution at which the antigen is still detectable. This method may also be used to identify fractions containing antigens of interest. In this case, individual wells of the plate are coated with different samples (e.g. samples of serum, fractions and column eluants and are subsequently incubated with identical samples of diluted antiserum.

**1.** Dissolve four filter papers (Millipore: RAWP 04700) in 10 ml of acetone. Coat the plates. Dry. This step is optional but may increase the level of response in cases where antigen binds poorly to the microtitre trays.

**2.** Coat the plates with 50 $\mu$l/well of antigen solution at 2 $\mu$g protein/ml in coating buffer, i.e. 0.1 $\mu$g protein/well, for 1–2 h room temperature or 37 °C.

**3.** Wash with PBS (no azide).

**4.** Fill up the wells with PBS containing 20% NS (heat-inactivated serum of the species in which the second antibody is raised) for 1–2 h at room temperature or 37 °C.

**5.** Wash with PBS.

**6.** Add $\simeq$ 50 $\mu$l of the supernatant (or dilutions of the serum etc.); run controls of culture medium, positive supernatant, etc.; incubate overnight at room temperature or for 2–3 h at 37 °C.

**7.** Wash with PBS.

**8.** Incubate with 50 $\mu$l of PBS containing 5% NS or BSA and 1:1000 diluted peroxidase-conjugated second antibody, $\simeq$ 2 h at 37 °C.

**9.** Wash with PBS and assay in the wells for peroxidase activity.

Assay for peroxidase with the following solution: dissolve 40 mg *o*-phenylene-diamine in 100 ml of phosphate-citrate buffer, pH 5.0. Add 40 $\mu$l of 30% $H_2O_2$. *Use at once* by adding 200 $\mu$l to each well. Stop the reaction after 20 min with 50 $\mu$l of 2.5 M $H_2SO_4$. Read either visually or in a spectrophotometer at 492 nm. An alternative assay solution which many authors prefer is given below (Section 7.5.4.3.).

*Other solutions required*

*Coating buffer: carbonate–bicarbonate, pH 9.6*

1.59 g of $Na_2CO_3$ + 2.93 g of $NaHCO_3$ + 0.2 g of $NaN_3$ made up to 1 l with $H_2O$. Do not store at room temperature longer than 2 weeks.

*Phosphate-citrate buffer, pH 5.0*

24.3 ml of 0.1 M citric acid (19.2 g/l) + 25.7 ml of 0.2 M phosphate (28.4 g $Na_2HPO_4$/l or 71.6 g $Na_2HPO_4.12H_2O$/l) + 50.0 ml of $H_2O$. Total volume is 100 ml.

NOTES

**1.** Do not include azide in the washing buffer (it can be replaced by 0.1 g/l thimerosal).

**2.** Wash extensively.

**3.** Inclusion of 0.5% Triton X-100 in the wash buffer reduces background staining; 2% low-fat milk powder in PBS is also very effective at blocking non-specific binding.

**4.** *o*-phenylenediamine is very poisonous.

**5.** Plates can be stored at various stages:
*(a)* coated with the nitrocellulose—dry;
*(b)* in PBS when coated with the antigen—wet chamber 4 °C;
*(c)* in PBS when saturated with serum—wet chamber 4 °C.

**6.** ELISA readers are available for direct assay of peroxidase activity in the microtitre wells (from Dynatec, Biorad). Interfaces with computers are also available.

### 7.5.4.2. Enzyme-linked immunoassays: quantitation of antigens in complex fractions by competition

REFERENCES: Engvall and Perlmann (1971); Engvall *et al.* (1971); Van Weeman and Schnuurs (1971a, b)

**1.** The wells are coated with pure antigen to provide immobilized antigen. Coat the wells of the microtitre plate with antigen as described in the previous method. Then wash and treat with 20% normal serum (heat-inactivated).

**2.** An antiserum dilution curve is obtained to determine the optimal antibody concentration for the assay. For affinity-purified antibodies vary the concentration from 2 $\mu$g/ml down to 5 ng/ml. Incubate overnight at room temperature. Then wash, and incubate with 1:1000 diluted peroxidase-conjugated second antibody for 2 h at 37 °C. Wash and assay in the wells for peroxidase activity. The *antibody* concentration is chosen such that the peroxidase reaction progresses in a linear fashion for 20 min. A curve such as that shown in Fig. 7.17 should be obtained. This dose–response curve shows a linear region, in this case from 280 to 35 ng/ml of antibody. For subsequent experiments, therefore, a concentration of 70 ng/ml, i.e. in the middle of the linear region, is chosen. When antisera are used, dilutions from 1:50 down to 1:250 000 are recommended.

**3.** A competition curve is established. Treat the wells of a plate as described above (1). Then incubate them with antibody at 70 ng/ml, *or* with samples containing mixtures of antibody at this concentration and pure antigen (1–300 ng). Competition of the antibodies for immobilized and soluble antigen results in decreased binding of antibody to the plate with increasing amount of soluble antigen. After overnight incubation, empty and wash the wells. The amount of bound antibody is then determined by

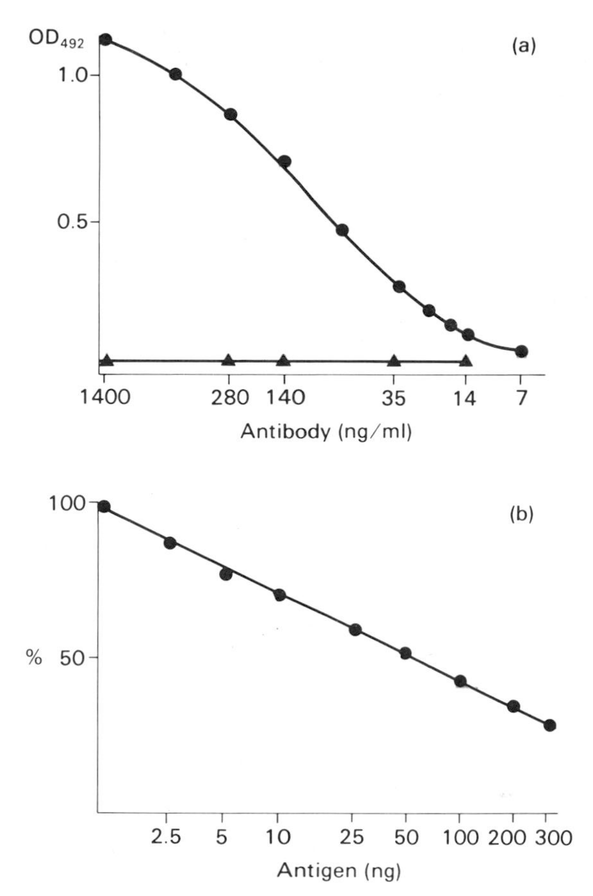


**Fig. 7.17** Enzyme-linked immune assay. Quantitation of the amount of an antigen by competition. (a) Determination of the optimal concentration of antibody for performing the assay. Serial dilutions of affinity-purified antibodies were incubated in the wells of an ELISA Tray which had previously been coated with antigen (39 ng bound per well). Peroxidase second antibody was used at 1:1000 dilution to detect bound antibody (●) or control IgG (▲). (b) Competition curve. Defined amounts of antigen (abscissa) were incubated with a fixed concentration of specific antibody (70 ng/ml). The ordinate indicates the colour intensity expressed as a percentage of the maximal value.

incubation with peroxidase-conjugated second antibody as described above. The competition curve should produce a straight line when absorbence is plotted against the $\log_{10}$ of the antigen amount present in the soluble phase.

Samples containing unknown amounts of antigen may obviously be processed at the same time as the standard curve. An example of a standard curve is shown in Fig. 7.17b.

NOTES: Clearly, this method can be used even if highly purified antigen is not available—provided that a monospecific antiserum is available. Similarly, if sufficiently highly purified antigen is available then the antiserum specificity is not critical.

### 7.5.4.3. ELISA notes

**1.** Peroxidase activity is normally detected with *o*-phenylenediamine (MacDonald *et al.* 1979) as described above. An alternative stain for peroxidase-conjugated antibodies is "ABTS", i.e. 2,2′-azino-di[3-ethyl-benz-thiazoline sulphonate] (Royston *et al.*, 1980; Porstmann *et al.*, 1981). Dissolve 0.165 g of ABTS in 10 ml of PBS, pH 7.4. Add 50 $\mu$l of 0.3% $H_2O_2$. Add 100 $\mu$l to each well and leave for 20–30 min. Stop the reaction with 10 $\mu$l of 1 M $NaN_3$. Read the absorbence at 414 nm. ABTS is reported to have many advantages over *o*-phenylenediamine, including the stability of solutions of the substrate and of the of the bright green reaction product.

**2.** Peroxidase-conjugated Protein A may be used for ELISAs instead of peroxidase-conjugated second antibody (Hawkes *et al.*, 1982b).

**3.** Streptavidin-linked reagents may also be used.

**4.** Alkaline phosphatase-conjugated antibodies may also be used for ELISAs (Moudallal *et al.*, 1984) and assayed with *p*-nitrophenyl phosphate (5.5 mM sodium salt in 50 mM glycine buffer, pH 10.5, containing 0.5 mM $MgCl_2$). Incubate for 30 min and stop the reaction with 100 $\mu$l of sodium

**Table 7.8** Methods for the preparation of immunoadsorbents

| Gel type | Activation with | Reference | Comments |
|---|---|---|---|
| Sepharose | Cyanogen bromide | Porath *et al.* (1973) | The activated gel is commercially available |
| Sepharose | Epichloro-hydrin | Porath and Sundberg (1970) | Gives highly cross-linked gel |
| Aminohexyl / Carboxyhexyl } Sepharose | Carbodiimide | Lowe and Dean (1974) | — |
| Aminoethyl-Biogel P-150 | Four simple steps | Fiddler and Gray (1978) | — |
| Cellulose | Azide | Miles and Hales (1968a, b) | — |

hydroxide solution (4 g per 500 ml of distilled water). Read the optical density at 405 or 410 nm.

### 7.5.5. Preparation of immunoadsorbents

Antibodies (and antigens) may be immobilized by a wide variety of methods. A selection of methods is presented in Table 7.8. In addition, a wide variety of other gel types are commercially available from Bio-Rad (e.g. Affigel 10, which is preactivated; Affigels 701 and 702, which have smaller beads than conventional matrices), Pharmacia, LKB and Pearse.

We give below details for preparing immunoadsorbents by coupling proteins to cyanogen-bromide-activated Sepharose and to Ultrogel.

#### 7.5.5.1. Preparation of cyanogen-bromide-activated Sepharose and coupling of immunoglobulins

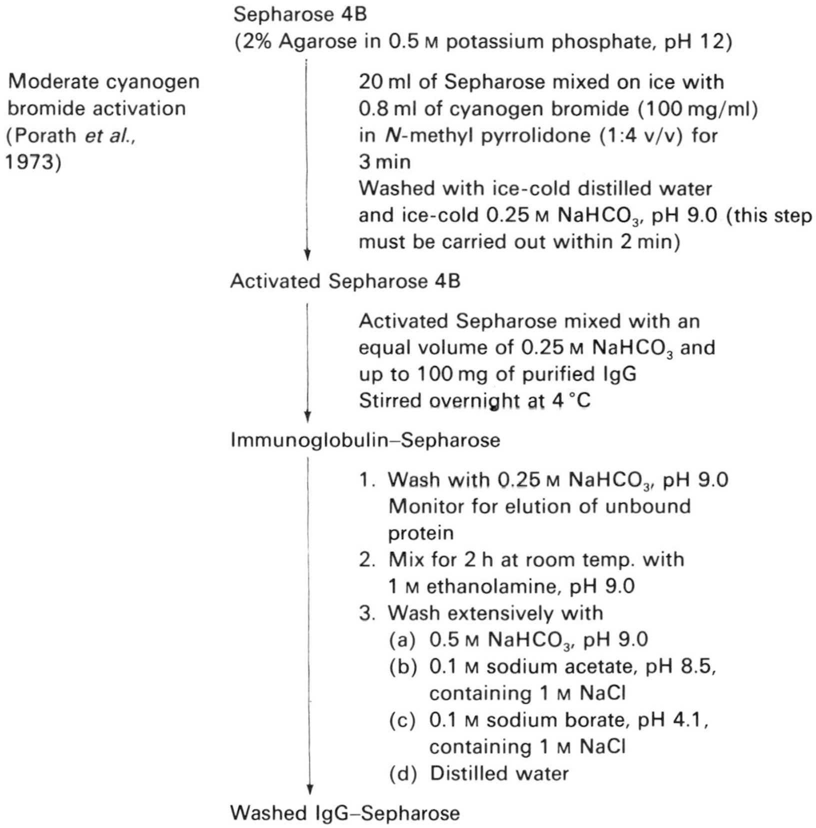

Sepharose 4B
(2% Agarose in 0.5 M potassium phosphate, pH 12)

Moderate cyanogen bromide activation (Porath *et al.*, 1973)

20 ml of Sepharose mixed on ice with 0.8 ml of cyanogen bromide (100 mg/ml) in *N*-methyl pyrrolidone (1:4 v/v) for 3 min
Washed with ice-cold distilled water and ice-cold 0.25 M $NaHCO_3$, pH 9.0 (this step must be carried out within 2 min)

↓

Activated Sepharose 4B

Activated Sepharose mixed with an equal volume of 0.25 M $NaHCO_3$ and up to 100 mg of purified IgG
Stirred overnight at 4 °C

↓

Immunoglobulin–Sepharose

1. Wash with 0.25 M $NaHCO_3$, pH 9.0
   Monitor for elution of unbound protein
2. Mix for 2 h at room temp. with 1 M ethanolamine, pH 9.0
3. Wash extensively with
   (a) 0.5 M $NaHCO_3$, pH 9.0
   (b) 0.1 M sodium acetate, pH 8.5, containing 1 M NaCl
   (c) 0.1 M sodium borate, pH 4.1, containing 1 M NaCl
   (d) Distilled water

↓

Washed IgG–Sepharose

**Fig. 7.18** Preparation of a Sepharose immunoadsorbent.

### 7.5.5.2. Ultrogel immunoadsorbent

REFERENCE: Ternynck and Avrameas (1976).

Glutaraldehyde-activated Ultrogel can bind 2–3 mg of protein per ml of gel.

Wash 20 ml of Ultrogel ACA22 with 10–12 volumes of double-distilled $H_2O$ on a glass Buchner funnel.

Wash with several volumes of 0.1 M potassium phosphate buffer (KPi), pH 7.4.

Activate: 20 parts of gel; 10 parts of 0.1 M KPi, pH 7.4; 10 parts of 50% glutaraldehyde (final concentration 5%); 50 parts of double-distilled $H_2O$.

Incubate at 37 °C for 18 h with rotation.

Remove the flask and let the contents settle.

Remove the fines.

Add 2 volumes of double-distilled $H_2O$, mix, allow to settle, and remove fines.

Repeat twice.

(It is very important to use double-distilled water.)

Keep washing until no smell of glutaraldehyde remains.

Add $NaN_3$ to 0.01% final concentration. The activated gel can keep at 4 °C for at least 1 month (longer if 0.5% glutaraldehyde is added).

Coupling proteins to the gel:

Wash with 0.1 M KPi, pH 7.4.

Add 1 volume of protein (2–4 mg/ml dialysed against 0.1 M KPi, pH 7.4) to 1 volume of gel.

Incubate at 37 °C for 18 h or 48 h at 4 °C with mixing.

Wash and collect the eluate for protein determination.

Quench the remaining aldehyde with 0.1 M lysine, pH 7.4, for at least 2 h at room temperature.

Wash with phosphate-buffered saline.

Pass the serum through a column at 10 ml/h. Recycle if the volume is large, or mix in a tube for 2 h at 37 °C or 18 h at 4 °C.

Wash the column until the $OD_{280}$ of the eluant is less than 0.02.

Transfer the column to the cold room and allow the column to become cold.

Elute bound antibodies with ice-cold 0.2 M HCl adjusted to pH 2.8 with 2 M glycine.

Add sufficient 1 M Tris base to tubes to neutralize the fractions collected of the glycine eluant. Include 0.2% gelatin.

Read the $OD_{280}$ (blanked on 0.2% gelatin).

Pool the fractions for which $OD_{280}$ is greater than 10% of the peak tube.

When $OD_{280}$ is less than 0.08, immediately neutralize the column with 1 M $K_2HPO_4$.

Wash the column with PBS, 0.01% $NaN_3$ and store in the cold.

Dialyse against PBS for 48 h with three changes of buffer.

(If a reduced yield is obtained with time it is usually possible to wash off bound antibody with 0.2 M HCl, pH 2, and glycine at 0 °C. Alternatively, guanidine-HCl may prove effective, although in this case the antibodies may require careful renaturation by gradual dialysis against decreasing concentrations of guanidine-HCl and finally against PBS.)

Immunoadsorbents may be prepared for purifying antigens by immobilizing antibodies in a similar way to the methods described above. In this case, it is recommended that immunoadsorbents are prepared from antisera obtained early in immunization schedules. In this way antibodies of relatively low affinity are immobilized and thus the antigen can be eluted from the column under less denaturing conditions. A variety of conditions may be used to elute bound antigen: electrophoresis (with or without denaturing agents, e.g. urea and sodium dodecyl sulphate); buffers at low pH (e.g. 0.2 M glycine-HCl, pH 2.8); strong salt solutions (e.g. 5 M magnesium chloride); chaotropic agents (e.g. 3 M lithium isothiocyanate); and strongly denaturing agents (e.g. urea and sodium dodecyl sulphate). It is recommended that the antigen is eluted from the immunoadsorbent as soon as possible after application of the sample.

## 7.6. Immunohistochemical Procedures

### 7.6.1. General

#### 7.6.1.1. Apparatus

*(a) Microtomes*

For paraffin sectioning and for cryostat sectioning.

*(b) Microscope*

For immunohistochemical studies you need a microscope with a *stable stand*. Optimal *illumination* for reflected light fluorescence is a 100 W d.c. mercury light source (HBO 100; Haaijman, 1983). Optimal *lenses* are (oil) immersion colour-corrected, flat-field and with high numerical apertures. Phase is often very useful but the phase rings in phase lenses cut down the amount of light passing through the lens. For fluorescence, care should be taken to choose filter blocks which produce a black background without seriously reducing fluorescent emission. The *camera* optimally has a shutter which produces mimimal vibration, and exposures are monitored to take into account fading of the image and reciprosity failure. For fluorescence microscopy some professional automatic cameras, especially the Olympus OM2, may provide excellent results, and may be a cheap alternative to the cameras designed specifically for microscope systems.

The following are some examples of good, relatively cheap, systems:

**1.** Zeiss standard 16, 50 W mercury light source, × 16/0.50 Plan-neofluor, oil; × 40/1.0 Planapochromat oil and × 63/1.4 Planapochromat oil (all phase objectives). Filter set 15, green excitation H′546 for rhodamine; filter set 16, blue excitation/H 485 for fluorescein. Photography either with the Zeiss automatic camera or with an Olympus OM2. Vibration due to the mirror and shutter can be a problem unless exposures are controlled, i.e. phase should be a couple of seconds.

**2.** Nikon Optiphot microscope, 100 W mercury light source, CF Fluor DL × 20/0.75, CF Fluor DL × 40, 0.85, dry phase objective. CF Fluor DL × 60/1.4 (oil, bright field); × 40/1.0 Planapomat oil, phase; × 100/1.35, F Planapomat (oil, phase). B2 excitation filter block/EF.D; G20 Excitation filter block/EF-D.

*(c) For immunohistochemistry*

Coplin jars, Columbia dishes for washing slides and cover slips respectively, and a moist box—this is a plastic box with lid containing wet tissue supporting a glass or plastic plate on which the slides or cover slips are supported.

### 7.6.1.2. Solutions

*(a) For washing specimens*

*Phosphate-buffered saline (PBS).* 10 mM sodium potassium phosphate, pH 7.2, 0.15 M sodium chloride. For 1 l, dissolve 1.48 g of $Na_2HPO_4$ (anhydrous), 0.43 g of $KH_2PO_4$ (anhydrous) and 7.2 g of NaCl in water and make up to volume. PBS2: for 1 l, 0.257 g $Na_2HPO_4.H_2O$ + 2.25 g $Na_2HPO_4.2H_2O$ + 8.767 g NaCl. Final pH is 7.4. Add $CaCl_2$ (0.9 mM final concentration) and $MgCl_2$ (0.5 mM final concentration) if desired. PBS3: 9.00 g $Na_2HPO_4$ $.2H_2O$ + 64.54 g $Na_2HPO_4.12H_2O$ + 160.00 g NaCl made up to 20 l with stirring for 12 h at room temperature.

*50 mM Tris-HCl, pH 7.6.* Dissolve 6.1 g of Tris in 50 ml of $H_2O$. Add 37 ml of 1 M HCl and dilute to 1 l with water.

*Tris-HCl buffered saline (TBS).* 10 mM Tris-HCl, pH 7.6 (at room temperature) containing 0.15 M sodium chloride. This suffers from considerable temperature-dependent variations in pH.

*Hepes buffered saline (HBS).* 10 mM Hepes, pH 7.4, with HCl and containing 0.15 M NaCl.

### (b) Fixative solutions

Optimally, formaldehyde should be prepared freshly from paraformaldehyde. Dissolve 3 g of paraformaldehyde in 100 ml of PBS2 by stirring at 80 °C *in a fume cupboard.* With constant stirring, add 10 μl of 1 M $CaCl_2$ and 10 μl of 1 M $MgCl_2$. Cool to room temperature, and adjust the pH to 7.4 if necessary. Suction filter through a 0.45 μm Millipore filter, aliquot and store frozen at −20 °C. Discard after thawing.

### (c) Adhesives

These are to help sections stick to the slides:

*Mayer's egg albumin.* Take the white of an egg and add an equal volume of glycerine. Mix well and filter through coarse filter paper. Add a crystal of thymol to prevent bacterial growth and store at 4 °C. Place a small drop on the slide and spread it evenly. Then dry in an oven at 60 °C for 30 min.

*Chrome-alum gelatin.* Add 0.3 g of gelatin to 100 ml of water at 60 °C. Stir until dissolved and then add 0.05 g of chrome alum (chromium potassium sulphate). Add a crystal of thymol and stir until dissolved. Filter the solution while hot and then coat slides by dipping. Blot the edge of the slide then dry in a hot oven.

*Poly(L-lysine)* (Huang *et al.*, 1983). Slides are treated with 0.5 mg/ml poly(L-lysine), mol. wt 300 000.

### (d) Mounting media for immunofluorescence microscopy

#### (i) Semipermanent mounting medium

REFERENCES: Rodriguez and Deinhardt (1960); Beug *et al.* (1979).

Poly(vinylalcohol), e.g. Gelvatol 20–30; Elvanol 51-OE (Dupont); Mowiol 4-88 (Hoechst).

Place 60 g of glycerol in a 50 ml disposable centrifuge tube. Add 2.4 g of poly(vinylalcohol) and stir thoroughly. Add 6 ml of distilled water and leave for 2 h at room temperature. Then add 12 ml of 0.2 M Tris-HCl, pH 8.5 (2.42 g Tris/100 ml $H_2O$, pH 8.5, with HCl). Incubate in a water bath at 50 °C for 10 min with occasional stirring. Spin at 5000 rpm for 15 min and aliquot in glass vials. Store at 4 °C.

Mount slides in a drop of medium on a cover slip. For immediate viewing, drops of nail varnish may be used to hold the cover slip in place. The medium gels in about 4 h at room temperature in the dark to yield a semipermanent specimen which can be stored at 4 °C for 9 months or more and be viewed repeatedly under the ultraviolet microscope.

*(ii) PBS-glycerine with fading retarder*

REFERENCE: e.g. Platt and Michael (1983). Something similar is available from Citifluor, The City of London University.

10 ml of phosphate-buffered saline (10 mM sodium phosphate, pH 7.4, containing 0.15 M NaCl) mixed with 90 ml of glycerol. Add *p*-phenylenediamine (0.1 g) to the phosphate-buffered saline prior to mixing with glycerol to intensify fluorescence and retard fading. Store the solution in aliquots at −20 °C. It acquires a brown discoloration with time although up to 3 months' storage did not affect its usefulness.

Fluorescein does not fluoresce optimally at pH 7.4 and therefore better results may be obtained with Tris-buffered saline, pH 8.5 (50 mM Tris-HCl, pH 8.5, containing 0.15 M sodium chloride). Use nail varnish to hold the cover slip in place.

*(iii) Permanent mounting for rhodamine-labelled second antibodies*

REFERENCE: Hanahan (1985).

Sections were taken through graded alcohol into xylene and mounted in Fluromount (Gurr) containing 2% diazabicyclooctane (DABCO, retards fading of fluorescence).

*(iv) Glycerine jelly*

REFERENCE: Woodhead (1892).

This mounting medium has been used since Victorian times and is still available commercially (Sigma). The medium consists of 30 parts gelatine, 70 parts water, 100 parts glycerine and 5 parts alcoholic solution of camphor. The gelatin is covered with water and allowed to swell for 12–24 h. The mixture is then boiled and strained through a warm filter. The glycerine and camphor are then added and quickly mixed. The mounting solution is stored in small bottles and used warm with the bottle immersed in warm water to keep the mounting medium fluid. It is also important to keep the slide warm until the cover glass is in position. As a concession to modern times the water and alcoholic solution of camphor may be replaced by phosphate-buffered saline containing 10 mM sodium azide. The cover glasses may be cemented in position with India rubber solution, or nail varnish.

## 7.6.2. Light microscopy

### 7.6.2.1. Immunofluorescence microscopy

*(a) Immunofluorescence microscopy on 10–20 μm cryostat sections*

REFERENCE: de Camilli *et al.* (1983).

*(i) Fixation*

Anaesthetize the animal with ether and perfuse transcardially for 5 min with ice-cold120 mM sodium phosphate buffer, pH 7.4, at a presence of 120 mm of Hg, followed by 4% (w/v) formaldehyde or 3% (w/v) formaldehyde/0.25% glutaraldehyde in phosphate buffer for 10–15 min. When perfusion fixation is not possible, simply remove 1–2 mm thick slabs of tissue as quickly as possible after death and place in fixative. With perfusion-fixed tissue, also cut 1–2 mm thick slabs of tissue and place in fixative for 3 h at 4 °C. Then wash to remove fixative with several changes of phosphate buffer.

*(ii) Cryoprotection to prevent ice-crystal damage to the tissue*

Transfer the tissue to 5% (w/v) sucrose in 120 mM sodium phosphate buffer. Then transfer it to 10% (w/v), 15% (w/v) and finally to 18% (w/v) sucrose in phosphate buffer. Allow 3–16 h for each step.

*(iii) Sectioning*

Place the fixed and cryoprotected sample on aluminium foil and immerse in isopentane chilled with liquid nitrogen. Place the isopentane in a shallow beaker which stands in a polystyrene container (e.g. an ice bucket) containing liquid nitrogen. When isopentane begins to solidify on the bottom of the beaker, immerse the sample for 1–2 min. If desired, the tissue slab may be frozen on to a disc of cork with TissueTech mounting fluid and the disc may then be mounted on to the metal specimen holder with more Tissue Tek or with water.

(*iv*) *Notes*

**1.** If optimal preservation is not required, samples of tissue may be prepared for cryostat sectioning without prior fixation or cryoprotection. Also some antigens may be lost by aldehyde fixation.

**2.** An alternative to liquid nitrogen/isopentane is a dry-ice/ethanol bath.

Place the specimen holder and sample in the cryostat and section at −20 °C or −25 °C. One of the most critical parts of the cryostat is the antiroll plate, which must be carefully adjusted. Clearly, a sharp knife is also essential, and must be set at the correct angle.

Transfer the sections on to cover slip by gently touching a room temperature cover slip on to the section, which then is brought out of the cryostat and allowed to air-dry. Sections stick best if one of the adhesives in the solutions section above is used (Mayer's egg albumin or chrome–alum–gelatin).

Sections may also be transferred on to slides by carefully manipulating the cold section on to a cold slide, using a cold paint brush and needle.

*Staining protocols.* Clearly, depending upon the antiserum, times and concentrations will vary. We assume here that a polyclonal antiserum is being used as serum at a concentration of 1:50–1:100, or affinity-purified polyclonal antibody at 5–50 μg/ml, or neat monoclonal antibody culture supernatant, or monoclonal ascites fluid, 1–10 μg IgG/ml.

*Permeabilization. (a)* Organic solvent extraction, e.g. for cytoskeletal antigens. Place the section into acetone at −20 °C for 5 min. Then transfer the section to buffered saline. Alternatively, use ethanol or methanol. Some authors perform this step at room temperature. *(b)* Triton X-100 extraction. Incubate the section in buffered saline containing 0.3% Triton X-100 for 60 min. Include Triton X-100 in all subsequent steps.

*Prevention of non-specific binding of antibodies. (a)*To quench unreacted aldehyde groups, use freshly prepared 1% sodium borohydride in water or 50 mM $NH_4Cl$ in buffered saline for 10 min or 30 min in 0.1% glycine (pH 7.4, with Tris base). *(b)* Also include 1% (w/v) bovine serum albumin in diluted antibody solutions and/or 5% (v/v) serum from the species used to prepare the fluorescently conjugated second antibody. *(c)* To prevent non-specific binding of antibody to nuclei, incubate antisera at 37 °C rather than 4 °C.

(*v*) *Incubation sequence with Triton extraction.*
Drops of solutions (10–100 μl, depending on section size) are placed on to the section and gently moved around with the side of a disposable plastic pipette tip to prevent concentration differences in different areas. Take care not to touch the section itself.

**1.** 30 min in 0.1 M glycine, pH 7.4, with Tris base. Drain off into filter paper and blot around the sample.

**2.** 30 min in Tris-buffered saline, pH 7.4, containing 3% (w/v) bovine serum albumin. The section may be treated with 0.3% Triton X-100 at the same time. Alternatively, permeabilize with acetone.

**3.** 60 min to 3 h in first antibody at room temperature or 37 °C. Drain.

**4.** Wash the sections in standard histology set-ups, e.g. Coplin jars for slides and Columbia jars for cover slips. Perform 5 × 5 washes in Tris-buffered saline. The courageous may like to try squirting buffer on to the slide above the section. The buffer flows over the section and washes it effectively—it may also wash it away. Several rinses may be sufficient.

**5.** 60 min in rhodamine- or fluorescein-conjugated second antibody diluted 1:50 in Tris-buffered saline containing 3% (w/v) bovine serum albumin.

**6.** Wash as (4) above.

**7.** Mount with a drop of mounting medium, preferably the semi-permanent mounting medium (see solutions).

**8.** View.

NOTE: Centrifuge first and second antibodies for 5 min in a microcentrifuge before use.

*(b) Immunofluorescence microscopy on 5–10 μm paraffin sections*

Fix tissue as described above for cryostat sections. Then *dehydrate* in an ethanol series. For a sample 4 × 2 × 1 mm, treat sequentially with 30%, 50%, 70%, 96% and 100% ethanol, changing each hour. Change the absolute ethanol three times. Then *clear* in toluene for 30 min, or in oil of wintergreen (methyl salicylate) or bergamot oil for 2 × 60 min or overnight. *Infiltrate* samples with paraffin or Paraplast by transferring the cleared sample into molten paraffin in an oven at 55–60 °C. Use three changes of paraffin during infiltration for 90 min each for a 1 mm thick sample. Optimally, the paraffin remains melted but does not rise in temperature more than 1 degree above its melting point. Finally, *embed* the samples in fresh paraffin. Place a plastic ice cube tray in the paraffin oven to warm it, then place paraffin into the tray and place the infiltrated specimen into the paraffin in the tray so that the face to be sectioned is flush with the bottom. A strip of paper with details of the sample may also be placed in the paraffin. Remove the ice cube tray from the oven and cool until a film forms on the paraffin. Then place the tray in the deep freeze (− 20 °C) for 30 min.

Remove blocks from the ice cube tray, mount on the specimen holder of a rotary microtome, and shape with a razor blade to produce a flat-topped pyramid with parallel top and bottom and tapering sides, as shown in Fig. 7.19.

Paraffin embedding can be performed with little more apparatus than an oven, paraffin, a few beakers and an ice cube tray. For greater convenience, special plastic specimen holders are available which provide a cage within which the specimen remains during all of the steps mentioned above. The major part of the plastic holder becomes the base on which the sample is mounted. These holders are available from histology suppliers (Raymond Lamb, Miles) and can be used as described above or used in automatic equipment.

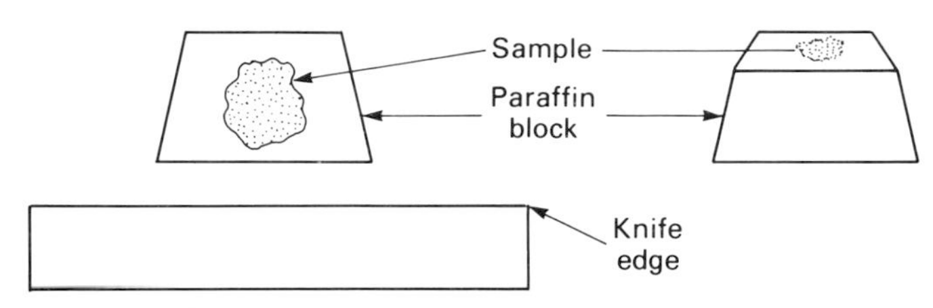


**Fig. 7.19** Sectioning paraffin-embedded tissue.

Cut sections on a rotary microtome at 5 $\mu$m, ribbons of sections can be obtained easily. Separate sections by means of a razor blade and transfer with a needle and/or a fine paint brush on to a few drops of water on the surface of a glass slide (pretreated with adhesive). The slide rests on a surface heated to 54–56 °C, 1 degree below the melting point of the paraffin. This enables the section to flatten. Once the section has spread out perfectly, remove the water it floats upon with filter paper and leave the slide to dry either on the spreading table or in a 37 °C oven overnight.

*For immunohistochemical staining.* The paraffin must first be removed by placing the slide in a glass Coplin jar containing xylene for 5 min and then repeating in fresh xylene. Remove the xylene with 100% ethanol for 5 min and then treat the slide with 50% ethanol for 5 min. After a subsequent 15 min in TBS the slide is ready for staining as described for cryostat sections.

NOTE: Samples may be fixed with a variety of fixatives (see Bibliography for general histology text). One very useful alternative to formaldehyde is Carnoy II, which penetrates tissue very rapidly and gives excellent preservation of nuclei. Since it destroys membranes, it does not give optimal preservation of cell structure, but the lack of anything which covalently modifies cellular proteins may be an advantage in certain circumstances. Carnoy II consists of 10% (v/v) glacial acetic acid, 60% (v/v) ethanol and

30% (v/v) chloroform. Tissue is placed into Carnoy II for 3–6 h, then washed for 2–3 h in 100% ethanol (several changes). The tissue is then cleared, infiltrated and embedded in paraffin.

*(c) Immunofluorescence microscopy on 1 μm plastic (Epon) sections*

REFERENCE: de Camilli *et al.* (1983).

Cut 1 μm sections of Epon-embedded tissue fixed as described for cryostat sections but then dehydrated and embedded in plastic (see Section 7.6.3.1.). Sections should be mounted on chrome–alum–gelatin treated slides. Remove plastic as described by Maxwell (1978). Add 2 g of KOH to 10 ml of methanol and 5 ml of propylene oxide. Stir for approximately 5 min until the KOH pellets disintegrate. Use the mixture to overlay the plastic sections for 2–5 min. Rinse with methanol and then with water. Then process exactly as described for cryostat sections.

*(d) Immunofluorescence microscopy on cells in culture*

Cells grown in culture on cover slips may be fixed and permeabilized in various ways.

*(i) Methanol procedure*

REFERENCE: Osborn and Weber (1982a,b).

Remove medium from the cells. Wash the cover slips twice in PBS at room temperature and then place them into methanol at −10 °C for 5 min. Wash the cover slip briefly with PBS and then treat with PBS containing 3% bovine serum albumin to prevent non-specific binding of the first antibody. Treat the samples with first antibody (5–50 μg/ml of affinity-purified antibody in PBS containing 0.5 mg/ml bovine serum albumin) for 30–60 min at 37 °C or at room temperature. After washing six times with PBS, treat the cells with second antibody (at the same concentration as first antibody for affinity-purified antibodies, or at 1:50 dilution for conjugated sera). After incubating for 30–60 min, wash, mount and view the cover slips.

*(ii) Formaldehyde/methanol procedure*

REFERENCE: Weber *et al.* (1975) originally for visualizing tubulin.

Fix cells for 10 min in 3.7% (w/v) formaldehyde in PBS at room temperature. Subsequently, place them in methanol at −10 °C for 5 min and then wash briefly with PBS before processing as described above.

### *(iii) Formaldehyde/Triton X-100 procedure*

REFERENCE: Heggeness *et al.*, (1977).

Fix cells for 10 min in 3.7% (w/v) formaldehyde in PBS at room temperature. Then treat them for 2 min at room temperature with 0.1% (v/v) Triton X-100 in PBS.

### *(iv) Preparation of cytoskeletons of cells in culture*

Wash the cover slips with PBS briefly, and then extract them in 10 mM Tris-HCl, pH 7.0, 40 mM KCl, 0.05% (v/v) Triton X-100 for 5 min at room temperature. Then fix them with 3.7% (w/v) formaldehyde in 10 mM Tris-HCl, pH 7.0, 40 mM KCl for 10 min at room temperature. Preservation of microtubules in cytoskeleton preparations requires the use of extraction buffers stabilizing these structures (Osborn and Weber, 1982b).

### *(v) Immunofluorescence microscopy on Saponin-permeabilized cells in culture*

REFERENCE: Willingham and Pastan (1985).

Saponin is a very mild detergent which causes minimum damage to cell ultrastructure. Saponin selectively removes cholesterol from membranes and must be included in all buffers to ensure permeability to antibodies.

Fix cells in 3.7% formaldehyde for 10 min at room temperature.

Wash with PBS.

Incubate with first antibody (10–100 μg/ml of affinity-purified or monoclonal antibody; or antiserum diluted 1:20 to 1:2000). Dilute the antibody in PBS containing 0.1% saponin, 4 mg/ml normal globulin (50% ammonium sulphate precipitate from serum dialysed against PBS, of the species used for the second antibody step, i.e. sheep if the second antibody is FITC sheep anti-rabbit IgG). Incubate for 15–30 min with gentle agitation.

Wash with PBS + 0.1% saponin five times over 10 min.

Incubate for 15 min at room temperature with second antibody (50–100 μg/ml of FITC or rhodamine-conjugated second antibody in PBS containing 0.1% saponin and 4 mg/ml normal globulin of the same species as the second antibody).

### *(vi) Notes on immunofluorescence*

#### I. IMMUNOFLUORESCENCE MICROSCOPY OF CELLS IN CULTURE

Many cover slips can be processed simultaneously by aliquoting antiserum on to parafilm.

Wet Whatman 3MM paper is placed in a plastic box and Parafilm is placed flat on top and numbered with a marker pen. Samples of antibodies (25 $\mu$l) are placed on the Parafilm near the numbers. Cover slips of cells which have been washed, fixed and treated with PBS 1% bovine serum albumin are drained and placed face down on the 25 $\mu$l drop of antibody. After incubation, the cover slips are picked up by pipetting 200 $\mu$l of PBS under the cover slip holder and washed in PBS. Incubation with second antibody is performed as for the first antibody using a fresh sheet of labelled Parafilm.

II. ALTERNATIVE STICKY-SITE BLOCKERS

**1.** Bovine serum albumin, 1–5% (w/v).

**2.** Heat-inactivated (56 °C × 30 min) serum diluted 5–10% (v/v) and obtained from species used to prepare the second antibody.

**3.** Carrageenan 0.25% (w/v) (Renfroe *et al.*, 1984).

### *(e) Immunofluorescence microscopy on cell suspension and whole mounts*

Some tissues can be homogenized to yield suspensions valuable for immunohistochemistry. Alternatively, thin tissues may be studied directly.

#### *(i) Small intestine epithelial cells*

REFERENCE: Bretscher and Weber (1978).

Small intestine (10–20 cm) is removed from the animal soon after death and its contents washed out with PBS. Both ends of the tube are clamped after filling the lumen with 3.7% (w/v) formaldehyde in PBS. The whole small intestine is immersed in fixative. After 10 min at room temperature the small intestine is cut open along its length and blotted with filter paper. Epithelial cells are scraped from the surface of the lumen with a microscope slide. The scrapings are suspended in PBS and passed several times through a 0.9 mm internal diameter syringe needle. Cells are then filtered through 100 $\mu$m mesh nylon net and spun down at 300*g* for 5 min. They are then resuspended in PBS, and samples are spread on to cover slips or slides and allowed to almost dry out at room temperature.

They are then treated for 5 min at −10 °C in methanol and rinsed in PBS before antibody incubation.

#### *(ii) Chicken breast muscle myofibrils*

REFERENCE: Herman and Pollard (1978).

Breast muscle from a freshly killed chicken is minced coarsely and then mixed with four volumes of 10 mM imidazole, pH 7.3, 0.15 M NaCl, 0.1 mM

EGTA and homogenized 2 × 10 s in a Polytron homogenizer. The myofibrils are centrifuged down and resuspended in the homogenization buffer containing 50% glycerol. The myofibrils can be stored at −20 °C for up to 1 year.

For indirect immunofluorescence the myofibrils are washed with PBS several times and then spread on to microscopic slides or cover slips. After drying at room temperature they are washed with PBS and treated with antibody.

*(iii) Muscle Z-discs*

REFERENCE: Granger and Lazarides (1979).

Muscle tissue can be processed to produce sheets of intact Z-discs.

*(iv) Cholinergic synapse preparations*

REFERENCES: Heuser and Salpeter (1979), Walker *et al.* (1985).

Gently homogenized *Torpedo* electric organ tissue produces large sheets of innervated membrane (see p. 89).

*(v) Whole mounts*

REFERENCE: Costa and Furness (1983).

Thin tissues such as mesentery, pericardium and iris are pinned on to balsa wood sheet with fine pins and then fixed in 4% (w/v) formaldehyde in PBS overnight at 4 °C. Hollow viscera such as large blood vessels, stomach, intestine, gall bladder and bladder are first cut open and stretched and pinned on to the balsa (≃1 cm × 1 cm). The balsa wood is then fixed tissue side down on a Petri dish containing fixative. Mesentery, pericadium and iris may be used directly for immunohistochemistry but the other tissues above must first be dissected into tissue layers. With fixed intestine the thin muscle layers with attached nerve plexuses peel off very easily. Small pieces of the peeled-off layer are incubated with antibodies in a routine manner, either keeping the tissue on a slide or for convenience placing it in the well of a microtitre plate. Tissue should be mounted as flat as possible when staining is completed. Clearly, the thickness of the sample may cause penetration problems. Costa and Furness (1983) recommend washing the fixed tissue in 80% ethanol (6 × 10 min) or 80% ethanol plus 95% ethanol (30 min), 100% ethanol (30 min), 100% xylene (30 min), 100%, 80%, and 50% ethanol (30 min each) and then PBS.

*(f) Counterstains for immunofluorescent samples*

Usually, specimens are viewed and photographed under phase contrast in addition to fluorescence. Other procedures can be used for bright-field/ fluorescence observation.

*Evans Blue* can be used to stain samples, and makes the cytoplasm fluoresce red. Stain with a drop of 0.5% dye in PBS for 1 min. Rinse in PBS, mount and view.

*Eriochrome black* (Difco: 1/10–1/500 dilution with PBS) can also be used to stain cell cytoplasms to fluoresce red. Stain for 1 min, rinse in water for 1 min, mount and view.

Nuclei can be made to fluoresce blue using the Hoechst dye bisbenzimide H33258 (Bainbridge and Macey, 1983; Hilwig and Gropp, 1973). Stain for 5 min in dye solution (0.5 mg/ml in PBS). Wash 5 × 10 min and view using Leitz filter pack A.

It may also be possible to use a combination of fluorescence and bright-field microscopy with samples counterstained with *haematoxylin* alone or with eosin. Photographs may be taken using a low level of illumination for the bright field. In this case sections must be stained with haematoxylin prior to the immunofluorescent staining (Weinstein and Lechago, 1977).

*(g) Photography of immunofluorescence*

*(i) For black and white photography*

**1.** Kodak Tri-X uprated to 800 or 1600 ASA and processed with Kodak HC-110 developer (1:7 dilution for 15 min for 1600 ASA).

**2.** Kodak Tri-X (800 ASA or 1600 ASA) processed with Acufine two-solution developer (Diafine Corporation, Chicago, USA).

**3.** Ilford HP5 (800 or 1600 ASA) processed with Ilford Microphen ($8\frac{1}{2}$ min at 20 °C for 800 ASA; 11 min at 20 °C for 1600 ASA).

NOTES

**1.** One should aim for the finest grain negative possible. If the samples are very bright, 400 ASA may work satisfactorily.

**2.** Rhodamine-stained preparations are normally overexposed on the negatives, and best results are obtained by altering the ASA setting on the camera. If, for example, you are using 400 ASA exposures for fluorescein, use 800, or better 1600, ASA setting for rhodamine and process the film as if all exposures were taken on 400 ASA.

*(ii) For colour photography*

Use 3M colour slide film (1000 ASA) or Kodak VR1000 colour print film or any conventional 400 ASA film balanced for daylight. Exposure times may need to be varied depending on how your camera responds to the fluorescent colours.

NOTE: To help correlate immunofluorescence with cell structures, it is often helpful to take photographs on immunofluorescence alone, phase alone, and a combination of both fluorescence and phase illumination. Phase illumination needs to be low so it does not swamp out the fluorescence.

### 7.6.2.2. Peroxidase-linked methods

*(a) Immunohistochemistry with peroxidase-conjugated second antibodies*

**1.** Cut 3–5 μm paraffin sections of formaldehyde-fixed tissue.

**2.** Remove paraffin 2 × 5 min in xylene.

**3.** Block endogenous peroxidase activity with 1% hydrogen peroxidase in absolute methanol for 30 min.

**4.** Wash in water 2 × 5 min.

**5.** Tris-buffered saline (TBS) 5 min.

**6.** Trypsin treatment if necessary—see 7.6.2.2. (*b*).6.

**7.** Wash 5 × 2 min with TBS in Coplin jars.

**8.** Drain and wipe around the sections.

**9.** Block "sticky" sites with 20% non-immune serum (NS) in TBS for 15 min using serum from the same species as that used to prepare the peroxidase-conjugated second antibody.

**10.** Remove the non-immune serum with a piece of filter paper and place the first antibody on to the section for 30 min. Use antisera at 1:10–1:500 dilutions in TBS containing 5% NS (TBSN) and affinity-purified IgG fractions or monoclonals at 1–10 μg/ml.

**11.** Wash 8 × 2 min in TBS in Coplin jars.

**12.** Incubate for 30 min with peroxidase-conjugated second antibody diluted 1:20 to 1:80 in TBSN.

**13.** Wash as step 11.

**14.** Stain for peroxidase activity using 3,3′-diaminobenzidine-HCl. (Dissolve 6 mg of DAB in 10 ml of 50 mM Tris-HCl, pH 7.6. Add 0.1 ml of 3% (v/v) $H_2O_2$. Filter.) Stain for 5–10 min.

**15.** Rinse in tap water.

**16.** Counterstain if necessary in haematoxylin for 30 s.

**17.** Wash in tap water.

**18.** Dehydrate in an ethanol series; 5 min in 50%; 5 min in 95%; 2 × 5 min in 100%.

**19.** Clear in xylene 2 × 5 min.

**20.** Mount in Fluoromount, D.P.X. or Eukitt (BDH, Raymond Lamb).

NOTE: Peroxidase staining with diaminobenzidine may be intensified if the stain solution contains 0.02% cobalt chloride in addition to the other components of the stain.

*(b) Immunohistochemistry at the light microscopic level with peroxidase–anti-peroxidase complexes*

REFERENCES: Sternberger (1979); Sternberger *et al.* (1970); Taylor (1976).

DAKO supply a useful manual of peroxidase staining methods for light microscopy.

For a paraffin section start at step (1).

For a cryostat section start at step (2), or at step (3) after 10 min in acetone at − 20 °C, or directly at step (3) depending upon the degree of permeabilization of the tissue required. Permeabilization can also be achieved by pretreating the section with 0.2% Triton X-100 for 60 min (see pp. 263–265).

**1.** Remove paraffin 2 × 5 min in xylene.

**2.** 5 min in absolute ethanol.

**3.** Block endogenous peroxidase activity with 1% hydrogen peroxide in absolute methanol for 30 min.

**4.** Wash in water 2 × 5 min.

**5.** Tris-buffered saline 5 min.

**6.** Trypsin treatment *if necessary*. Incubate for 30 min in freshly prepared 0.1% (w/v) trypsin in TBS containing 0.1% calcium chloride. Trypsin treatment may be necessary in heavily fixed samples. It is easiest first to try the protocol without trypsinization, and only trypsinize if an antiserum known to be positive, e.g. by immunoblotting, fails to result in staining.

**7.** Wash 5 × 2 min with TBS.

**8.** Drain and wipe around the sections.

**9.** Block sticky sites with 20% non-immune serum (NS) in TBS for

15 min using serum from the same species as that used to prepare the "linker" antibody.

**10.** Remove the non-immune serum with a piece of filter paper and place first antibody on to the section for 30 min. The first antibody should be tested over a wide range of dilutions (1:10 to 1:1000) to optimize the staining obtained.

**11.** Wash 8 × 2 min in TBS in Coplin jars.

**12.** Incubate for 15–30 min with "linker" antibody diluted 1:50 in TBS containing 5% NS (TBSN). For increased sensitivity the linker antibody may be peroxidase-conjugated.

**13.** Wash as at step (11).

**14.** Incubate in PAP complex diluted 1:50 in TBSN for 15–30 min.

**15.** Wash as at step (11).

**16.** Stain for peroxidase activity using 3,3′-diaminobenzidine-HCl. (Dissolve 6 mg of DAB in 10 ml of 50 mM Tris-HCl, pH 7.6. Add 0.1 ml of 3% (v/v) $H_2O_2$. Filter.) Stain for 5–10 min.

**17.** Wash in tap water.

**18.** Stain with haematoxylin, if required, for 30 s.

**19.** Wash in tap water.

**20.** Dehydrate in an ethanol series, 5 min in 50%, 5 min in 95%, 2 × 5 min in 100%.

**21.** Clear in xylene, 2 × 5 min.

**22.** Mount in Fluoromount, DPX or Eukitt (BDH, Raymond Lamb).

NOTE: The diaminobenzidine reaction product can be intensified by rinsing the section with water after step (16) and osmicating for 10 s with a few drops of a 1% solution of osmium tetroxide in 0.1 M sodium phosphate buffer, pH 7.4 (Busachi *et al.*, 1978). Alternatively, cobalt intensification may be used as described above (7.6.2.2.(a)) or with the very sensitive silver staining kit from Amersham International. Various intensification procedures for diaminobenzidine have been examined by Scopsi and Larssen (1986). The most useful involve (1) the addition of imidazole to the reaction mixture, and (2) intensification with silver and gold after pretreatment with nickel sulphate.

*(c) Alternative stains for peroxidase activity*

The stains given below are soluble in organic solvents.

*WARNING: Do not* try to dehydrate, clear and mount in Eukitt or DPX.

*(i) 3-amino-9-ethylcarbazole*

REFERENCE: Graham *et al.* (1965).

Follow protocol (a) above up to step 13 or protocol (b) to step 15. Then:

**1.** Wash with water 2 × 30 s.

**2.** Rinse for 30 s with 50 mM sodium acetate buffer, pH 5.0.

**3.** Incubate for 5–10 min in a filtered solution containing 0.02% (w/v) 3-amino-9-ethylcarbazole, and 0.02% hydrogen peroxide in 50 mM sodium acetate buffer, pH 5.0.

**4.** Rinse in TBS and mount in glycerol:buffered saline (1:9) or in glycerol:gelatin (Sigma; Raymond Lamb; do *not* try to dehydrate, clear, and mount in Eukitt or DPX.)

*(ii) 4-chloro-1-naphthol* (4CN)

After step 15 of protocol 7.6.2.2(a) or step 15 of protocol 7.6.2.2(b):

**1.** Incubate in stain solution for 5–15 min at room temperature. To prepare the stain, dissolve 30–40 mg of 4CN in 0.2–0.5 ml of absolute ethanol. Add, with stirring, 100 ml of 50 mM Tris-HCl, pH 7.6, containing 50–100 μl of 30% $H_2O_2$. Filter off the white precipitate.

**2.** Rinse in water and mount in glycerol:buffered saline (1:9) or in glycerol:gelatin.

### 7.6.2.3. Biotin/(strept)avidin methods

REFERENCES: Amersham protocols booklet and Hsu, S. *et al.*, (1981).

Biotin-labelled antibodies provide versatile reagents which can be coupled to a variety of detection systems based on the egg-white protein avidin or a similar bacterial protein, streptavidin. Streptavidin is to be preferred since it shows less non-specific binding. Streptavidin can be obtained (Sigma, Amersham) labelled with fluorescein, rhodamine, various enzymes, including peroxidase, and $^{125}I$. Alternatively, streptavidin, being multivalent, may be used to provide a bridge between biotin-labelled antibody and biotin-conjugated enzymes.

Finally, a third possibility is the use of complexes formed by mixing avidin or streptavidin with biotin-labelled enzymes in proportions which ensure an excess of biotin-binding sites in the formed complex. Such preformed complexes may be obtained commercially (Amersham) or prepared from commercially available kits (Vectastain, Vector Laboratories, also available from DAKO) and give much stronger staining.

The basic procedure is essentially identical to those described above. Recommended dilutions:

*(a)* Biotin-labelled second antibody, 1–5 $\mu$g/ml of affinity-purified antibody for 60 min at room temperature.

Then either:

*(b)* Streptavidin–biotin enzyme complexes (Amersham), 1:100 for $\beta$-galactosidase, 1:200 for peroxidase, for 60 min;

or

*(c)* Streptavidin peroxidase, 1:300 (Amersham);

or

*(d)* Streptavidin 2.5–5.0 $\mu$g/ml for 15 min at room temperature and then biotin-labelled peroxidase, diluted 1:500 (Amersham), 5.0 $\mu$g/ml for 30 min.

Colloidal gold-labelled streptavidin is also available commercially (Amersham), thus enabling the streptavidin method to be linked to the very sensitive silver-enhancement technique associated with gold-labelled antibodies.

#### 7.6.2.4. Localization of antigens at the light-microscopic level with gold-labelled antibodies

REFERENCES: Extensive technical information is available from Janssen Pharmaceuticals; Hacker *et al.* (1985).

Cut sections of paraffin-embedded tissue, deparaffinize and take through ethanol to PBS as described in the protocols above. Block sticky sites with 5% normal serum of the same species as the second antibody diluted in PBS for 20 min. Incubate with the first antibody for 1 h at various dilutions (1:100 to 1:10 000?). High dilutions of antiserum can often be used. Wash eight times with PBS containing 0.1% BSA. Then incubate with 5 nm gold-labelled second antibody at dilutions of 1:20 to 1:160. Longer incubation times are necessary for higher dilutions and, in general, longer times are needed for gold reagents than for immunofluorescence and enzyme-linked procedures.

Incubation times of 1 h to overnight may be used.

Rinse and wash the specimens again and then intensify the gold label by silver enhancement.

Labelled antigens are stained red, and may be intensified by precipitation of silver on to gold. For the intensification procedure use double-distilled, deionized water.

*Silver-enhancement solutions* (light-sensitive)

*Solution 1.* 2 M sodium citrate buffer, pH 3.85, 23.5 g trisodium citrate ($2H_2O$) + 25.5 g citric acid ($1H_2O$) in 100 ml of water.

*Solution 2.* Dilute 50 ml of solution 1 to 500 ml.

*Solution 3.* Developer—prepare immediately before use. 77 mM hydroquinone, 5.5 mM sodium lactate in 200 mM citrate buffer at pH 3.85. Make the solutions up as follows in containers entirely covered in aluminium foil. Mix 10 ml of solution 1 with 60 ml of water (solution A). Dissolve 0.85 g of hydroquinone in 15 ml of water (Solution B). Dissolve 0.11 g of silver lactate in 15 ml of water (Solution C). Add B to A. Add C and mix well.

*Solution 4.* Fixer.

*Silver enhancement procedure*

*Step 1.* Wash the slide 2 × 10 min in water.

*Step 2.* 2 min in solution 2.

*Step 3.* 5–15 min in solution 3—developer in a Petri dish entirely covered with aluminium foil.

*Step 4.* 5 min in fixative (standard photography fixer).

*Step 5.* Wash in water 3 × 5 min and air-dry.

NOTE: A more convenient silver enhancement kit is available which is not light-sensitive (Intense from Janssen Pharmaceuticals).

#### 7.6.2.5. References to other enzyme-linked antibody methods

Alkaline phosphatase conjugates (Ponder and Wilkinson, 1981) and complexes of alkaline phosphatase and monoclonal antialkaline phosphatase (Cordell *et al.*, 1984) may be used in immunohistochemistry. Endogenous alkaline phosphatases may need to be inhibited prior to the immune staining procedure (Ponder and Wilkinson, 1981).

### 7.6.3. Localization of antigens by electron microscopy

The reader especially interested in EM localization of antigens is referred to the extensive references given in Chapter 4. Also, the Bibliography lists several textbooks devoted to immunohistochemistry. Finally, Janssen Pharmaceuticals supply extensive details of various protocols for the localization of antigens at the EM level using gold-labelled markers.

#### 7.6.3.1. Pre-embedding immunoperoxidase histochemistry at the EM level

REFERENCE: Fiedler and Walker (1985).

Preliminary experiments should be performed with immunofluorescence to determine the maximal concentration of glutaraldehyde which can be used without destroying the antigen. Parallel electron microscopy will indicate whether sufficient ultrastructure remains for the experiment to be of value.

Immunohistochemical electron microscopy is then performed as described below. Fix blocks of tissue in 4% formaldehyde plus 0.02% glutaraldehyde in 0.15 M sodium cacodylate buffer, pH 7.4, at 4 °C for 2 h. Cut 200 $\mu$m thick sections with a tissue sectioner (Sorvall TC-2) and embed in 1% agarose. This prevents sections sticking together and works as an additional filter for the diaminobenzidine (DAB) (see below). Then perform the following incubation steps in phosphate-buffered saline (PBS; 10 mM sodium phosphate buffer, pH 7.4, containing 0.15 M sodium chloride) at room temperature unless otherwise stated:

(1) 0.1% Triton X-100 plus 20% normal sheep serum (NSS) for 60 min.

(2) 5 min in PBS.

(3) Incubate in the specific test serum or the control serum at a concentration of 1:50 plus 2% NSS overnight at 4 °C.

(4) Wash 4 × 20 min.

(5) Incubate for 2 h in peroxidase-labelled goat anti-rabbit IgG (e.g. Miles) 1:100 plus 10% NSS.

(6) Wash 3 × 20 min in PBS and 2 × 20 min in 0.05 M Tris-HCl, pH 7.4.

(7) Preincubate for 15 min in 0.05% 3,3′-diaminobenzidine tetra-HCl (Sigma) in 0.05 M Tris-HCl followed by 10 min 0.01% $H_2O_2$ in the same buffer. Filter DAB through a 0.2 $\mu$m membrane filter before use.

(8) Wash 2 × 15 min in Tris-HCl and 1 × 15 min in 0.1 M phosphate buffer (PB), pH 7.4.

(9) Treat with 1% $OsO_4$ in PB for 1 h.

(10) Wash 3 × 10 min in PB.

Then dehydrate the sections and embed in Epon 812 or Durcopan (see protocol below). Make thin sections in the transverse direction to the 200 $\mu$m section. As a control, absorb the antiserum with antigen. Determine the titre by an enzyme-linked immunosorbent assay (ELISA)-test. All sera should be used as an IgG-fraction. Pre-absorb the peroxidase-labelled goat anti-rabbit IgG with tissue acetone powder or with a membrane fraction. Contrast the thin sections with uranyl acetate and lead citrate.

This protocol may also be used with 50 μm cryostat sections and with vibratome sections (for details see "Immunohistochemistry" edited by Cuello, 1983).

*Preparation of samples for electron microscopy using Durcopan* (Fluka)

Perfusion-fix the animal or, immediately after death, fix tissues in 4% glutaraldehyde (EM grade) in 0.1 M sodium cacodylate buffer, pH 7.0, for 2 h. Rinse with buffer and post-fix in 1% osmium tetroxide for 2 h. Embed as follows:

**1.** Dehydration: 30–50–70–96% alcohols, absolute alcohol/propylene-oxide, 10 min each. When samples are in 70% alcohol start to prepare the resin mixture.

**2.** Resin mixture: 10 g Durcopan A + 10 g Durcopan B + 0.3 g Durcopan D + 0.4–0.5 g Durcopan C (in this order!). Mix thoroughly but not too vigorously. It is enough to use a glass rod. Place the mixture into a 60 °C oven for 10–15 min until air bubbles leave.

**3.** Impregnation: transfer samples from propylene-oxide to 2:1, 1:1 and 1:2 mixtures of propylene-oxide and resin mixture, 30 min in each.

**4.** Embedding: transfer samples into pure resin mixture and place them in capsules.

**5.** Polymerization: put capsules into a 60 °C oven and leave them to polymerize for 16–24 h.

The Durcopan Kit (Fluka) contains all the necessary ingredients.

### 7.6.3.2. Immunogold, pre-embedding staining of cells in culture

REFERENCES: Langanger *et al.* (1984a,b); DeMey *et al.* (1981); Janssen Pharmaceutical Technical Data.

Grow cells on cover slips. Plastic cover slips which can be sectioned may be advantageous. Alternatively, perform the entire procedure directly in the Petri dish. Fix using 0.3–1% highly purified glutaraldehyde in buffered saline for 10 min. Alternatively, use 2–4% freshly prepared paraformaldehyde containing 0.05–0.1% glutaraldehyde. Permeabilize cells with 0.2–0.5% Triton X-100 for 5–30 min. Treat 2 × 10 min with freshly prepared $NaBH_4$

(1 mg/ml) in buffered saline. Then wash with buffered saline. Then treat the cells as follows.

**1.** 5% heat-inactivated normal serum (NS) in TBS containing 0.1% BSA (TBSA) for 20 min.

**2.** Primary antibody 1–10 μg/ml in 1% NS in TBSA for 1–12 hours.

**3.** Wash 3 × 10 min in TBSA.

**4.** Gold-labelled second antibody diluted in TBSA (1:20 to 1:100) for 30 min to several hours.

**5.** Wash as at step 3.

**6.** Wash twice with Sorensen's phosphate buffer, pH 7.2 (P).

**7.** Post-fix with 1% glutaraldehyde, 0.2% tannic acid in P for 30–60 min.

**8.** Rewash in P.

**9.** Post-fix in 0.5% osmium tetroxide in P buffer for 10 min at 4 °C.

**10.** Rewash 3 × 5 min in P.

**11.** Dehydrate in 70% ethanol (several changes).

**12.** Impregnate with 0.5% uranyl acetate 1% phosphotungstic acid in 70% ethanol for 30 min.

**13.** Dehydrate completely and embed in EPON. Embed cells grown on cover slips either directly in the tissue-culture dish (using culture dishes resistant to acetone and epoxy resins) or by supporting cover slips with labelled cells over microscope slides with spacers made from slivers of cover slip. A drop of EPON is placed on the microscope slide and the sandwich is hardened for 3 days at 50 °C.

Treatment of the cover slip and slide with release agent (e.g. MS-123 from Miller-Stephenson Chemical Co. Inc.) will facilitate removal of the EPON subsequently. Then separate the EPON from the slide and cover slip by alternately dipping it in boiling water and liquid nitrogen!

Sorensen's buffer: 36 ml of 0.2 M $Na_2HPO_4$ + 14 ml of 0.2 M $Na_2HPO_4$ made up to 100 ml with $H_2O$.

### 7.6.3.3. Post-embedding EM localization of antigens using ultrathin sections of Lowicryl K4M embedded tissue

REFERENCES: Valentine *et al.* (1985); Roth *et al.* (1981). Extensive details are available from the manufacturer.

Preferably, animals are anaesthetized and perfusion-fixed. Perfuse rats through the heart with 100 ml of 0.9% sodium chloride followed by 250 ml of

4% paraformaldehyde, 0.1% glutaraldehyde, 0.2 mM $CaCl_2$ in 0.1 M sodium cacodylate buffer, pH 7.4. Allow fixation to proceed for 2 h *in situ* and then remove the tissue of interest and place in the same fixative without glutaraldehyde overnight at 4 °C. Dissect out blocks of 1 $mm^3$, and place in 0.1 M sodium cacodylate buffer, pH 7.4, for 1 h with several changes.

Then quench aldehyde with 50 mM $NH_4Cl$ in 150 mM NaCl, 100 mM sodium phosphate, pH 7.4. Stain in block with 2% aqueous uranyl acetate for 1–2 h at 4 °C. Dehydrate for 10 min each in 50%, 70% and 95% ethanol on ice. Mix Lowicryl K4M (Polysciences): 2.7 g of the cross-linker A + 17.3 g of monomer B + 0.1 g of benzoin ethyl ether for room-temperature polymerization, or 0.1 g of initiator C for low-temperature embedding (less than 0 °C). Mix under nitrogen. Then infiltrate the tissue: 10 min in a mixture of 2 parts ethanol and 1 part Lowicryl; 15 min in a mixture of 1 part ethanol and 1 part Lowicryl; 20 min in 100% Lowicryl; 25 min in Lowicryl—all at room temperature on a rotator (Altman, 1984).

Then place the tissue in BEEM capsules (Polyscience) filled with fresh Lowicryl and cap. Perform polymerization with a UV light source (2 × 15 W lamps) (long wave 400 $\mu W/cm^2$) 30 cm from the capsules. After 45 min at 4 °C, continue the UV polymerization for 24–48 h at room temperature. Alternatively, perform the polymerization at −30 °C for 24 h. The BEEM capsules can be suspended over the UV source in a cardboard box lined with aluminium foil. Then section the blocks.

Semithin sections may be collected and used for LM immunohistochemistry. Ultrathin sections (silver-gold) are collected on Formvar-coated nickel grids. The face of the block must be kept dry during sectioning. The sections are then stained. Drops of solutions are placed on dental wax in a moist environment, e.g. a Petri dish containing moistened filter paper. The grids are then transferred from one drop to another. The procedure is as follows:

**1.** Rinse the grid on drops of water.

**2.** Etch with saturated aqueous sodium metaperiodate for 10–60 min (Bendayan and Zollinger, 1983). This step is not always necessary.

**3.** Rinse on drops of water for 5 min.

**4.** Incubate for 30 min on drops of first antibody (2–25 $\mu$g of affinity-purified IgG/ml or more than 500-fold diluted serum) diluted in PBS containing 0.3% BSA, gelatin or ovalbumin.

**5.** Wash with PBS containing 0.3% BSA. If necessary 0.05% Tween 20 and 0.5 M NaCl may be added to reduce background staining (Craig and Goodchild, 1982). Wash the grids 2 × 10 min.

**6.** Incubate the grids for 30 min on drops of gold-labelled Protein A diluted 1:10 to 1:200 with PBS + 1% BSA prior to use.

**7.** Repeat washing step 5.

**8.** Wash with distilled water 2 × 5 min and then contrast with 2% aqueous osmium tetroxide for 10 min and 2% aqueous uranyl acetate for 5 min.

NOTE 1

Optimal preservation of antigens may require embedding at low temperature. In this case samples are processed as follows.

Dehydrate and infiltrate as follows:

30 min in ethanol (30%) at 0 °C; 60 min in 50% ethanol at −20 °C; 60 min in 70% ethanol at −35 °C; 60 min in 95% ethanol at −35 °C; 2 × 60 min in 100% ethanol at −35 °C; 60 min in a 1:1 mixture of K4M (2.7 g crosslinker A + 17.3 g monomer B + 0.1 g initiator C) and ethanol at −35 °C; 60 min in a 2:1 mix of K4M and ethanol at −35 °C; 60 min in pure K4M resin at −35 °C; and overnight in fresh K4M resin at −35 °C.

Polymerization is then achieved by means of ultraviolet light.

NOTE 2

Problems may be encountered in sectioning Lowicryl, which can be rather soft when polymerized at low temperatures. Refer to the manufacturers' instructions (and Lowicryl Letters) for details of the optimal sectioning procedures.

## 7.7. Miscellaneous

### 7.7.1. Protein assay protocols

#### 7.7.1.1. Lowry-Folin

REFERENCES: Lowry *et al.* (1951), as modified by Dully and Grieve (1975) and Peterson (1977).

*(a) Solutions*

**A.** *2% $Na_2CO_3$ in 0.1 M $NaOH$ + 0.5% SDS*: 10 g $Na_2CO_3$ + 2.5 g SDS + 50 ml of 1 M NaOH in 500 ml.

**B.** *0.1 M NaOH*: 2 g NaOH in 500 ml.

**C.** *0.5% (w/v) $CuSO_4 . 5H_5O$*: 0.5 g in 100 ml.

**D.** *1% NaK-tartrate*: 1.0 g in 100 ml.

**X.** Freshly made A:C:D in the ratio 98:1:1.

*(b) Procedure:*

For samples in the SDS sample buffer (or 1–10 mg/ml):

**1.** Take 10 $\mu$l or 25 $\mu$l of sample.

**2.** Add 1.0 ml of 10% TCA and leave for 15 min at room temperature.

**3.** Spin in Eppendorf for 15 min.

**4.** Remove the supernatant with a Pasteur pipette and discard.

**5.** Dissolve the pellet in 0.25 ml of 0.1 M NaOH, i.e. solution B.

**6.** Transfer to a test-tube and add 2.5 ml of X. Mix and leave for 10 min.

**7.** Add 0.25 ml of Lowry–Folin–Ciocaeulteau reagent, 1:2 diluted with $H_2O$ (i.e. 1 volume of LFC + 2 volumes of $H_2O$). Leave for 20 min.

**8.** Read OD at 750 nm.

*(c) Standards in triplicate*

| | | |
|---|---|---|
| 0 $\mu$l BSA | | + 250 $\mu$l 0.1 M NaOH |
| 10 $\mu$l BSA | 1 mg/ml | + 240 $\mu$l 0.1 M NaOH |
| 25 $\mu$l BSA | 1 mg/ml | + 225 $\mu$l 0.1 M NaOH |
| 50 $\mu$l BSA | 1 mg/ml | + 200 $\mu$l 0.1 M NaOH |
| 75 $\mu$l BSA | 1 mg/ml | + 175 $\mu$l 0.1 M NaOH |

A recent sensitive alternative to the Lowry procedure which shows very little sensitivity to interfering substances has been described (Smith *et al.*, 1985) and is available from the Pierce Chemical Co. The method can be used for extremely dilute protein solutions (0.5–10 $\mu$g/ml) but is rather expensive.

#### 7.7.1.2. Bradford protein assay

A very convenient assay has been described for soluble proteins (Bradford, 1976). This method suffers from interference by Triton X-100 and sodium dodecyl sulphate but is otherwise extremely fast.

Prepare the assay solution by dissolving 100 mg of Coomassie brilliant blue G-250 in 50 ml of 95% ethanol, adding 100 ml of 85% phosphoric acid, diluting to 1 l with water and then filtering.

For 5–100 $\mu$g of protein, add 5 ml of assay solution to 0.1 ml of protein sample.

For 1–10 $\mu$g of protein, add 1 ml of assay solution to 0.1 ml of protein sample.

Read the OD at 595 nm after 2 min and before 2 h. This method can be performed very conveniently on a microscale using the Biorad ELISA reader to measure the optical densities of samples and to produce a printed record of the results.

#### 7.7.1.3. Schaffner and Weissman (1973)

These authors have described a protein assay which enables large volumes of dilute samples (0.75 $\mu$g/ml) to be assayed for protein with minimum interference. Make samples 10% with trichloroacetic acid and leave for 2 min at room temperature. Then suck the samples on to a nitrocellulose filter (e.g. Millipore 0.45 $\mu$m) and stain with 0.1% amido black in methanol:glacial acetic acid:$H_2O$, 45:10:45, for 2–3 min.

After 30 s in $H_2O$, destain the background in methanol:glacial acetic acid:$H_2O$, 90:2:8. Then wash the filters with $H_2O$ for 1–2 min and finally elute with 25 mM NaOH, 5 mM EDTA in 50% aq. ethanol.

Measure the OD at 630 nm.

#### 7.7.1.4. Fluorescamine

This can be used as a very sensitive assay for proteins and primary amines in the picomolar range (Udenfriend *etal.*, 1972).

### 7.7.2. Subfractionation of membrane proteins with Triton X-114

REFERENCE: Bordier (1981).

Triton X-114 extracts of cells or cell membranes separate into detergent and aqueous phases at 30 °C. Many integral membrane proteins partition into the detergent phase, while soluble proteins and peripheral membrane proteins partition into the aqueous phase.

Samples are extracted at 4 °C with 0.5% Triton X-114 in PBS containing protease inhibitors (PMSF 40 $\mu$g/ml, iodoacetamide 10 mM and Trasylol (Bayer) 1:1000 dilution).

For extracts of cells in culture, centrifugation for 3 min in a microcentrifuge at 4 °C pellets the nuclei.

Transfer the supernatant to a new microcentrifuge tube at 30 °C and incubate for 4 min.

Then spin for 4 min at room temperature in an Eppendorf centrifuge (3000 rpm).

The detergent phase forms an oily pellet which dissolves when mixed with ice-cold PBS. The supernatant may be "re-extracted" for membrane proteins by adding 7.5% Triton X-114 to return the detergent concentration to 0.5%.

For SDS-gel electrophoresis samples may be precipitated with 10% TCA on ice for 60 min. After centrifugation the pellets may be washed at room temperature with ethanol, ethanol ether (1:1) and then ether. After brief drying, the pellets may be dissolved in sample buffer.

# Bibliography


## General

*Methods in Enzymology*. Academic Press. Especially Volumes 70, 73, 74, 84, 92, 93 and 108 for antibody methods and Volume LVIII for cell culture.
*Laboratory Techniques in Zoology*. R. Mahoney. Butterworths, London (1975).
*Methods in Molecular Biology, 1. Proteins*. Ed. by J. M. Walker, Humana Press, Clifton, N.J. (1984).

## Biochemical Methods

*A Biologist's Guide to Principles and Techniques of Practical Biochemistry*. Ed. by B. Williams & K. Wilson. Edward Arnold (1978).
*The Tools of Biochemstry*. T. G. Cooper. Wiley Interscience, New York, London (1977).
*Protein Purification: Principles and Practice*. R. K. Scopes. Springer-Verlag, New York (1982).
*Gel Electrophoresis of Proteins*. Ed. by B. D. Hames & D. Rickwood. IRL Press, Oxford, England (1981).

## Immunochemical Methods

*Practical Immunology*. L. Hudson & F. C. Hay. Blackwell Scientific Publishers, London (1976).
*Immunochemistry in Practice*. Johnstone and Thorpe. Blackwell Scientific Publishers, London (1982).
*An Introduction to Radioimmunoassay and Related Techniques: Laboratory Techniques in Biochemistry and Molecular Biology*. T. Chard. Elsevier Science Publishers, Amsterdam, The Netherlands (1986).
*Practice and Theory of Enzyme Immunoassays: Laboratory Techniques in Biochemistry and Molecular Biology*. P. Tijssen. Elsevier Science Publishers, Amsterdam, The Netherlands (1985).

## Cell Culture and Monoclonal Antibody Production

*Cell Culture for Biochemists: Laboratory Techniques in Biochemistry and Molecular Biology*. Adams. Elsevier Science Publishers, Amersterdam, The Netherlands (1985).
*Animal Cell Culture*. Freshney. IRL Press (1986).
*Monoclonal Antibodies*. J. W. Goding. Academic Press, New York & London (1984).
*Monoclonal Antibody Technology: Laboratory Techniques in Biochemistry and Molecular Biology*. A. M. Campbell. Elsevier Science Publishers, Amsterdam, The Netherlands (1984).

## Histology

*The Rat Brain in Stereotoxic Coordinates*. Ed. by G. Paxinos & C. Watson (1976).
*The Cell—An Atlas of Ultrastructure*. D. W. Fawcett. Saunders (1966).
*Histology, Cell and Tissue Biology*. Ed. by L. Weiss. Macmillan (1983).
*Theory and Practice of Histological Techniques*. Ed. by J. D. Bancroft & A. Stevens. Churchill Livingstone, London (1982).
*Ultrastructure of the Mammalian Cell*. R. V. Krstic. Springer-Verlag, New York, Berlin (1983).
*Practical Methods in Electron Microscopy*. Series of several volumes. North Holland.

## Immunocytochemistry

*Immunocytochemistry*. L. A. Sternberger. John Wiley, New York (1979).
*Immunohistochemistry*. Ed. by A. C. Cuello. John Wiley (1983).
*Techniques in Immunocytochemistry*. Several volumes. Ed. by G. R. Bullock & P. Petrusz, Academic Press, Orlando, Florida, USA (1982–1985).
*Immunolabelling for Electron Microscopy*. Ed. by J. M. Polak & I. M. Varndell. Elsevier, Holland (1985).
*An Atlas of Immunofluorescence in Cultured Cells*. M. C. Willingham & I. Pastan. Academic Press, London (1985).

# Manufacturers' Addresses

Manufacturers are listed in alphabetical order. Numbers appearing in parentheses after the manufacturer's name correspond to the following categories:

(1) Laboratory equipment.
(2) Chemicals.
(3) Biochemicals and Enzymes.
(4) Antibodies.
(5) Microscopy.
(6) Radioisotopes.
(7) Cell Cultures.

Amersham International plc (3,4,6)
Lincoln Place, Green End, Aylesbury, Bucks, HP20 2TP.

BCL (Boehringer) (1,2,3,4)
Bell Lane, Lewes, East Sussex, BN7 1LG.

BDH (1,2,3)
Broom Road, Poole, Dorset, BH12 4NN.

Beckman (1)
Progress Road, Sands Industrial Estate, High Wycombe, Bucks.

Bio-Rad Laboratories (1,2,3,4)
Caxton Way, Watford Business Park, Watford, Herts, WD1 8RP.

Dynatech Laboratories (1)
Daux Road, Billingshurst, West Sussex, RH14 9SJ.

F. T. Scientific Instruments (1)
Station Industrial Estate, Brendon, Nr Tewkesbury, Glos. GL20 7HH.

Gibco (2,3,7)
P.O. Box 35, Trident House, Renfrew Road, Paisley, PA3 4EF.

Heraeus Equipment (1,7)
9 Wales Way, Brentwood, Essex, CM15 9TB.

Janssen Life Sciences Products (2,4,5)
Janssen Pharmaceutical Ltd, Grove, Wantage, Oxon, OX12 0DQ.

E. Leitz Ltd (5)
48 Park Street, Luton, Bedfordshire.

LKB Produkter AB (1,2,3,5)
232 Addington Road, Selsdon, South Croydon, Surrey, CR2 8YD.

Nikon UK (5)
Haybrook, Halesfield 9, Telford, Shropshire, TF7 4EW.

Miles Laboratories Ltd (1,3,4,5)
Stoke Court, Stoke Poges, Buckinghamshire.

Olympus Optical Co. (5)
2–8 Honduras Street, London, EC1 0TX.

Pharmacia (1,2,3)
Pharmacia House, Midsummer Boulevard, Milton Keynes, Bucks, MK9 3HP.

Raymond A. Lamb (5)
6 Sunbeam Road, North Acton, London, NW10 6JL.

Shandon Southern Products (1)
Chadwick Road, Astmoor, Runcorn, Cheshire, WA7 1PR.

Sera-Lab Ltd (4)
Crawley Down, Sussex RH10 4FF.

Sigma-Aldrich (2,3,4,7)
Fancy Road, Poole, Dorset, BH17 7NH.

Whatmam Biochemicals Ltd (1,2)
Springfield Mill, Maidstone, Kent.

Carl Zeiss (Oberkochen) (5)
PO Box 78, Woodfield Road, Welwyn Garden City, Herts, AL7 1LU.

# References


Abuchowski, A., Van Es, T., Palczuk, N. C. & Davis, F. F. (1977a). *J. Biol. Chem.* **252,** 3578–3581.

Abuchowski, A., McCoy, J. R., Palczuk, N. C., Van Es, T. & Davis, F. F. (1977b). *J. Biol. Chem.* **252,** 3583–3586.

Ackerstrom, B., Brodin, T. Reis, K. & Bjorck, L. (1985). *J. Immunol.* **135,** 2589–2592.

Adair, W. S., Jurvich, D. & Goodenough, U. W. (1978). *J. Cell Biol.* **79**, 281–285.

Adams, J. M., Jepeson, P. G. N., Sanger, F. & Barrell, B. G. (1969). *Nature* **223,** 1009–1014.

Addison, G. M. & Hales, C. N. (1971). *Horm. Metab. Res.* **3,** 59–60.

Adolf, G. R., Hartter, E., Ruis, M. & Swetly, P. (1980). *Biochem. Biophys. Res. Commun.* **95,** 350–356.

Alberts, A. W., Strauss, A. W., Hennessy, S. & Vagelos, P. R. (1975). *Proc. Natl Acad. Sci USA* **72,** 3956–3960.

Algranati, D. I., Milstein, C. & Ziegler, A. (1980). *Eur. J. Biochem.* **103,** 197–207.

Alkan, S. S., Nitecki, D. E. & Goodman, J. W. (1971). *J. Immunol.* **107,** 353–358.

Allington, W. B., Cordry, A. L., McCullough, G. A., Mitchell, D. E. & Nelson, I. W. (1978). *Anal. Biochem.* **85,** 188–196.

Allsop, D., Landon, M. L., Kidd, M., Lowe, J. S., Reynolds, G. P. & Gardener, A. (1986). *Neuroscience Lett.* **68,** 252–256.

Al-Sarraj, K., White, D. A. & Mayer, R. J. (1978). *Biochem. J.* **173,** 877–883.

Al-Sarraj, K., Newbury, J., White, D. A. & Mayer, R. J. (1979). *Biochem. J.* **182,** 837–845.

Altman, L. G. (1984). *J. Histochem. Cytochem.* **32,** 1217–1223.

Ames, B. N. (1966). *Methods Enzymol.* **8,** 115–117.

Ames, G. F. & Nikaido, K. (1976). *Biochemistry* **15,** 616–623.

Amit, T., Barkey, R. J., Gavish, M. & Youdim, M. B. H. (1986). *Endocrinol.* **118,** 835–843.

Amos, W. B. (1976). *Anal. Biochem.* **70,** 612–615.

Anderson, C. W., Baum, P. R. & Gesteland, R. F. (1973). *J. Virol.* **12,** 241–252.

Anderson, N. G. & Anderson, N. L. (1978). *Anal. Biochem.* **85,** 331–340.

Anderson, R. G. W., Vasile, E., Mello, R. J., Brown, M. S. & Goldstein, J. L. (1978). *Cell* **15,** 919–933.

Anderton, B. H., Breinburg, D., Downes, M. J., Green, P. J., Tomlinson, B. E. Ulrich, J., Wood, J. N. & Kahn, J. (1982). *Nature* **298,** 84–86.

Anglister, L., Rogozinski, S. & Silman, I. (1976). *FEBS Lett.* **69,** 129–132.

Anglister, L., Tarrab-Hazdai, R., Fuchs, S. & Silman, I. (1979). *Eur. J. Biochem.* **94,** 25–29.

Anglister, J., Frey, T. McConnell, H. M. (1984). *Biochemistry* **23,** 5372–5375.

Anker, H. S. (1970). *FEBS Lett.* **7,** 293.

Armbruster, B. L., Gravito, R. M. & Kellenberger, E. (1983). *J. Histochem. Cytochem.* **31,** 1380–1384.

Arnon, R. (1971). *In* "Current Topics in Microbiology and Immunology", Vol. 54, pp. 47–93. Springer-Verlag, Berlin.

Arnon, R. (1973). *In* "The Antigens", Vol. 1 (M. Sela, Ed.), pp. 87–159. Academic Press, New York and London.

Avrameas, S. (1969). *Immunochemistry* **6,** 43–52.
Avrameas, S. & Ternynck, T. (1969). *Immunochemistry* **6,** 53–66.
Avrameas, S. & Ternynck, T. (1971). *Immunochemistry* **8,** 1175–1179.
Axelsen, N. H., Bock, E. & Krøll, J. (1973). *In* "A Manual of Quantitative Immunoelectrophoresis" (N. H. Axelsen, J. Krøll & B. Weeke, Eds), pp. 91–94. Universitetsforlaget, Oslo.
Bainbridge, D. R. & Macey, M. M. (1983). *J. Immunol. Methods* **62,** 193–195.
Baines, A. J. & Bennett, V. (1985). *Nature* **315,** 410–413.
Ballard, D. W., Lynn, S. P., Gardner, J. F. & Voss, E. W. Jr. (1984). *J. Biol. Chem.* **259,** 3492–3498.
Bang, B. E., Hurme, M., Juntunen, K. & Makela, O. (1981). *Scand. J. Cli. Lab. Invest.* **41,** 75–78.
Bankert, R. B. (1982). *Sea Notes.* FMC Corporation, Marine Colloids Division, Rochland, Maine, Autumn.
Barger, B. D., White, F. C., Pace, J. L., Kemper, D. L. & Ragland, W. G. (1976). *Anal. Biochem.* **70,** 327–335.
Bartlett, P. F., Noble, M. D., Pruss, R. M., Raff, M. C., Rattray, S. & Williams, C. A. (1981). *Brain Res.* **204,** 339–351.
Basbaum, C. B., Mann, J. K., Chow, A. W. & Finkbeiner, W. E. (1984). *Proc. Natl Acad. Sci.* **81,** 4419–4423.
Batteiger, B., Newhall, W. J. & Jones, R. B. (1982). *J. Immunol. Methods* **55,** 297–307.
Bechtel, P. J., Beavo, J. A. & Krebs, E. G. (1977). *J. Biol. Chem.* **252,** 2691–2697.
Beeley, J. G. (1976). *Biochem. J.* **155,** 345–351.
Beesley, J. E. (1985). *Proceedings Royal Microscopy Society* **20,** 1872–1896.
Beisiegel, V., Schneider, W. J., Goldstein, J. L., Anderson, R. G. W. & Brown, M. S. (1981). *J. Biol. Chem.* **256,** 11923–11931.
Belin, M. & Boulanger, P. (1985). *Exp. Cell Res.* **160,** 356–370.
Bell, M. L. & Engvall, E. (1982). *Anal. Biochem.* **123,** 329–335.
Benchimol, S., Matlashewski, G. & Crawford, L. (1984). *Biochem. Soc. Trans.* **12,** 708–711.
Benda, P., Tsuji, S., Daussant, J. & Changeux, J. P. (1970). *Nature* **225,** 1149–1150.
Bendayan, M. & Zollinger, M. (1983). *J. Histochem. Cytochem.* **31,** 101–109.
Berson, S. A. & Yalow, R. S. (1959). *J. Clin. Invest.* **38,** 1996–2016.
Berzins, K., Lando, P., Raftell, M. & Blomberg, F. (1977). *Biochim. Biophys. Acta* **497,** 337–348.
Berzofsky, J. A., Hicks, G., Fedorko, J. & Minna, J. (1980). *J. Biol. Chem.* **255,** 11188–11191.
Betts, S. A. & Mayer, R. J. (1975). *Biochem. J.* **151,** 263–270.
Betts, S. A. & Mayer, R. J. (1977). *Biochim. Biophys. Acta* **496,** 302–311.
Beug, H., van Kirchbach, A., Doderlein, G., Conscience, J. F. & Graf, T. (1979). *Cell* **18,** 375–390.
Billett, E. E. & Mayer, R. J. (1986). *Biochem. J.* **235,** 257–263.
Billett, E. E., Gunn, B. & Mayer, R. J. (1984). *Biochem. J.* **221,** 765–776.
Billingsley, M. L., Pennypacker, K. R., Hoover, C. G., Brigati, D.J. & Kincaid, R. L. (1985). *Proc. Natl Acad. Sci. USA* **82,** 7585–7589.
Binder, M., Tourmente, S., Roth, J., Renaud, M. & Gehring, W. J. (1986). *J. Cell Biol.* **102,** 1646–1653.
Birrer, M. J., Bloom, B. R. & Udem, S. (1981). *Virology* **108,** 381–390.
Bischoff, R., Eisert, R. M., Schedel, I., Vienken, J. & Zimmerman, U. (1982). *FEBS Lett.* **147,** 64–68.

Bjercke, R. J., Hale, P. M., Mercer, W. D. & McQuire, W. L. (1981). *Biochem. Biophys. Res. Commun.* **99,** 550–556.
Bjerrum, O. J. & Bøg-Hansen, T. C. (1976a). *Biochim. Biophys. Acta* **455,** 66–89.
Bjerrum, O. J. & Bøg-Hansen, T. C. (1976b). *In* "Biochemical Analysis of Membranes" (A. H. Maddy, Ed.), pp. 378–426. Chapman & Hall, London.
Bjerrum, O. J., Lundahl, P., Brogren, C-H. & Hjerten, S. (1975). *Biochim. Biophys. Acta* **394,** 173–181.
Bjorck, L. & Kronvall, G. (1984). *J. Immunol.* **133,** 969–974.
Blake, M. S. (1984). *Anal. Biochem.* **136,** 175–179.
Blessing, W. W., Costa, M., Geffen, L. B., Rush, R. A. & Fink, G. (1977). *Nature* **267,** 368–369.
Blobel, G. (1977). *In* "Gene Expression" (B. F. C. Clark, H. Klenow & J. Zeuthen, Eds), pp. 99–108. Pergamon Press, Oxford.
Blomberg, F. & Berzins, K. (1975). *Eur. J. Biochem.* **56,** 319–326.
Blomberg, F. & Perlmann, P. (1971a). *Exp. Cell Res.* **66,** 104–112.
Blomberg, F. & Perlmann, P. (1971b). *Biochim. Biophys. Acta* **233,** 53–60.
Blomberg, F., Cohen, R. S. & Siekevitz, P. (1977). *J. Cell Biol.* **74,** 204–255.
Bloom, W. S., Fields, K. L., Haver, K., Schook, W. & Puszkin, S. (1980). *Proc. Natl Acad. Sci. USA* **77,** 5520–5524.
Blythman, H. E., Casellas, P., Gros, O., Gros, P., Jansen, F. K., Paolucci, F., Pau, B. & Vidal, H. (1981). *Nature* **290,** 145–146.
Bock, E. (1978). *J. Neurochem.* **30,** 7–14.
Bock, K. W. & Matern, S. (1973). *Eur. J. Biochem.* **38,** 20–24.
Bolton, A. E. & Hunter, W. M. (1973a). *Biochem. J.* **133,** 529–538.
Bolton, A. E. & Hunter, W. M. (1973b). *Biochim. Biophys. Acta* **329,** 318–330.
Bordier, C. (1981). *J. Biol. Chem.* **256,** 1604–1607.
Borgese, N. & Meldolesi, J. (1976). *FEBS Lett.* **63,** 231–234.
Borrebaeck, C. A. K. & Etzler, M. E. (1981). *J. Biol. Chem.* **256,** 4723–4725.
Bouma, H. & Fuller, G. M. (1975). *J. Biol. Chem.* **250,** 4678–4683.
Boyd, J. B. & Mitchell, H. B. (1965). *Anal. Biochem.* **13,** 28–42.
Bradford, M. M. (1976). *Anal. Biochem.* **72,** 248–254.
Brandon, C. & Wu, J.-Y. (1978). *J. Neurochem.* **30,** 791–797.
Brandt, J., Elde, R. P. & Goldstein, M. (1979). *Neuroscience* **4,** 249–270.
Brecker, W. (1969). *Immunochem.* **6,** 539–546.
Breschkin, A. M., Ahern, J. & White, D. O. (1981). *Virology* **113,** 130–140.
Bretscher, A. (1984). *J. Biol. Chem.* **259,** 12873–12880.
Bretscher, A. & Weber, K. (1978). *J. Cell Biol.* **79,** 839–845.
Brockhaus, M., Magnani, J. L., Merlyn, M., Blaszczyk, M., Steplewski, Z., Koprowski, M. & Ginsburg, V. (1982). *Arch. Biochem. Biophys.* **217,** 614–651.
Brogren, C. H. & Bøg-Hansen, T. C. (1975). *Scand. J. Immunology* **4,** Supplement 2, 35–51.
Brower, D. L., Smith, R. J. & Wilcox, M. (1980). *Nature* **285,** 403–405.
Brown, J. P., Seight, P. W., Hart, C. E., Woodbury, R. G., Hellstrom, K. E. & Hellstrom, I. (1980). *J. Biol. Chem.* **255,** 4980–4983.
Brown, J. P., Hellstrom, K. E. & Hellstrom, I. (1981a). *Clin. Chem.* **27,** 1592–1596.
Brown, J. P., Woodbury, R. G., Hart, C. E., Hellstrom, I. & Hellstrom, K. E. (1981b). *Proc. Natl Acad. Sci.* **78,** 539–543.
Brown, J. P., Hellstrom, K. E. & Hellstrom, I. (1982). *Clin. Chem.* **27,** 1592–1596.
Brown, A., Feizi, T., Gooi, H. C., Embleton, K. J., Picard, J. K. & Baldwin, R. W. (1983). *Biosci. Rep.* **3,** 163–170.

Brownsey, R. W., Hughes, W. A., Denton, R. M. & Mayer, R. J. (1977). *Biochem. J.* **168,** 441–445.
Bruck, C., Portetelle, D., Glineur, C. & Bollen, A. (1982). *J. Immunol. Methods* **53,** 313–319.
Brulet, P., Babinet, C., Kemler, R. & Jacob, F. (1980). *Proc. Natl Acad. Sci.* **77,** 4113–4117.
Buckley, K. M. & Kelly, R. B. (1985). *J. Cell Biol.* **100,** 1284–1294.
Bundesen, P. G., Drake, R. G., Kelly, K., Worsley, I. G., Friesen, H. G. & Sehon, A. H. (1980). *J. Clin. Endocrin. Metab.* **51,** 1472–1474.
Burchiel, S. W., Billman, J. R. & Alber, T. R. (1983). *J. Immunol. Methods* **69,** 33–42.
Burgess, W. H., Watterson, D. M. & van Eldik, L. J. (1984). *J. Cell Biol.* **99,** 550–557.
Burke, B., Walter, C., Griffiths, G. & Warren, G. (1983). *Eur. J. Cell Biol.* **31,** 315–324.
Burnette, W. N. (1981). *Anal. Biochem.* **112,** 195–203.
Burridge, K. (1978). *In* "Methods in Enzymology", Vol. L (V. Ginsburg, Ed.), pp. 54–64. Academic Press, New York and London.
Busachi, C. A., Ray, M. B. & Desmet, V. J. (1978). *J. Immunol. Methods* **19,** 95–99.
Cahill, A. L. & Morris, S. J. (1979). *J. Neurochem.* **32,** 855–867.
Cailla, H. L., Vannier, C. J. & Delaage, M. A. (1976). *Anal. Biochem.* **70,** 195–202.
Campbell, K. P., MacLennan, D. H. & Jorgensen, A. O. (1983). *J. Biol. Chem.* **258,** 11267–11273.
Catt, K. & Tregear, G. W. (1967). *Science* **158,** 1570–1572.
Chaiet, L. & Wolf, F. L. (1964). *Arch. Biochem. Biophys.* **106,** 1.
Chan, S. H. P. & Tracy, R. P. (1978). *Eur. J. Biochem.* **89,** 595–606.
Chang, T. H., Steplewski, Z. & Koprowski, M. (1980). *J. Immunol. Methods* **39,** 369–375.
Chao, L. P. (1975). *J. Neurochem.* **25,** 261–266.
Chao, L. P., Wolfgram, F. J. & Eng. L. F. (1977). *Neurochem. Res.* **2,** 323–325.
Chiorazzi, N. (1986). *Biotechnology* **4,** 210–218.
Cho-Chung, Y. S. & Pitot, H. C. (1968). *Eur. J. Biochem.* **3,** 401–406.
Choo, K. H., Myer, J., Cotton, R. G. H., Camakaris, J. & Danks, D. M. (1980). *Biochem. J.* **191,** 665–668.
Christie, W. W. (1973). *Lipid Analysis.* Pergamon Press, Oxford.
Chua, N. H. & Blomberg, F. (1979). *J. Biol. Chem.* **254,** 215–233.
Chun, L. L. U., Patterson, P. H. & Cantor, H. (1980). *J. Exp. Biol.* **89,** 73–83.
Cinader, B. (1983). *Ann. NY Acad. Sci.* **103,** 495–548.
Cinader, B. (1967) *In* "Antibodies to Biologically Active Molecules" (B. Cinader, Ed.), pp. 85–137. Pergamon, Oxford.
Cinader, B. (1977). *In* "Methods in Immunology and Immunochemistry", Vol. IV (C. A. Williams & M. W. Chase, Eds), pp. 313–375. Academic Press, New York and London.
Cinader, B. & Lafferty, K. J. (1964). *Immunol.* **7,** 342–362.
Clark, W. A., Everett, A. W., Fitch, F. W., Frogner, K. S., Jakovcic, S., Rabinowitz, M., Warner, A. M. & Zak, R. (1980). *Biochem. Biophys. Res. Commun.* **95,** 1680–1686.
Clarke, M. F., Gelmann, E. P. & Reitz, M. S. Jr. (1983). *Nature* **305,** 60–62.
Clausen, J. (1971). *In* "Laboratory Techniques in Biochemistry and Molecular Biology", pp. 443–444. North Holland, Amsterdam.
Cleveland, D. W., Fischer, S. G., Kirschner, M. W. & Laemmli, U. K. (1977). *J. Biol. Chem.* **252,** 1102–1106.

Cleveland, W. L., Wassermann, N. H., Sarangarajan, R., Penn, A. S. & Erlanger, B. F. (1983). *Nature* **305,** 56–57.
Click, R. E., Benck, L. & Alter, B. J. (1972). *Cell Immunol.* **3,** 264–269.
Cohen, J. & Selvendran, S. Y. (1981). *Nature* **291,** 421–423.
Colcher, D., Horanhand, P., Teramoto, Y. A., Wunderlieh, D. & Schlom, J. (1981). *Cancer Res.* **41,** 1451–1459.
Collins, W. P. & Hennam, J. F. (1976). *In* "Molecular Aspects of Medicine", Vol. 1 (H. Baum & J. Gergely, Eds), pp. 3–128. Pergamon, Oxford.
Collot, M., Louvard, D. & Singer, S. J. (1984). *Proc. Natl Acad. Sci. USA* **81,** 788–792.
Colover, J. (1960). *Biochem. J.* **78,** 5P.
Comoglio, P. M., Di Renzo, M. F., Tarone, G., Giancotti, F. G., Naldini, L. & Marchisio, P. C. (1984). *EMBO J.* **3,** 483–489.
Converse, C. A. & Papermaster, D. S. (1975). *Science* **189,** 469–472.
Coons, A. H., Creech, H. J., Jones, R. N. & Berliner, E. (1942). *J. Immunol.* **45,** 150–170.
Cordell, J. L., Pael, N. T. & Chubb, I. W. (1984). *J. Histochem. Cytochem.* **32,** 219–229.
Costa, M. & Furness, J. B. (1983). *In* "Immunohistochemistry" (A. C. Cuello, Ed.), pp. 373–397. John Wiley, New York.
Costa, M., Rush, R. A., Furness, J. B. & Geffen, L. B. (1976). *Neurosci. Lett.* **3,** 201–207.
Costa, M., Furness, J. B., Geffen, L. B., Lewis, S. Y. & Rush, R. A. (1978). *J. Anat.* **126,** 652.
Craig, S. & Goodchild, D. J. (1982). *Eur. J. Cell Biol.* **28,** 251.
Creighton, W. D., Lambert, P. H. & Miescher, P. A. (1973). *J. Immunol.* **111,** 1219–1227.
Croce, C. M., Linnebhach, A., Hall, W., Stepewski, Z. & Koprowski, M. (1980). *Nature* **288,** 488–489.
Croy, R. R. D., Gatehouse, J. A., Tyler, M. (1980). *Biochem. J.* **191,** 509–516.
Crumpton, M. J. & Parkhouse, R. M. E. (1972). *FEBS Lett.* **22,** 210–222.
Cuello, A. C., Priestley, J. V. & Milstein, C. (1982). *Proc. Natl Acad, Sci.* **79,** 665–669.
Cutting, J. A. & Roth, T. F. (1973). *Anal. Biochem.* **54,** 386–394.
Dahlberg, A. E., Dingman, C. W. & Peacock, A. C. (1969). *J. Mol. Biol.* **41,** 139–147.
Dale, G. & Latner, A. L. (1969). *Clin. Chim. Acta.* **24,** 61–68.
Danscher, G. (1981). *Histochemistry* **71,** 81–88.
Davis, B. J. (1964). *Ann. NY Acad. Sci.* **121,** Part 2, 404–427.
Davis, J. & Bennett, V. (1983). *J. Biol. Chem.* **258,** 7757–7766.
Dean, B. (1979). *Anal. Biochem.* **99,** 105–111.
Dean, P. D. G., Brown, P., Leyland, M. J., Watson, D. H., Angal, S. & Harvey, M. J. (1977). *Biochem. Soc. Trans.* **5,** 1111–1113.
De Camilli, P., Cameron, R. & Greengard, P. (1983). *J. Cell Biol.* **96,** 1337–1354.
Denney, R. M., Fritz, R. R., Pael, N. T. & Abell, C. W. (1982). *Science* **215,** 1400–1403.
Dennick, R. G. & Mayer, R. J. (1977). *Biochem. J.* **161,** 167–174.
DeMey, J. (1981). *Cell Biol. Int. Reports* **5,** 889.
DeMey, J. (1983). *In* "Immunohistochemistry" (A. C. Cuello, Ed.), IBRO Handbook Series, Vol. 3, pp. 348–372, John Wiley, Chichester, New York.
DeMey, J. (1984). *EMSA Bulletin* **14,** 54–66.

DeMey, J., Moeremans, M., Geuens, G., Nuydens, R. & DeBrabander, M. (1981). *Cell Biol. Int. Rep.* **5,** 889–899.
De Potter, W. P. & Chubb, I. W. (1977). *Neuroscience* **2,** 167–174.
DeSavigny, D. & Voller, A. (1980). *Immunoenzymatic Assay Techniques*. Martinus Nijhoff, The Hague.
Deschamps, J. R., Hildreth, J. E., Derr, D. & August, J. T. (1985). *Anal. Biochem.* **147,** 451–454.
de St Groth, S. F. & Scheidegger, D. (1980). *J. Immunol. Methods* **35,** 1–21.
Dewald, B., Dulaney, J. T. & Touster, O. (1974). *Methods in Enzymol.* **32,** 982–91.
Dippold, W. G., Jay, G., DeLeo, A. B., Khoury, G. & Old, L. J. (1981). *Proc. Natl Acad. Sci.* **78,** 1695–1699.
Don, M. & Masters, C. J. (1975). *Biochim. Biophys. Acta* **384,** 25–36.
Donnelly, D., Mihovilovic, M., Gonzalez-Ros, J. M., Ferragut, J. A., Richman, D. & Martinez-Carrion, M. (1984). *Proc. Natl Acad. Sci.* **81,** 7999–8003.
Dorsey-Stuart, W., Bishop, J. G., Carson, H. L. & Frank, M. B. (1981). *Proc. Natl Acad. Sci.* **78,** 3751–3754.
Dulley, J. R. & Grieve, P. A. (1975). *Anal. Biochem.* **64,** 136–141.
Dyson, H. J., Cross, K. J., Ostresh, J., Houghton, R. A., Wilson, I. A., Wright, P. E. & Lerner, R. A. (1986). *In* "Ciba Foundation Symposium 119, Synthetic Peptides as Antigens", pp. 58–75. Wiley & Sons, Chichester.
Edelman, G. M. & Rutishauser, U. (1974). *In* "Methods in Enzymology", Vol. 34 (W. B. Jakoby & M. Wilchek, Eds), pp. 195–225. Academic Press, New York and London.
Edwards, P. A. W. (1980). *Biochem. Soc. Trans.* **8,** 334–335.
Edwards, P. A. W. (1981). *Biochem. J.* **200,** 1–10.
Edwards, P. A. W., Smith, C. M., Neville, A. M. & O'Hare, M. J. (1982). *Eur. J. Immunol.* **12,** 641–648.
Eichner, R., Bonitz, P. & Sun, T. (1984). *J. Cell Biol.* **98,** 1388–1396.
Eisenbarth, G. S. (1981). *Anal. Biochem.* **111,** 1–16.
Eisenbarth, G. S., Oie, H., Gazdar, A., Chick, W., Schultz, J. A. & Scearce, R. M. (1981). *Diabetes* **30,** 226–230.
Ek, B. & Heldin, C. (1984). *J. Biol. Chem.* **259,** 11145–11152.
Ekwall, K., Soderholm, J. & Wadstrom, T. (1975). *J. Immunol. Methods.* **12,** 103–115.
Elkon, K. B. (1984). *J. Immunol. Methods* **66,** 313–321.
Eng, L. F., Ueda, C. T., Chao, L. P. & Wolfgram, F. (1974). *Nature* **250,** 243–245.
Engvall, E. & Perlmann, P. (1971). *Immunochemistry* **8,** 871–874.
Engvall, E. & Perlmann, P. (1972). *J. Immunol.* **109,** 129–135.
Engvall, E., Jonsson, K. & Perlmann, P. (1971). *Biochim. Biophys. Acta* **251,** 427–434.
Evan, G. I. (1984). *J. Immunol. Methods* **73,** 427–435.
Evan, G. I., Hancock, D. C., Littlewood, T. & Panza, C. D. (1986). *In* "Ciba Foundation Symposium 119, Synthetic Peptides as Antigens", pp. 245–263. Wiley & Sons Chichester.
Faulk, W. P. & Taylor, G. M. (1971). *Immunochemistry* **8,** 1081–1083.
Ferretti, P. & Borroni, E. (1986). *J. Neurochem.* **46,** 1888–1894.
Fiddler, M. B. & Gray, G. R. (1978). *Anal. Biochem.* **86,** 716–724.
Fiedler, W. & Walker, J. H. (1985). *Eur. J. Cell Biol.* **38,** 34–41.
Fillenz, M., Gagnon, C., Stoeckel, K. & Thoenen, H. (1976). *Brain Res.* **114,** 293–303.
Fisher, A. G. & Brown, G. (1980). *J. Immunol. Methods* **39,** 377–385.
Forsgren, A. & Sjoquist, J. (1966). *J. Immunol.* **97,** 822–827.
Frackleton, A. R. & Rotman, B. (1980). *J. Biol. Chem.* **255,** 5286–5290.
Fraser, C. M. & Venter, J. C. (1980). *Proc. Natl Acad. Sci.* **77,** 7034–7038.

Frosch, M., Gorgen, I., Boulnois, G. J., Timmis, K. N. & Bitter-Suermann, D. (1985). *Proc. Natl Acad. Sci.* **82,** 1194–1198.

Fuxe, K., Hokfelt, T., Eneroth, P., Gustafsson, J.-A. & Skett, P. (1977). *Science* **196,** 899–900.

Galfre, G. & Milstein, C. (1981). *Methods Enzymol.* **73,** 1–45.

Galfre, G., Howe, S. C., Milstein, C., Butcher, G. W. & Howard, J. C. (1977). *Nature* **266,** 550–552.

Galfre, G., Milstein, C. & Wright, B. (1979). *Nature* **277,** 131–133.

Garson, J. A., Beverley, P. C. L., Coakham, H. B. & Harper, E. I. (1982). *Nature* **298,** 375–377.

Gee, N. S., Matsas, R. & Kenny, A. J. (1983). *Biochem. J.* **214,** 377–386.

Gershoni, J. M. (1985). *TIBS* **10,** 103–106.

Gershoni, J. M. & Palade, G. E. (1982). *Anal. Biochem.* **124,** 396–405.

Gershoni, J. M. & Palade, G. E. (1983). *Anal. Biochem.* **131,** 1–15.

Geuze, H. J., Slot, J. W., van der Ley, P. A. & Scheffer, R. C. T. (1981). *J. Cell Biol.* **89,** 653–665.

Geysen, H. M., Meloen, R. H. & Barteling, S. J. (1984). *Proc. Natl Acad. Sci.* **81,** 3998–4002.

Geysen, H. M., Barteling, S. J. & Meloen, R. H. (1985). *Proc. Natl Acad. Sci.* **82,** 178–182.

Ghangas, G. S. & Milman, G. (1977). *Science* **196,** 1119–1120.

Gilbert, D. (1978). *Nature* **272,** 577–578.

Gintzler, A. R., Gersham, M. D. & Spector, S. (1978). *Science* **199,** 447–448.

Girardet, C., Ladisch, S., Heunan, D., Mach, J-P. & Carrel, S. (1983). *Int. J. Cancer* **32,** 177–183.

Gisiger, V., Vigny, M., Gautron, J. & Rieger, F. (1978). *J. Neurochem.* **30,** 501–516.

Glenney, J. R. & Weber, K. (1980). *J. Biol. Chem.* **255,** 10551–10554.

Göding, J. N. (1980). *J. Immunol. Methods* **39,** 285–308.

Goodfellow, P., Banting, G., Levy, R., Povey, S. & McMichael, A. (1980). *Somatic Cell Genet.* **6,** 777–787.

Gopalkinshun, R. & Anderson, W. B. (1982). *BBRC* **104,** 830–836.

Gordon, A. S., Davis, C. G. & Diamond, I. (1977a). *Proc. Natl Acad. Sci. USA* **74,** 263–267.

Gordon, A. S., Davies, C. G., Milfay, D. & Diamond, I. (1977b). *Nature* **267,** 539–540.

Gordon, L. K. (1981). *J. Immunol. Methods* **44,** 241–245.

Goridis, C., Hirn, M., Santoni, M.-J., Gennarini, G., Deagostini-Bazin, H., Jordan, B. R., Kiefer, M. & Steinmetz, M. (1985). *EMBO J.* **4,** 631–635.

Grabar, P. & Williams, C. A. (1953). *Biochim. Biophys. Acta* **10,** 193–194.

Graessmann, M. & Graessmann, A. (1976). *Proc. Natl Acad. Sci.* **73,** 366–370.

Graham, R. C., Lundholm, U., Karnovsky, M. J. (1965). *J. Histochem. Cytochem.* **13,** 150–152.

Granger, B. L. & Lazarides, L. (1978). *Cell* **15,** 1253–1268.

Granger, B. L. & Lazarides, L. (1979). *Cell* **18,** 1053–1063.

Greene, G. L., Nolan, C., Engler, J. P. & Jensen, E. V. (1980). *Proc. Natl Acad. Sci.* **77,** 5115–5119.

Greengard, P. (1976). *Nature* **260,** 101–108.

Greengard, P., McAfee, D. A. & Kebabian, J. W. (1972). *In* "Advances in Cyclic Nucleotide Research", Vol. I (P. Greengard, R. Paoletti & G. A. Robinson, Eds), pp. 337–355. Raven Press, New York.

Greenwood, F. C., Hunter, W. M. & Glover, J. S. (1963). *Biochem. J.* **89,** 114–123.

Griesser, G. H. (1978). *Neuroscience* **3,** 301–306.
Griffiths, G., Brands, R., Burke, B., Louvard, D. & Warren, G. (1982). *J. Cell Biol.* **317,** 248–250.
Groopman, J. D., Trudel, L. J., Donahue, P. R., Marshak-Rothstein, A. & Wogan, G. N. (1984). *Proc. Natl Acad. Sci.* **81,** 7728–7731.
Guillet, J. G., Kaveri, S. V., Durieu, O., Delavier, C., Hoebeke, J. & Strosberg, A. D. (1985). *Proc. Natl Acad. Sci.* **82,** 1781–1784.
Gzanna, R., Morrison, J. H., Coyle, J. T. & Molliver, M. E. (1977). *Neurosci. Lett.* **4,** 127–134.
Haaiman, J. J. (1983). *In* "Immunohistochemistry" (A. C. Cuello, Ed.) pp. 47–86. John Wiley, New York.
Haase, W., Schafer, A., Murer, H. & Kinne, R. (1978). *Biochem. J.* **172,** 57–62.
Hackenbrock, C. R. & Hammon, K. M. (1975). *J. Biol.* **250,** 9185–9197.
Hacker, G. W., Springall, D. R., Grimelius, L. & Polak, J. M. (1985). *Virchows Arch. [Pathol. Anat.]* **406,** 449–461.
Hackett, C. J., Askonas, B. A., Webster, R. G. & van Wyke, K. (1980). *J. Gen. Virol.* **47,** 497–501.
Haeuptle, M. T., Aubert, M. L., Djiane, J. & Kraehenbuhl, J. P. (1983). *J. Biol. Chem.* **258,** 305–314.
Hamaguchi, Y., Kato, K., Fukui, H., Shirakawa, I., Okawa, S., Ishikawa, E., Kobayashi, K. & Katunuma, N. (1976a). *Eur. J. Biochem.* **71,** 459–467.
Hamaguchi, Y., Kato, K., Ishikawa, E., Kobayashi, K. & Katunuma, N. (1976b). *FEBS Lett.* **69,** 11–14.
Hanahan, D. (1985). *Nature* **315,** 115–122.
Hance, A. J., Robin, E. D., Simon, L. M., Alexander, S., Herzenberg, L. A. & Theodore, J. (1980). *J. Clin. Invest.* **66,** 1258–1264.
Hancock, K. & Tsang, V. C. W. (1983). *Anal. Biochem.* **133,** 157–163.
Handschin, H. E. & Ritschard, W. J. (1976). *Anal. Biochem.* **71,** 143.
Hansen, J. N. (1977). *Anal. Biochem.* **76,** 37–44.
Hansen, R. S. & Beavo, J. A. (1982). *Proc. Natl Acad. Sci.* **79,** 2788–2792.
Harboe, N. & Ingild, A. (1973). *In* "A Manual of Quantitative Immunoelectrophoresis" (N. H. Axelsen, J. Krøll & B. Weeke, Eds), pp. 161–164. Universitetsforlaget, Oslo.
Hardwicke, P. M. D. (1976). *Biochim. Biophys. Acta* **422,** 357–364.
Hartman, B. C., Zide, D. & Udenfriend, S. (1972). *Proc. Natl Acad. Sci. USA* **69,** 2722–2726.
Haspel, M. V., Onodera, T., Prabhakar, B. S., McClintock, P. R., Essani, K., Ray, V. R., Yagihashi, S. & Notkins, A. L. (1983). *Nature* **3094,** 73–76.
Hatta, K., Okada, T. S. & Takeichi, M. (1985). *Proc. Natl Acad. Sci.* **82,** 2789–2793.
Haugen, A., Groopman, J. D., Hsu, I-C., Goodrich, G. R., Wogan, G. N. & Harris, C. C. (1981). *Proc. Natl Acad. Sci.* **78,** 4124–4127.
Hauri, H-P., Quaroni, A. & Isselbacher, K. J. (1980). *Proc. Natl Acad Sci.* **77,** 6629–6633.
Haustein, D. & Warr, G. W. (1976). *J. Immunol. Meth.* **12,** 323–336.
Hawkes, R., Niday, E. & Gordon, J. (1982a). *Anal. Biochem.* **119,** 142–147.
Hawkes, R., Nielay, E. & Matus, A. (1982b). *Cell* **28,** 253–258.
Hayman, E. G., Engvall, E., Green, T. J. & Harrison, R. (1982). *J. Cell Biol.* **95,** 20–23.
Heggeness, M. H., Wang, K. & Singer, S. J. (1977). *Proc. Natl Acad. Sci. USA* **74,** 3883–3887.

Heidelberger, M. & Kendall, F. E. (1929). *J. Exp. Med.* **50,** 809–823.
Heidmann, T. & Changeaux, J. P. (1978). *Annu. Rev. Biochem.* **47,** 317–357.
Heilbronn, E. & Bartfai, T. (1978). *Prog. Neurob.* **11,** 171–188.
Heilbronn, E. & Mattson, C. (1974). *J. Neurochem.* **22,** 315–317.
Heilbronn, E. & Stalberg, E. (1978). *J. Neurochem.* **31,** 5–11.
Helenius, A. & Simons, K. (1977). *Proc. Natl Acad. Sci. USA* **74,** 529–532.
Helle, K. B., Fillenz, M., Stanford, C., Pihl, K. E. & Srebro, B. (1979). *J. Neurochem.* **32,** 1351–1355.
Henderson, D. & Weber, K. (1981). *Exp. Cell Res.* **132,** 297–311.
Herbert, W. J. (1978). *In* "Handbook of Experimental Immunology" (D. M. Weir, Ed.). Blackwell Publishers, London.
Herlyn, M., Steplewiski, Z., Herlyn, D. & Koprowski, H. (1979). *Proc. Natl Acad. Sci.* **76,** 1438–1442.
Herman, I. M. & Pollard, T. D. (1978). *Exp. Cell Res.* **114,** 15–25.
Hertzog, P. J., Shaw, A., Lindsay-Smith, J. R. & Garner, R. C. (1983). *J. Immunol. Methods* **62,** 49–58.
Herzenberg, L. A. & Herzenberg, L. A. (1978). *In* "Handbook of Experimental Immunology", 3rd edn, Vol. 1 (D. M. Weir, Ed.). Blackwell Scientific Publications, Oxford.
Heuser, J. E. & Salpeter, S. R. (1979). *J. Cell Biol.* **82**, 150–173.
Higgins, R. C. & Dahmus, M. E. (1979). *Anal. Biochem.* **93**, 257–260.
Hilwig, I. & Gropp, A. (1973). *Exp. Cell Res.* **81**, 474–477.
Hino, Y., Asano, A. & Sato, R. (1978). *J. Biochem.* **83,** 925–934.
Hirn, M., Pierres, M., Deagostini-Bazin, H., Hirsch, M. & Goridis, C. (1981). *Brain Res.* **214,** 433–439.
Hizi, A. & Yagil, G. (1974). *Eur. J. Biochem.* **45,** 211–221.
Hochkeppel, H. K., Menge, U. & Collins, J. (1981). *Nature* **291,** 500–501.
Hoeldtke, R. & Kaufman, S. (1977). *J. Biol. Chem.* **252,** 3160–3169.
Hoffman, G. J., Lazarowitz, S. G. & Hayward, S. D. (1980). *Proc. Natl Acad. Sci.* **77,** 2979–2983.
Hogg, N., Slvsarenko, M., Cohen, J. & Reiser, J. (1981). *Cell* **24,** 875–884.
Hokfelt, T., Fuxe, K. & Goldstein, M. (1973). *Brain Res.* **62,** 461–469.
Hokfelt, T., Fuxe, K., Goldstein, M. & Johansson, O. (1974). *Brain Res.* **66,** 235–251.
Hokfelt, T., Elde, R., Johansson, O., Luft, R. & Arimura, A. (1975). *Neurosci. L.* **1,** 231–235.
Holgate, C. S., Jackson, P., Cowen, P. N. & Bird, C. C. (1983). *J. Histochem. Cytochem.* **31,** 938–944.
Holland, P. C. & MacLennan, D. H. (1976). *J. Biol. Chem.* **251,** 2030–2036.
Hopff, W. H., Riggio, G. & Waser, P. G. (1975). *In* "Cholinergic Mechanisms" (P. G. Waser, Ed.). Raven Press, New York.
Hopgood, M. F., Ballard, F. J., Resherf, L. & Hanson, R. W. (1973). *Biochem. J.* **134,** 445–453.
Hortnagl, H., Winkler, H. & Lochs, H. (1973). *J. Neurochem.* **20,** 977–985.
Houdebine, L.-M. (1976). *Eur. J. Biochem.* **68,** 219–225.
Houdebine, L.-M. & Gaye, P. (1976). *Eur. J. Biochem.* **63,** 9–14.
Houslay, M. D. & Tipton, K. F. (1973). *Biochem. J.* **135,** 173–186.
Howard, G. C., Abmayr, S. M., Shinefield, L. A., Sato, V. L. & Elgin, S. C. R. (1981). *J. Cell Biol.* **88,** 219–225.
Hsu, D., Hoffman, P. & Mashburn, T. A. (1972). *Anal. Biochem.* **46,** 156–163.
Hsu, I-C., Yolken, R. M. & Harris, C. C. (1981). *Methods Enzymol.* **73,** 383–394.

Hsu, S. (1984). *Anal. Biochem.* **142,** 221–225.
Hsu, S., Raine, L. & Fanger, H. (1981). *J. Histochem. Cytochem.* **29,** 577–580.
Huang, K. S., Wallner, B. P., Mattaliano, R. J., Tizard, R. *et al.* (1986). *Cell* **46,** 19–199.
Huang, W. M., Gibson, S. J., Facer, P., Gu, J. & Polak, J. M. (1983). *Histochemistry* **77,** 289–302.
Hubbard, A. L., Bartles, J. R. & Braiterman, L. T. (1985). *J. Cell Biol.* **100,** 1115–1125.
Hubbard, A. & Cohn, F. C. (1975). *J. Cell Biol.* **64,** 438–460.
Hudson, L. & Hay, F. C. (1976). "Practical Immunology". Blackwell, Oxford, England.
Hudson, L. & Hay, F. C. (1980). "Practical Immunology", 2nd edn. Blackwell Scientific Publications, Oxford.
Hughes, E. N. & August, J. T. (1981). *J. Biol. Chem.* **256,** 664–671.
Hunkapillar, S. (1983). *Methods Enzymol.* **91,** 227–236.
Hunter, W. M. (1967). *In* "Handbook of Experimental Immunology" (D. M. Weir, Ed.), pp. 608–654. Blackwell, Oxford.
Huttner, W. B., Schiebler, W., Greengard, P. & DeCamilli, P. (1983). *J. Cell Biol.* **196,** 1374–1389.
Hyden, H. (1973). *In* "Macromolecules and Behaviour" (G. B. Ansell & P. D. Bradley, Eds), pp. 3–26. Macmillan, London.
Ikehara, Y., Takahashi, K., Mansho, K., Eto, S. & Kato, K. (1977). *Biochim. Biophys. Acta* **470,** 202–211.
Iscove, N. N. & Melchers, F. (1978). *J. Exp. Med.* **147,** 922–933.
Izant, J. G. & McIntosh, J. R. (1980). *Proc. Natl Acad. Sci.* **77**, 4741–4745.
Izant, J. G., Weatherbee, J. A. & McIntosh, J. R. (1982). *Nature* **295,** 245–250.
Jacob, L., Tron, F., Bach, J-F. & Louvard, D. (1984). *Proc. Natl Acad. Sci.* **81,** 3843–3845.
Jean, D. H. & Albers, R. W. (1976). *Biochim. Biophys. Acta* **452,** 219–226.
Jefferis, R., Deverill, I. & Steansgaard, J. (1981). *Biochem. Soc. Trans.* **9,** 116–117.
Jensen, K. (1959). Thesis. Munksgaard, Copenhagen.
Jerne, N. K. (1974). *Ann. Immunol. (Inst. Pasteur)* **125C,** 373–389.
Johnson, D. A., Gautsch, J. W., Sportsman, J. R. & Elder, J. H. (1984). *Gene Anal. Techn.* **1,** 3–8.
Johnson, G. D. & Araujo, G. M. (1981). *J. Immunol. Methods* **43,** 349–350.
Johnson, L. V., Walsch, M. L. & Chen, L. B. (1980). *PNAS* **77,** 990–994.
Johnson, L. V., Walsch, M. L., Bockus, B. J. & Chen, L. B. (1981). *J. Cell Biol.* **88,** 526–535.
Johnston, J. P. (1968). *Biochem. Pharmacol.* **17,** 1285–1297.
Jolley, M. E., Wang, C. J., Ekenberg, S. J., Luelke, M. S. & Kelso, D. M. (1984). *J. Immunol. Methods.* **67,** 21–35.
Jones, M. I., Massingham, W. E. & Spragg, S. P. (1980). *Anal. Biochem.* **106,** 446–449.
Jonsson, S. & Kronvall, G. (1974). *Eur. J. Immunol.* **4,** 29–33.
Jørgensen, A. O., Subrahmanyan, L., Turnbull, C. & Kalnins, V. I. (1976). *Proc. Natl Acad. Sci. USA* **73,** 3129–3196.
Jørgensen, O. S. (1976). *Science J. Neurochem.* **27,** 1223–1227.
Jørgensen, O. S. (1977). *FEBS Lett.* **79,** 42–44.
Julien, J. P. & Mushynski, W. E. (1982). *J. Biol. Chem.* **257,** 10467–10470.
Just, W. W. (1980). *Anal. Biochem.* **102,** 134–144.
Kabat, E. A. (1967). *In* "Methods in Immunology and Immunochemistry", Vol. 1

(C. A. Williams & M. W. Chase, Eds), pp. 335–339. Academic Press, New York and London.

Kabat, E. A. (1971). "Kabat and Mayers Experimental Immunochemistry", 3rd ed., pp. 89–90. Charles C. Thomas, Springfield, Ill.

Kaltschmidt, E. & Wittmann, H. G. (1970). *Anal. Biochem.* **36,** 401–412.

Kan, K. S. K., Chao, L. P. & Eng, L. F. (1978). *Brain Res.* **146,** 221–229.

Kao, I. & Drachman, D. B. (1977). *Science* **196,** 527–529.

Kaplan, G., Unkeless, J. C. & Cotin, Z. A. (1979). *Proc. Natl Acad. Sci. USA* **76,** 3824–3828.

Kaplan, H. & Olsson, L. (1980). *Proc. Natl Acad. Sci.* **77,** 5429–5431.

Karlin, A., Holtzman, E., Valderrama, R., Damle, V., Hsu, K. & Reyers, F. (1978). *J. Cell Biol.* **76,** 577–592.

Karlsson, E., Heilbronn, E. & Widlund, L. (1972). *FEBS Lett.* **28,** 107–111.

Kartenbeck, J., Schwecheimer, K., Moll, R. & Franke, W. W. (1984). *J. Cell Biol.* **98,** 1072–1082.

Katan, M. B., Van Harten-Loosebroek, N. & Groot, G. S. P. (1976). *Eur. J. Biochem.* **70,** 409–417.

Kay, R. E., Walwick, E. R. & Gifford, C. K. (1964). *J. Phys. Chem.* **68,** 1896–1906.

Kearney, J. F., Radbruch, A., Liesegany, B. & Rajewsky, K. (1979). *J. Immunol.* **123,** 1548–1550.

Keller, G., Tokhyasu, K. T., Dutton, A. H. & Singer, S. J. (1984). *Proc. Natl Acad. Sci. USA* **81,** 5744.

Kemp, D. J. & Cowman, A. F. (1981). *Proc. Natl. Acad. Sci.* **78,** 4520–4524.

Kennett, R. M., McKearn, T. J. & Bechtol, U. B. (Eds) (1981). "Monoclonal Antibodies, Hybridomas: A New Dimension in Biological Analyses". Plenum Press, New York and London.

Kessler, S. W. (1975). *J. Immunol.* **115,** 1617–1624.

Kessler, S. W. (1976). *J. Immunol.* **117,** 1482–1490.

Khyse-Andersen, J. (1984). *J. Biophys. Biochem. Methods* **10,** 203–209.

Klausner, A. (1986). *Biotechnology* **4,** 185–194.

Koch, C. & Nielsen, H. E. (1975). *Scand. J. Immunol.* **4,** Suppl. 2, 121–124.

Kohen, F., Kim, J. B., Lindner, H. R., Eshhar, Z. & Green, B. (1980). *FEBS Lett.* **111,** 427–430.

Köhler, G. & Milstein, C. (1975). *Nature* **256,** 495–497.

Köhler, G. & Milstein, C. (1976). *Eur. J. Immunol.* **6,** 511–519.

Korman, A. J., Knudsen, P. J., Kaufman, J. F. & Strominger, J. L. (1982). *Proc. Natl Acad. Sci USA* **79,** 1844–1848.

Kostner, G. & Holasek, A. (1972). *Anal. Biochem.* **46,** 680–683.

Kraehenbuhl, J. P., de Grandi, P. B. & Campiche, M. A. (1971). *J. Cell Biol.* **50,** 432–445.

Kraehenbuhl, J. P., Galardy, R. E. & Jamieson, J. D. (1974). *J. Exp. Med.* **139,** 208–223.

Kramer, A., Haars, R., Kabisch, R., Will, H., Bantz, F. A. & Bantz, E. K. F. (1980). *Mol. Gen. Genet.* **180.** 193–199.

Kristiansen, T. (1976). *In* "Immunoadsorbents in Protein Purification" (E. Ruoslahti, Ed.), pp. 19–27. Universitetsforlaget, Oslo.

Krohn, K. A., Knight, L. C., Harwig, J. F. & Welch, M. J. (1977). *Biochim Biophys. Acta* **490,** 497–505.

Krøll, J. (1973a). *In* "A Manual of Quantitative Immunoelectrophoresis" (N. H. Axelsen, J. Krøll & B. Weeke, Eds), pp. 57–59. Universitetsforlaget, Oslo.

Krøll, J. (1973b). *In* "A Manual of Quantitative Immunoelectrophoresis" (N. H. Axelsen, J. Krøll & B. Weeke, Eds), pp. 79–81. Universitetsforlaget, Oslo.
Kronvall, G. & Frommel, D. (1970). *Immonochem.* **7**, 124–127.
Kronvall, G., Grey, H. M. & Williams, R. C. (1970). *J. Immunol.* **105**, 1116–1123.
Kronvall, G., Seal, U. S., Finstad, J. & Williams, R. C. Jr. (1970). *J. Immunol.* **104**, 140–147.
Kuma, F., Prough, R. A. & Masters, B. S. S. (1976). *Arch. Biochem. Biophys.* **172**, 600–607.
Kwapinski, J. B. G. (1972). "Methodology of Immunochemical and Immunological Research", pp. 286–306. Wiley-Interscience, New York.
Kyhse-Andersen, J. (1984), *J. Biochim. Biophys. Methods* **10**, 203–209.
Kyte, J. (1976). *J. Cell Biol.* **68**, 287–303.
Lachmann, P. J., Oldroyd, R. G., Milstein, C. & Wright, B. W. (1980). *Immunology* **41**, 503–515.
Laemmli, U. K. (1970). *Nature* **227**, 680–685.
Lagercrantz, H. (1976). *Neuroscience* **1**, 81–92.
Lamberts, R. & Goldsmith, P. C. (1985). *J. Histochem. Cytochem.* **33**, 499–507.
Lamoyi, E. & Nisonoff, A. (1983). *J. Immunol. Methods* **56**, 235–243.
Landon, J., Livanon, J. & Greenwood, F. C. (1967). *Biochem, J.* **105**, 1075–1083.
Langanger, G., DeMey, J., Moeremans, M. & Small, J. V. (1984a). *J. Cell Biol.* **99**, 1324–1334.
Langanger, G., DeMey, J., Moeremans, M., Small, J. V. & DeBrabander, M. (1984b). *J. Submicrosc. Cytol.* **16**, 43–45.
Lanhonr, J. J. (1980). *Methods in Enzymol.* **70**, 221–247.
Lanzerotti, R. H. & Gullino, P. M. (1972). *Anal. Biochem.* **50**, 344–353.
Larsson, L. I., Fahrenkrug, J. & Schaffalitzky de Muckadell, O. B. (1977). *Science* **197**, 1374–1375.
Laurell, C.-B. (1966). *Anal. Biochem.* **15**, 45–52.
Lazarides, E. (1976). *J. Cell Biol.* **68**, 202–219.
Lazarides, E. & Weber, K. (1974). *Proc. Natl Acad. Sci. USA* **71**, 92–107.
Leary, J. J., Brigati, D. J. & Ward, D. C. (1983). *Proc. Natl Acad. Sci. USA* **80**, 4045–4049.
Lee, H. U. & Kaufmann, S. J. (1981). *Dev. Biol.* **B81**, 81–95.
Lee, J. S., Lewis, J. R., Morgan, A. R., Mosmann, T. R. & Singh, B. (1981). *Nucleic Acids Res.* **9**, 1707–1721.
Lee, L. D., Baden, H. P., & Cheng, C. K. (1978). *J. Immunol. Methods* **24**, 155–162.
Lee, T.-C., Baker, R. C., Stephens, N. & Snyder, F. (1977). *Biochim. Biophys. Acta.* **489**, 25–31.
Leek, A. E. & Chard, T. (1974). *In* "L'Alpha Foetoproteine" (R. Masseyeff, Ed.). INSERM, Paris.
Lehto, V. P., Virtanen, I. & Kurki, P. (1978). *Nature* **272**, 175–177.
Leiber, D., Harbon, S., Guilet, J-G., Andre, C. & Strosberg, A. D. (1984). *Proc. Natl Acad. Sci.* **81**, 4331–4334.
Lennon, V. A., Thompson, M. & Chen, J. (1980). *J. Biol. Chem.* **255**, 4395–4398.
LePeoq, J. B. (1971). *Methods of Biochem. Anal.* **20**, 41–86.
Lerner, E. A., Lerner, M. R., Janeway, C. A. & Steitz, J. A. (1981). *Proc. Natl Acad. Sci.* **78**, 2737–2741.
Leserman, L. D., Barbet, J., Kourilsky, F. & Weinstein, J. W. (1980). *Nature* **288**, 602–604.
Levine, W., Lu, A. Y. H., Thomas, P. E., Ryan, D., Kizer, D. E. & Griffin, M. J. (1978). *Proc. Natl Acad. Sci.* **75**, 3240–3243.

Levine, L. (1967). *In* "The Neurosciences" (G. C. Quarton, T. Melneckuk & F. O. Schmitt, Eds). Rockefeller Universal Res., New York.
Lin, J. J-C. (1981). *Proc. Natl Acad. Sci.* **78**, 2335–2339.
Lin, L.-F. Y., Clejan, L. & Beattie, D. S. (1978). *Eur. J. Biochem.* **87,** 171–179.
Lind, I. & Mansa, B. (1968). *Acta. Pathol. Microbiol. Scand. (B)* **73,** 637–645.
Lind, I., Live, I. & Mansa, B. (1970). *Acta. Pathol. Microbiol. Scand. (B)* **78,** 673–682.
Lindmark, R., Thoren-Tolling, K. & Sjoquist, J. (1983). *J. Immunol. Methods* **62,** 1–13.
Ling, C. M. & Overby, L. R. (1972). *J. Immunol.* **109,** 834–841.
Linsenmayer, T. R. & Hendrix, M. J. C. (1980). *Biochem. Biophys. Res. Commun.* **92,** 440–446.
Lipsich, L. A., Lewis, A. J. & Brugge, J. S. (1983). *J. Virol.* **48,** 352–360.
Lipsky, N. G. & Pagano, R. E. (1985). *Science* **228,** 745–747.
Littlefield, J. W. (1964). *Science* **145,** 709–710.
Loft, H. (1975). *Scand. J. Immunol.* **4,** Suppl. 2, 115–119.
Loh, Y. P. (1979). *Proc. Natl Acad. Sci.* **76,** 796–800.
Louvard, D., Semeriva, M. & Maroux, S. (1976a). *J. Mol. Biol.* **106,** 1023–1035.
Louvard, D., Vannier, C. H., Maroux, S., Pages, J.-M. & Ladunski, C. (1976b). *Anal. Biochem.* **76,** 83–94.
Louvard, D., Reggio, H. & Warren, G. (1982). *J. Cell Biol.* **92,** 92–107.
Lowe, C. R. & Dean, P. D. G. (1974). "Affinity Chromatography". Wiley, London.
Lowry, O. H., Roslebrough, N. J., Farr, A. L. & Randall, R. J. (1951). *J. Biol. Chem.* **193,** 265–275.
Lubeck, M. D. & Gerhard, W. (1981). *Virology* **113,** 64–72.
Luben, R. A., Brazeau, P., Bohlen, P. & Guillemin, R. (1982). *Science* **218,** 887–889.
Lundahl, P. & Liljas, L. (1975). *Anal. Biochem.* **65,** 50–59.
Lundblad, A., Schroer, K. & Zopf, D. (1984). *J. Immunol. Methods* **68,** 227–234.
Luzio, J. P., Newby, A. C. and Hales, C. N. (1976). *Biochem. J.* **154,** 11–21.
MacDonald, D. J., Belfield, A., Steele, C. J., Mack, D. S. & Shah, M. M. (1979). *Clin. Chem. Acta* **94,** 41–49.
Maizel, J. R. (1971). *Methods in Virology* **5,** 179–245.
Malthe-Sørenssen, D., Eskeland, T. & Fonnum, F. (1973). *Brain Res.* **62,** 517–522.
Malthe-Sørenssen, D., Lea, T., Fonnum, F. & Eskeland, T. (1978). *J. Neurochem.* **30,** 35–46.
Mancini, G., Carbonara, A. O. & Hereman, S. J. F. (1965). *Immunochemistry* **2,** 235–254.
Mangeat, P. H. & Burridge, K. (1984). *J. Cell Biol.* **98,** 1363–1377.
Mann, D. L., Popovic, M., Sarin, P., Murray, C., Reitz, M. S., Strong, D. M., Hayness, B. F., Gallo, R. C. & Blattner, W. A. (1983). *Nature* **305,** 58–60.
Manning, R., Dils, R. & Mayer, R. J. (1976). *Biochem. J.* **153,** 463–468.
Manningley, C. & Roth, J. (1985). *J. Histochem. Cytochem.* **33,** 1247–1251.
Marchalonis, J. J. (1969). *Biochem. J.* **113,** 299–305.
Marchesi, V. T. & Furthmayr, H. (1976). *Ann. Rev. Biochem.* **45,** 667–698.
Mardian, J. K. W. & Isenberg, I. (1978). *Anal. Biochem.* **91,** 1–12.
Marengo, T. S., Harrison, R., Lunt, G. G. & Behan, P. O. (1979). *Lancet* **1,** 442.
Markwell, M. A. K. (1982). *Anal. Biochem.* **125,** 427–432.
Martinez, P. & McCauley, R. (1977). *Biochim. Biophys. Acta* **497,** 437–446.
Mason, T. L., Poyton, R. O., Wharton, D. C. & Schatz, G. (1973). *J. Biol. Chem.* **248,** 1346–1354.

Mather, I. H., Nace, C. S., Johnson, V. G. & Goldsby, R. A. (1980). *Biochem. J.* **188,** 925–928.

Matsuidaira, P. T. & Burgess, D. R. (1978). *Anal. Biochem.* **87,** 386–396.

Matsura, S., Fujii-Kuriyama, Y. & Toshiro, Y. (1978). *J. Cell Biol.* **78,** 503–519.

Matthew, W. D., Tsavaler, L. & Reichardt, L. F. (1981). *J. Cell Biol.* **91,** 257–269.

Maurer, P. M. (1971). *In* "Methods in Immunology and Immunochemistry", Vol. III (C. A. Williams & M. W. Chase, Eds), pp. 1–58. Academic Press, New York and London.

Maxam, A. M. & Gilbert, W. (1977). *Proc. Natl Acad. Sci.* **74,** 560–564.

Maxwell, M. H. (1978). *J. Microsc. (London)* **112,** 253–255.

Mayer, R. J. & Walker, J. H. (1978). *In* "Techniques in the Life Sciences", Vol. B1/I, "Protein and Enzyme Biochemistry", B119 (K. F. Tipton, Ed.). Elsevier, Amsterdam.

Mayer, R. J. & Walker, J. H. (1980). "Immunochemical Methods in the Biological Sciences: Enzymes and Proteins". Academic Press, London.

Mayer, R. J. & Billett, E. (1985). *In* "Techniques in the Life Sciences, Techniques in Protein and Enzyme Biochemistry", Part II, Supplement (K. F. Tipton, Ed.). Elsevier, Amsterdam.

McCans, J. L., Lane, L. K., Lindenmayer, G. E., Butler, V. P. Jr. & Schwartz, A. (1974). *Proc. Natl Acad. Sci.* **71,** 2249–2452.

McCans, J. L., Lindenmayer, G. E., Pitts, B. J. R., Ray, M. W., Rayner, B. D., Butler, V. P. J. & Schwartz, A. (1975). *J. Biol. Chem.* **250,** 7257–7265.

McCauley, R. & Racker, E. (1973). *Mol. Cell Biochem.* **1,** 73–81.

McDonnell, M. W., Simon, M. N. & Studier, F. W. (1977). *J. Mol. Biol.* **110,** 119–129.

McEver, R. P., Baenziger, N. L. & Majerus, P. W. (1980). *J. Clin. Invest.* **66,** 1311–1318.

McKay, R. O. G. (1980). *FEBS Lett.* **118,** 219–224.

McKeon, F., Kirschner, M. & Caput, D. (1986). *Nature* **319,** 463–468.

McKinney, R. M. & Spillane, J. T. (1975). *Ann. NY Acad. Sci.* **254,** 55–64.

McLaughlin, B. J., Wood, J. G., Saito, K., Roberts, E. & Wu, J. (1975). *Brain Res.* **85,** 355–371.

McNeil, T. H. & Sladek, J. R. (1978). *Science* **200,** 72–74.

Mellman, I. S. & Unkeless, J. C. (1980). *J. Exp. Med.* **152,** 1048–1069.

Merisko, E. M., Farquhar, M. G. & Palade, G. E. (1982). *J. Mol. Biol.* **92,** 846–857.

Merril, C. R. (1981). *Science* **211,** 1437–1438.

Merril, C. R., Switzer, R. C. & VanKeuren, M. L. (1979). *PNAS* **76,** 4335–4339.

Metzger, D. W. S., Miller, A. & Sercorz, E. E. (1980). *Nature* **287,** 540–542.

Miles, L. E. M. & Hales, C. N. (1968a). *Nature* **219,** 186–189.

Miles, L. E. M. & Hales, C. N. (1986b). *Biochem. J.* **108,** 611–618.

Milne, R. W., Douste-Blazy, P., Marcel, Y. L. & Retegui, L. (1981). *J. Clin. Invest.* **68,** 111–117.

Milstein, C. & Cuello, A. C. (1983). *Nature* **305,** 537–540.

Minamuira, N. & Yasunobu, K. T. (1978). *Arch. Biochem. Biophys.* **189,** 481–489.

Mishell, B. B. & Shiigi, S. M. (Eds) (1980). "Selected Methods in Immunology", Chapter 12. Freeman, CA.

Mitchison, N. A. (1968). *Immunology* **15,** 509–530.

Moeremans, M., Daniels, G. & DeMey, J. (1985). *Anal. Biochem.* **145,** 315–321.

Molday, R. S. & MacKenzie, D. (1983). *Biochemistry* **22,** 653–660.

Momoi, M., Kennett, R. H. & Glick, M. C. (1980). *J. Biol. Chem.* **255,** 914–921.

Moody, G. J. (1976). *Lab. Pract.* **25,** 575–581.
Moorman, A. F. M., Grivell, L. A., Lamie, F. & Smits, H. L. (1978). *Biochim. Biophys. Acta* **518,** 351–365.
Moradi-Ameli, M. & Godinot, C. (1983). *Proc. Natl Acad. Sci.* **80,** 6167–6171.
Morgan, J. L., Rodkey, L. S. & Spooner, B. S. (1977). *Science* **197,** 578–580.
Morimoto, T., Matsura, S., Sasaki, S., Tashiro, Y. & Omura, T. (1976). *J. Cell Biol.* **68,** 189–201.
Moudallal, Z. A., Altschuh, D., Briand, J. P. & van Regenmortel, M. H. V. (1984). *J. Immunol. Meth.* **68,** 35–43.
Mukasa, H., Shimaraura, A. & Tsumori, H. (1982). *Anal. Biochem.* **123,** 276–284.
Munro, S. & Pelham, H. R. B. (1984). *EMBO J.* **3,** 3087–3093.
Murphy, M. J. (1976). *Biochem. J.* **159,** 287–292.
Muruyama, K., Mikawa, T. & Ebashi, S. (1984). *J. Biochem. (Tokyo)* **95,** 511–519.
Nagatsu, I. & Kando, Y. (1975). *Act. Hist. Cyt.* **8,** 279–287.
Naiem, M., Gerdes, J., Abdulaziz, Z., Sunderland, C. A., Allington, M. J., Stein, H. & Mason, D. Y. (1982). *J. Immunol. Methods* **50,** 145–160.
Nelles, L. P. & Bamburg, J. R. (1976). *Anal. Biochem.* **73,** 522–531.
Nelson, N., Deters, D. W., Nelson, H. & Racker, E. (1973). *J. Biol. Chem.* **248,** 2049–2055.
Neville, D. M. (1971). *J. Biol. Chem.* **246,** 6328–6334.
Newman, G. R., Jasani, B. & Williams, E. D. (1983). *Histochem. J.* **15,** 543–555.
Newman, J. (1984). *Histochem. J.* **15,** 543–550.
Newman, P. J., Kahn, R. A. & Hines, A. (1981). *J. Cell Biol.* **90,** 249–253.
Nicklin, M. G. & Stephen, J. (1974). *Immunochem.* **11,** 35–40.
Niemi, W. D., Nastuk, W. L., Chang, H. W., Penn, A. S. & Rosenberg, T. L. (1979). *Exp. Neurol.* **63,** 1–27.
Nigg, E. A., Cooper, J. A. & Hunter, T. (1983). *J. Cell Biol.* **96,** 1601–1609.
Nilsson, G., Said, S. & Goldstein, M. (1978). *Brain Res.* **155,** 239–248.
Niman, H. L., Thompson, A. M. H., Yu, A., Markman, M., Willems, J. J., Herwig, K. R., Habib, N. A., Wood, C. R., Houghton, R. A. & Lerner, R. A. (1985). *Proc. Natl Acad. Sci.* **82,** 7924–7928.
Nisonoff, A. & Palmer, J. L. (1964). *Science* **143,** 376–379.
Noonan, K. D., Starling, J. J. & Seitz, T. L. (1981). *In* "Monoclonal Antibodies in Endocrine Research" (R. E. Fellows & G. S. Eisenbarth, Eds). Raven Press, New York.
Norrild, B., Bjerrum, O. J. & Vestergaard, B. F. (1977). *Anal. Biochem.* **81,** 432–441.
Noshiro, M. & Omura, T. (1978). *J. Biochem.* **83,** 61–77.
Noteboom, W. D., Knurr, K. E., Kim, H. S., Richmond, W. G., Martin, A. P. & Vorbeck, M. L. (1984). *J. Immunol. Methods* **75,** 141–148.
Nudelman, E., Hakamori, S. I., Knowles, B. B., Sotter, D., Nowinski, R. C., Tam, M. R. & Young, W. W. (1980). *Biochem. Biophys. Res. Commun.* **97,** 443–451.
Nunberg, J. H., Rodgergs, G., Gilbert, J. H. & Snead, R. M. (1984). *Proc. Natl Acad. Sci.* **81,** 3675–3679.
Nussenzweig, M. C., Steinman, R. M., Witmer, M. D. & Gutchinov, B. (1982). *Proc. Natl Acad. Sci.* **79,** 161–165.
Oakley, B. R., Kersh, D. R. & Morris, N. R. (1980). *Anal. Biochem.* **105,** 361–363.
O'Donnell, C. M. & Suffin, S. C. (1979). *Anal. Biochem.* **51,** 33A.
Oesch, F. & Bentley, P. (1976). *Nature* **259,** 53–55.
O'Farrell, P. H. (1975). *J. Biol. Chem.* **250,** 4007–4021.
O'Farrell, P. Z., Goodman, H. M. & O'Farrell, P. H. (1977). *Cell* **12,** 1133–1142.

Ogata, K., Arakawa, M., Kasahara, T., Shoiri-Nakano, K. & Hiraoka, K. (1983). *J. Immun. Methods* **65,** 75–82.
Oi, V. T. & Herzenberg, L. A. (1980). *In* "Selected Methods in Cellular Immunology" (B. B. Mishell & S. M. Shiigi, Eds), pp. 351–371. Freeman, San Francisco.
Okayasu, T., Onon, T. & Shinojima, K. (1977). *Lipids* **12,** 267–271.
Okret, S., Wikstrom, A-C., Wrange, O., Andersson, B. & Gustafsson, J-A. (1984). *Proc. Natl Acad. Sci.* **81,** 1609–1613.
Olden, K. & Yamada, K. M. (1977). *Anal. Biochem.* **78,** 483–490.
Olmstead, J. B. (1981). *J. Biol. Chem.* **256,** 11955–11957.
Olsen, R. W., Meunier, J.-C. & Changeux, J.-P. (1972). *FEBS Lett.* **28,** 96–100.
Olsson, L. & Kaplan, H. S. (1980). *Proc. Natl Acad. Sci.* **77,** 5429–5431.
Orci, L., Ravazzola, M., Amhardt, M., Louvard, D. & Perrelet, A. (1985). *Proc. Natl Acad. Sci. USA* **82,** 5385–5389.
Orlov, G. E. & Gurvich, A. E. (1971). *In* "Kongr. Mikrobiol. Mater. Kongr. Microbiol. Bulg. 2nd 1969" (I. Pashev, Ed.), Vol. 1, pp. 225–229.
Ornstein, L. (1964). *Ann. NY Acad. Sci.* **121,** Part 2, 404–427.
Osborn, M. (1981). *In* "Techniques in the Life Sciences", Vol. P1/I, "Cellular Physiology", P107 (P. F. Baker, Ed.). Elsevier, Amsterdam.
Osborn, M. & Weber, K. (1982a). *Cell* **31,** 303–306.
Osborn, M. & Weber, K. (1982b). "Methods in Cell Biology", Vol. 24A, pp. 97–132. Academic Press, New York.
Osborn, M. & Weber, K. (1983). *Lab. Invest.* **48,** 372–394.
Osborn, M., Debus, E. & Weber, K. (1984). *Eur. J. Cell Biol.* **34,** 137–143.
Osborn, M., Altmannsberger, M., Debus, E. & Weber, K. (1984). *Cancer Cells, Cold Spring Harbour* **1,** 191–200.
Otto, J. J. (1983). *J. Cell Biol.* **97,** 1283–1287.
Ouchterlony, O. (1968). *In* "Handbook of Immunodiffusion and Immunoelectrophoresis". Ann-Arbor Science Publ., Ann Arbor.
Owen, P. & Smith, C. J. (1977). *In* "Immunochemistry of Enzymes and their Antibodies" (M. R. J. Salton, Ed.). Wiley, London.
Pages, J.-M., Varenne, S. & Lazdunski, C. (1976). *Eur. J. Biochem.* **67,** 145–153.
Palacios, R., Palmiter, R. D. & Schimke, R. T. (1972). *J. Biol. Chem.* **247,** 2316–2321.
Papermaster, B. W., Sordahl, L. A. & Stene, H. L. (1976). *In* "Non Isotopic Immunoassays", Conference Handbook. Robert S. First, Inc.
Pardue, R. L., Brady, R. C., Perry, G. W. & Dedman, J. R. (1983). *J. Cell Biol.* **96,** 1149–1154.
Park, S. S., Persson, A. V., Mihor, J., Coon, M. J. & Gelboin, H. V. (1980). *FEBS Lett.* **116,** 231–234.
Paskin, N. & Mayer, R. J. (1976). *Biochem. J.* **159,** 181–184.
Patrick, J. & Lindstrom, J. (1973). *Science* **180,** 871–872.
Patton, J. G., Alleyh, M. C. & Mao, S. J. T. (1982). *J. Immunol. Methods* **55,** 193–203.
Peacock, A. C. & Dingman, C. W. (1967). *Biochemistry* **6,** 1818–1827.
Peacock, A. C. & Dingman, C. W. (1968). *Biochemistry* **7,** 659–667.
Peavy, D. E. & Hansen, R. J. (1975). *Biochem. Biophys. Res. Commun.* **66,** 1106–1111.
Peferoen, M., Huybrechts, R. & De Loof, A. (1982). *FEBS Lett.* **145,** 369–372.
Peltz, G., Spudich, J. A. & Parham, P. (1985). *J. Cell Biol.* **100,** 1016–1023.
Pepinsky, R. B. & Sinclair, L. K. (1986). *Nature* **321,** 81–84.
Perrin, D. & Aunis, D. (1985). *Nature* **15,** 589–592.
Petersen, E., Dingle, J. T. & Smith, A. (1982). *FEBS Lett.* **145,** 369–372.
Peterson, G. L. (1977). *Anal. Biochem.* **83,** 346–356.

Petrucci, T. C., Thomas, C. & Bray, D. (1983). *J. Neurochem.* **40,** 1507–1516.
Philippidis, H., Hansen, R. W., Reshef, L., Hopgood, M. F. & Ballard, F. J. (1972). *Biochem. J.* **126,** 1127–1134.
Piazzi, S. E. (1969). *Anal. Biochem.* **27,** 281–284.
Pickel, V. M., Tong, H. J. & Reis, D. J. (1975). *Proc. Natl Acad. Sci. USA* **72,** 659–663.
Pinnas, J. L., Northway, J. D. & Tan, E. M. (1973). *J. Immunol.* **111,** 996–1004.
Plaisancie, H., Alexandre, Y., Uzan, G., Besmond, C., Benarous, R., Frain, M., Trepat, J. S., Dreyfus, J-C. & Kahn, A. (1984). *Anal. Biochem.* **142,** 271–276.
Platt, J. L. & Michael, A. F. (1983). *J. Histochem. Cytochem.* **31,** 840–842.
Ponder, B. A. & Wilkinson, M. M. (1981). *J. Histochem. Cytochem.* **29,** 981–984.
Poole, A. R. (1974). "Immunological Methods for the study of the Cellular Localization of Proteins in Biochemistry, Vol 4 (E. Reid, Ed.), Subcellular Studies". Longman, London.
Porath, J. & Sundberg, L. (1970). *Protides Biol. Fluids* **18,** 401–407.
Porath, J., Aspberg, K., Drevin, H. & Axen, R. (1973). *J. Chromatogr.* **86,** 53–56.
Porstmann, B. (1981). *J. Clin. Chem. Clin. Biochem.* **19,** 435–439.
Poyton, R. O. & Schatz, G. (1975). *J. Biol. Chem.* **250,** 762–766.
Prescott, L. (1983). *Biotechnology* (April), 157–161.
Price, M. R. & Baldwin, R. W. (1977). "Cell Surface Reviews", Vol. 1 (G. Post & G. Nicholson, Eds). Elsevier, Amsterdam.
Priestley, J. V. & Cuello, A. C. (1983). *In* "Immunohistochemistry" (A. C. Cuello, Ed.), pp. 273. IBRO, John Wiley, Chichester.
Prough, R. A. & Ziegler, D. M. (1977). *Arch. Biochem. Biophys.* **180,** 363–373.
Pryde, J. G. (1986). *Trends in Biochem. Sci.* **11,** 160–163.
Radola, B. J. (1973). *Biochim. Biophys. Acta.* **295,** 412–428.
Raftell, M., Berzins, K. & Blomberg, F. (1977). *Arch. Biochem. Biophys.* **181,** 534–541.
Rammohan, K. W., McFarland, H. F. & McFarlin, D. E. (1981). *Nature* **290,** 588–589.
Ravdin, P. & Axelrod, D. (1977). *Anal. Biochem.* **80,** 585–592.
Readhead, C., Addison, G. M., Hales, C. W. & Letimann, H. (1973). *J. Endocr.* **59,** 313–323.
Reed, K., Vandlen, R. L., Bode, J., Duguid, J. K. & Raftery, M. A. (1975). *Arch. Biochem. Biophys.* **167,** 138–144.
Reid, M. S. & Bieleski, R. L. (1968). *Anal. Biochem.* **22,** 374–381.
Reis, D. J., Pickel, V. M., Shikimi, T. & Joh, T. H. (1975). *Trans. Amer. Soc. Neurochem.* **6,** 155.
Remacle, J., Fowler, S., Beaufay, H. & Berthet, J. (1974). *J. Cell Biol.* **65,** 237–240.
Remacle, J., Fowler, S., Beaufay, H., Amarcostesec, A. & Berthet, J. (1976). *J. Cell Biol.* **71,** 551–564.
Remy, M. H. & Poznansky, M. J. (1978). *Lancet*, July 68–70.
Renfroe, J. B., Chronister, R. B., Haycock, J. W. & Waymire, J. C. (1984). *Brain Res. Bull.* **13,** 109–126.
Rennert, O. M. (1967). *Nature* **213,** 1133.
Rhoads, R. E., McKnight, G. S. & Schmike, R. T. (1973). *J. Biol. Chem.* **248,** 2031–2039.
Rice, R. H. & Means, G. E. (1971). *J. Biol. Chem.* **246,** 831–832.
Richards, E. G., Coll, J. A. & Gratzer, W. B. (1965). *Anal. Biochem.* **12,** 452–471.

Richardson, P. J., Walker, J. H., Jones, R. T. & Whittaker, V. P. (1982). *J. Neurochem.* **38,** 1605–1614.
Richman, D. P., Gomez, C. M., Berman, P. W., Burres, S. A., Fitch, F. W. & Arnason, B. G. W. (1980) *Nature* **286,** 738–739.
Rieger, F., Bon, S., Massoulie, J., Cartaud, J., Picard, B. & Benda, P. (1976). *Eur. J. Biochem.* **68,** 513–521.
Rindler, M. J., Ivanov, I. E., Plesken, H. & Sabatini, D. D. (1985). *J. Cell Biol.* **100,** 136–151.
Risau, W., Saumweber, H. & Symmonds, P. (1981). *Exp. Cell Res.* **133,** 47–54.
Riva, M., Memet, S., Micouin, J-Y., Huet, J., Treich, I., Dassa, J., Young, R., Buhler, J-M., Sentenac, A. & Fromageot, P. (1986). *Proc. Natl Acad. Sci.* **83,** 1554–1558.
Roberts, R. & Painter, A. (1977). *Biochim. Biophys. Acta* **480,** 521–526.
Rodriguez, J. & Deinhardt, F. (1960). *Virology* **12,** 316–317.
Roisen, F., Inczedy-Marcsek, M., Hsu, L & Yorke, W. (1978). *Science* **199,** 1445–1448.
Rosen, J. M., Woo, S. L. C. & Comstock, J. P. (1975). *Biochemistry* **14,** 2895–2903.
Rosenberry, T. L. & Richardson, J. M. (1977). *Biochemistry* **16,** 3550–3558.
Ross, E. & Schatz, G. (1976). *J. Biol. Chem.* **251,** 1997–2004.
Ross, M. E., Reis, D. J. & Joh, T. H. (1981). *Brain Res.* **208,** 493–498.
Ross, R. A., Joh, T. H. & Reis, D. J. (1978). *J. Neurochem.* **31,** 1491–1500.
Rossier, J. (1975). *Brain Res.* **98,** 619–622.
Rossier, J. (1976a). *J. Neurochem.* **26,** 543–548.
Rossier, J. (1976b). *J. Neurochem.* **26,** 549–553.
Rossier, J., Bauman, A., Rieger, F. & Benda, P. (1975). *In* "Cholinergic Mechanisms" (P. G. Waser, Ed.), pp. 283–292. Raven Press, New York.
Roth, J. (1982). *J. Histochem. Cytochem.* **30,** 691–696.
Roth, J. (1983). *Immunocytochemistry* **2,** 217–284.
Roth, J. (1984). Lowicryl Letters No. 2, C. Available from Chemicho Wern Lowri, P.O. Box 1660, D-8264 Waldkraiburg, West Germany.
Roth, J. (1984). *J. Cell Biol.* **98,** 399–406.
Roth, J., Bendayan, M., Carkmalm, E. & Kwan, J. (1981). *J. Histochem. Cytochem.* **29,** 663–671.
Rothman, J. E. & Lenard, J. (1977). *Science* **195,** 743–753.
Royston, I., Majda, J. A., Baird, S. M., Meserve, B. L. & Griffiths, J. C. (1980). *J. Immunol.* **125,** 725–731.
Ruoslahti, E. (1976). *In* "Immunoadsorbents in Protein Purification" (E. Ruoslahti, Ed.), pp. 3–7. Universitetsforlaget, Oslo.
Rush, R. A., Costa, M., Furness, J. B. & Geffen, L. B. (1976). *Neurosci. Lett.* **3,** 209–213.
Russell, S., Davey, J. & Mayer, R. J. (1978a). *Proc. Europ. Soc. Neurochem.* **1,** 544.
Russell, S., Davey, J. & Mayer, R. J. (1978b). *Proc. Europ. Soc. Neurochem.* **1,** 545.
Rye, D. B., Saper, C. B. & Wainer, B. H. (1984). *J. Histochem. Cytochem.* **32,** 1145–1153.
Ryan, D. E., Thomas, P. E. & Levin, W. (1977). *Mol. Pharmacol.* **13,** 521–532.
Saito, K. (1978). *Fol. Pharm. J.* **74,** 427–440.
Salvaterra, P. M. & Mahler, H. R. (1976). *J. Biol. Chem.* **251,** 6327–6334.
Sandler, M. & Youdim, M. B. H. (1972). *Pharmacol. Rev.* **24,** 331–348.
Sargent, J. R. & George, S. G. (1975). "Methods in Zone Electrophoresis". BDH Chemicals, Poole, England.
Schaffner, W. & Weissmann, C. (1973). *Anal. Biochem.* **56,** 502–514.

Schecter, I. (1973). *Proc. Natl Acad. Sci.* **46,** 4898–4903.
Schenk, D. B. & Leffert, H. (1983). *Proc. Natl Acad. Sci.* **80,** 5281–5285.
Schimke, R. T., Sweeney, E. W. & Berlin, C. M. (1965). *J. Biol. Chem.* **240,** 322–331.
Schlaepfer, W. W. (1977). *J. Cell Biol.* **74,** 226–240.
Schmidt, J. & Raftery, M. A. (1973). *Biochemistry* **12,** 852–856.
Schmitt, M., Rittinghaus, K., Scheurich, P., Schwulera, U. & Dose, K. (1978). *Biochim. Biophys. Acta* **509,** 410–418.
Schneider, M. D. & Eisenbarth, G. S. (1979). *J. Immunol. Methods* **24,** 331–342.
Schreiber, A. B., Couraud, P. O., Andre, C., Vray, B. & Strosberg, A. D. (1980). *Proc. Natl Acad. Sci.* **77,** 7385–7389.
Schulman, M., Wilde, C. D. & Kohler, G. (1976). *Nature* **276,** 269–270.
Schultzberg, M., Dreyfus, C. F., Gershon, M. D., Hokfelt, T., Elde, R. P., Nilsson, G., Said, S. & Goldstein, M. (1978). *Brain Res.* **155,** 239–248.
Schultzberg, M., Hokfelt, T., Terenius, L., Brandt, J., Elde, R. P. & Goldstein, M. (1979). *Neuroscience* **4,** 249–270.
Schutz, G. S., Kieval, B., Groner, A. E., Sippel, D. T., Kurz, M. and Fergelson, P. (1977). *Nucleic Acids Res.* **256,** 1495–1503.
Schwartz, A. L., Marshak-Rothstein, A., Rup, D. & Lodish, H. F. (1981). *Proc. Natl Acad. Sci.* **78,** 3348–3352.
Scopsi, L. & Larsson, L. I. (1986). *Histochemistry* **84,** 221–230.
Scott Hausen, R. & Beavo, J. A. (1982). *Proc. Natl Acad. Sci.* **79,** 2788–2792.
Secher, D. S. & Burke, D. C. (1980). *Nature* **285,** 446–449.
Sege, K. & Peterson, P. A. (1978). *Proc. Natl Acad. Sci.* **75,** 2443–2447.
Sewell, M. M. H. (1967). *Science Tools* **14,** 11–12.
Shaper, N. L., Shaper, J. H., Meuth, J. L., Fox, J. L., Chang, H., Kirsch, I. R. & Hollis, G. F. (1986). *Proc. Natl Acad. Sci.* **83,** 1573–1577.
Shapiro, S. Z. & Young, J. R. (1981). *J. Biol. Chem.* **256,** 1495–1498.
Shaw, G., Osborn, M. & Weber, K. (1981). *Eur. J. Cell Biol.* **26,** 68–82.
Sherline, P. & Schiavone, K. (1977). *Science* **198,** 1038–1040.
Shuster, L. & O'Toole, C. (1974). *Life Sci.* **15,** 645–656.
Siddle, K. & Soos, M. (1981). *Biochem. Soc. Trans.* **9,** 142.
Siegelman, M., Bond, M. W., Gallatin, W. M., St John, T., Smith, H. T., Fried, V. A. & Weissman, I. L. (1986). *Science* **231,** 845–850.
Sigel, M. B., Sinha, Y. N. & Van der Laan, W. P. (1983). *Methods Enzymol.* **93,** 3–12.
Sikora, K. & Neville, A. M. (1982). *Nature* **300,** 316–317.
Singer, S. J., Ash, J. F., Bourguignon, L. Y. W., Heggeness, M. H. & Louvard, D. (1978). *J. Supram. Struct.* **9,** 373–389.
Singh, V. K. & McGreer, P. L. (1974). *Life Sci.* **15,** 901–913.
Slaughter, C. A., Coseo, M. C., Cancro, M. P. & Harris, H. (1981). *Proc. Natl Acad. Sci.* **78,** 1124–1128.
Slot, J. W. & Geuze, H. J. (1983). *In* "Immunochemistry" (A. C. Cuello, Ed.), pp. 323–346. John Wiley and Sons, Chichester.
Slot, J. W. & Geuze, H. J. (1985). *Eur. J. Cell Biol.* **44,** 1401–1407.
Small, J. V. & Celis, J. E. (1978). *J. Cell, Sci.* **31,** 393–409.
Smith, J. A., Hurrell, J. G. R. & Leach, S. J. (1978). *Anal. Biochem.* **87,** 299–305.
Smith, P. K., Krohn, R. I., & Hermanson, G. T. (1985). *Anal. Biochem.* **150,** 76–85.
Snabes, M. C., Boyd, A. E. & Bryan, J. (1981). *J. Cell Biol.* **97,** 1283–1287.
Snabes, M.C., Boyd, A. E., Pardue, R. L. & Bryan, J. (1981). *J. Biol. Chem.* **256,** 6291–6295.
Sobel, A., Weber, M. & Changeux, J.-P. (1977). *Eur. J. Biochem.* **80,** 215–224.

Sobel, J. H., Ehrlich, P. H., Birken, S., Saffran, A. J. & Canfield, R. E. (1983). *Biochemistry* **22,** 4175–4183.
Sobieszek, A. & Bremel, R. D. (1975). *Eur. J. Biochem.* **55,** 49–60.
Sobieszek, A. & Small, J. V. (1977). *J. Mol. Biol.* **112,** 559–576.
Soderholm, C. J., Smyth, C. J. & Wadstrom, T. (1975). *Scand. J. Immunol.* **4,** Suppl. 2, 107–113.
Soini, E. & Hemmila, I. (1979). *Clin. Chem.* **25,** 353–361.
Sommer, I. & Schachner, M. (1981). *Dev. Biol.* **83,** 311–327.
Southern, E. M. (1975). *J. Mol. Biol.* **98,** 503–517.
Speake, B. K., Dils, R. & Mayer, R. J. (1975). *Biochem. J.* **148,** 309–320.
Speake, B. K., Dils, R. & Mayer, R. J. (1976). *Biochem. J.* **154,** 359–370.
Spector, S., Felix, A., Semenuk, G. & Finberg, J. P. M. (1978). *J. Neurochem.* **30,** 685–689.
Speth, M., Alejandro, R. & Lee, E. Y. C. (1984). *J. Biol. Chem.* **259,** 3475–3481.
Springall, D. R., Hocker, G. W., Grimeliush, L. & Polak, J. M. (1984). *Histochemistry* **81,** 603–608.
Springer, T. A. (1981). *In* "Monoclonal Antibodies Hybridomas: a New Dimension in Biological Analyses" (R. H. Kennet, T. J. McKearn, & K. B. Bechtol, Eds). pp. 185–212. Plenum Press, New York & London.
Spudick, A. W. & Watt, S. (1971). *J. Biol. Chem.* **246,** 4806–4871.
Stadler, H. & Tashiro, T. (1978). *Proc. Eur. Soc. Neurochem.* **1,** 580.
Stadler, H. & Whittaker, V. P. (1978). *Brain Res.* **153,** 408–413.
Stahli, C., Staehelin, T., Miggiano, V., Schmidt, J. & Haring, P. (1980). *J. Immunol. Methods* **32,** 297–304.
Staines, N. A. & Lew, A. M. (1980). *Immunology* **40,** 287–293.
Stallcup, K. C., Springer, T. A. & Mescher, M. F. (1981). *J. Immunol.* **127,** 923–930.
Stanley, E. F. & Drachman, D. B. (1978). *Science* **200,** 1285–1287.
Stansbie, D., Denton, R. M., Bridges, B. J., Pask, H. T. & Randle, P. J. (1976). *Biochem. J.* **154,** 225–236.
Stein, S., Chang, C. H., Bohlen, P., Imai, K. & Udenfriend, S. (1974). *Anal. Biochem.* **60,** 272.
Steiner, A. L. (1974). *Methods Enzymol.* **38,** 96–105.
Stephens, R. E. (1975). *Anal. Biochem.* **65,** 369–379.
Sternberger, L. A. (1974). "Immunocytochemistry". Prentice Hall, New Jersey.
Sternberger, L. A. (1979). "Immunocytochemistry", 2nd Edition. John Wiley, New York.
Sternberger, L. A., Hardy, P. H., Cuculis, J. J. & Meyer, H. G. (1970). *J. Histochem. Cytochem.* **18,** 315–333.
Sternberger, L. A., Harwell, L. W. & Sternberger, N. H. (1982). *Proc. Natl Acad. Sci.* **79,** 1326–1330.
Stevenson, G. T.(1974). *Nature* **247,** 477–478.
Stocker, J. W. & Heusser, C. N. (1979). *J. Immunol. Methods* **26,** 87–95.
Stott, D. I., McLearie, J. & Marsden, H. S. (1985). *Anal. Biochem.* **149,** 454–460.
Strand, M. (1980). *Proc. Natl Acad. Sci.* **77,** 3234–3348.
Strange, P. G. (1978). *Biochem. J.* **176,** 583–590.
Strauss, A. W., Alberts, A. W., Hennessy, S. & Vagelos, P. R. (1975). *Proc. Natl Acad. Sci. USA* **72,** 4366–4370.
Strosberg, A. D. (1983). *Springer Semin. Immunopathol.* **6,** 67–78.
Studier, F. W. (1973). *J. Mol. Biol.* **79,** 237–248.
Sudhof, T. C., Ebbecke, M., Walker, J. H., Fritsche, U. & Boustead, C. M. (1985). *Biochemistry* **23,** 1103–1109.

Sugiyama, H., Benda, P., Meunier, J.-C. & Changeux, J.-P. (1973). *FEBS Lett.* **35,** 124–128.

Surek, B. & Latzko, E. (1984). *BBRC* **121,** 284–289.

Surolia, A., Pain, D. & Khan, M. I. (1982). *TIBS* (February), 74–76.

Suttie, J. W., Carlisle, T. L. & Cranfield, L. (1977). *In* "Calcium-Binding Proteins and Calcium Function" (R. H. Kretsinger, D. H. MacLennan & F. L. Siegel, Eds). N. Holland Publ. Co., New York.

Svendsen, P. J. (1973). *Scand. J. Immunol.* **4,** Suppl. 1, 69–70.

Swaab, D. F. & Fisser, B. (1977). *Neuroscience Lett.* **7,** 313–317.

Takesue, Y. & Nishi, Y. (1976). *J. Biochem.* **79,** 479–488.

Takesue, Y. & Nishi, Y. (1978). *J. Membrane Biol.* **39,** 285–296.

Takesue, Y. & Omura, T. (1970). *Biochem. Biophys. Res. Commun.* **40,** 369–377.

Talian, J. C., Olmsted, J. B. & Goldman, R. D. (1983). *J. Cell Biol.* **97,** 1277–1282.

Talbot, D. N. & Yaphantis, P. A. (1971). *Anal. Biochem.* **44,** 246–253.

Tashiro, T. & Stadler, H. (1978). *Eur. J. Biochem.* **90,** 479–487.

Taylor, C. R. (1976). *Eur. J. Cancer* **12,** 61–75.

Teichberg, V. I., Sobel, A. & Changeux, J.-P. (1977). *Nature* **267,** 540–542.

Terasaki, M., Song, J., Wong, U. R., Weiss, M. J. & Chen, L. B. (1984). *Cell* **38,** 101–108.

Ternynck, T., & Avrameas, S. (1976). *In* "Immunoadsorbents in Protein Purification" (E. Ruoslahti, Ed.), pp. 29–35. Universitetsforlaget, Oslo.

The, T. H. & Feltkamp, T. E. W. (1970). *Immunology* **18,** 875–881.

Thomas, P. E., Lu, A. Y. H., Ryan, D., West, S. B., Kawalek, J. & Levin, W. (1976). *Mol. Pharmacol.* **12,** 746–758.

Thomas, P. E., Lu, A. Y. H., West, S. B., Ryan, D., Miwa, G. T. & Levin, W. (1977). *Mol. Pharmacol.* **13,** 819–831.

Thomas, P. E., Reidy, J., Reik, L. M., Ryan, D. E., Koop, D. R. & Levin, W. (1984). *Arch. Biochem. Biophys.* **235,** 239–253.

Thornton, J. M. & Sibanda, B. L. (1983). *J. Mol. Biol.* **167,** 443–460.

Thorpe, R., Bird, C. R. & Spitz, M. (1984). *J. Immunol. Methods* **73,** 259–265.

Tijssen, P. & Kurstak, E. (1983). *Anal. Biochem.* **128,** 26–35.

Tipton, K. F., Houslay, M. D. & Mantle, T. J. (1976). *In* "Monoamine Oxidase and its Inhibition", Ciba Found. Sym. **39** (New Series), pp. 5–33. Elsevier, North Holland, Amsterdam.

Tokayasu, K. T. (1983). *J. Histochem. Cytochem.* **31,** 164–167.

Towbin, H. & Gordon, J. (1984). *J. Immunol. Methods* **72,** 313–340.

Towbin, H., Staehlin, T. & Gordon, J. (1979). *Proc. Natl Acad. Sci. USA* **76,** 4350–4354.

Towbin, M., Ranjoue, H-P., Unster, H., Liverant, D. & Godwin, J. (1982). *J. Biol. Chem.* **257,** 12 709–12 715.

Tripathi, R. K. & O'Brien, R. D. (1977). *Biochim. Biophys. Acta* **480,** 382–389.

Trowbridge, I. S. & Lopez, F. (1982). *Proc. Natl Acad. Sci.* **79,** 1175–1179.

Tsang, V., Peralta, J. & Simons, A. R. (1983). *Methods Enzymol.* **92,** 377–391.

Tsuji, S., Rieger, F., Peltre, G., Massoulie, J. & Benda, P. (1972). *J. Neurochem.* **19,** 989–997.

Turner, B. M. (1983). *J. Immunol. Methods* **63,** 1–6.

Tzartos, S. J. & Lindstrom, J. M. (1980). *Proc. Natl Acad. Sci.* **77,** 755–759.

Tzartos, S. J., Rand, D. E., Einarson, B. L. & Lindstrom, J. M. (1981). *J. Biol. Chem.* **256,** 8635–8645.

Udenfriend, S., Stein, S., Bohlen, P. L., Dairman, W., Leingruben, W. & Weigele, M. (1972). *Science* **178,** 871–872.

Uhl, G. R., Goodman, R. R., Kuhar, M. J., Childers, S. R. & Snyder, S. H. (1979). *Brain Res.* **166,** 75–94.
Ungewickell, E. (1985). *EMBO Journal* **4,** 3385–3391.
Unsicker, K., Drenkhahn, D., Groschel-Stewart, U., Schumacher, U. (1978). *Cell Tissue Res.* **188,** 341–344.
Updyke, T. W. & Nicholson, G. L. (1984). *J. Immunol. Methods* **73,** 83–95.
Uriel, J. (1964). *In* "Immunoelectrophoretic Analysis" (P. Grabar & J. P. Burtin, Eds). Elsevier, Holland.
Valderrama, R., Weill, C. L., McNamee, M. & Karlin, A. (1976). *Ann. NY Acad. Sci.* **274,** 108–115.
Valdivia, M. M. & Brinkley, B. R. (1985). *J. Cell Biol.* **101,** 1124–1134.
Valentine, K. L., Crumrine, D. A. & Reichardt, L. F. (1985). *J. Histochem. Cytochem.* **33,** 969–973.
Valle, G., Jones, E. A. & Colman, A. (1982). *Nature* **300,** 71–74.
van Ness, J., Laemmli, U. K. & Pettijohn, D. E. (1984). *Proc. Natl Acad. Sci.* **81,** 7879–7901.
Vannier, O. H., Louvard, D., Maroux, S. & Desnuelle, P. (1976). *Biochim. Biophys. Acta* **455,** 185–199.
van Weemen, B. K. & Schnuurs, A. H. W. M. (1971a). *FEBS Lett.* **15,** 232–236.
van Weemen, B. K. & Schnuurs, A. H. W. M. (1971b). *FEBS Lett.* **24,** 77–81.
van Wyck, K. L., Hinshaw, V. S., Bean, W. J. Jr. & Webster, R. G. (1980). *J. Virol.* **35,** 24–30.
Virtanen, I., Ekblom, P. & Laurila, P. (1980). *J. Cell Biol.* **85,** 429–434.
Voller, A., Bidwell, P. E. & Bartlett, A. (1979). *In* "The Enzyme Linked Immunoadsorbent Assay (ELISA)". Dynatec, Nuffield Laboratories of Comparative Medicine, The Zoological Society of London.
Vora, S. & Francke, V. (1981). *Proc. Natl Acad. Sci.* **78,** 3738–3742.
Walker, J. H. & Mayer, R. J. (1976). *Biochem. Soc. Trans.* **4,** 342–344.
Walker, J. H. & Mayer, R. J. (1977). *Biochem. Soc. Trans.* **5,** 1101–1103.
Walker, J. H., Betts, S. A., Manning, R. & Mayer, R. J. (1976). *Biochem. J.* **159,** 355–362.
Walker, J. H., Obrocki, J. & Zimmermann, C. W. (1983). *J. Neurochem.* **41,** 209–216.
Walker, J. H., Stadler, H. & Witzemann, V. (1984a). *J. Neurochem.* **42,** 314–320.
Walker, J. H., Boustead, C. M. & Witzemann, V. (1984b). *EMBO J.* **3,** 2287–2290.
Walker, J. H., Boustead, C. M., Witzemann, V., Shaw, G., Weber, K. & Osborn, M. (1985). *Eur. J. Cell Biol.* **38,** 123–133.
Walker, J. H., Kristiansen, G. I. & Stadler, H. (1986). *J. Neurochem.* **46,** 875–881.
Wallace, R. W., Yu, P. H., Dieckart, J. P. & Dieckart, J. W. (1974). *Anal. Biochem.* **61,** 86.
Walsh, F. S., Moore, S. E. & Dhut, S. (1981). *Dev. Biol.* **84,** 121–132.
Wardi, A. H. & Michos, G. A. (1972). *Anal. Biochem.* **49,** 607–609.
Watson, S. J., Richard, C. W. & Barchas, J. D. (1978). *Science* **200,** 1180–1182.
Webb, K. S., Mickey, D. D., Stove, K. R. & Paulson, D. F. (1977). *J. Immun. Methods.* **14,** 343–353.
Weber, K. & Groschel-Steward, U. (1974). *Proc. Natl Acad. Sci. USA* **71,** 4561–4564.
Weber, K. & Osborn, M. (1969). *J. Biol. Chem.* **244,** 4406–4412.
Weber, K., Pollack, R. & Bibring, T. (1975). *Proc. Natl Acad. Sci. USA* **72,** 459–463.
Webster, R. E., Anderson, D., Osborn, M. & Weber, K. (1978). *PNAS* **75,** 5511–5515.
Weeke, B. (1973a). *In* "A Manual of Quantitative Immunoelectrophoresis", (N. H. Axelsen, J. Krøll & B. Weeke, Eds), pp. 37–46. Universitetsforlaget, Oslo.

Weeke, B. (1973b) *In* "A Manual of Quantitative Immunoelectrophoresis" (N. H. Axelsen, J. Krøll & B. Weeke, Eds), pp. 47–56. Universitetsforlaget, Oslo.
Wehland, J. & Willingham, M. C. (1983). *J. Cell Biol.* **97,** 1476–1490.
Weidenmann, B. & Franke, W. W. (1985). *Cell* **41,** 1017–1028.
Weiler, E. W. & Zenk, M. (1981). *Methods Enzymol.* **73,** 395–406.
Weinstein, W. M. & Lechago, J. (1977). *J. Immunol. Methods* **17,** 375–378.
Werner, S. (1974). *Eur. J. Biochem.* **43,** 39–48.
Werner, S. & Machleidt, W. (1978). *Eur. J. Biochem.* **90,** 99–105.
Whitnall, M. H., Mezey, E. & Gainer, H. (1985). *Nature* **317,** 248–250.
Whittaker, V. P. (1977). *Naturwissenschaften* **64,** 606–611.
Willard, M., Simon, C., Baitinger, C., Levine, J. & Skene, P. (1980). *J. Cell Biol.* **85,** 587–596.
Williams, A. F., Galfre, G. & Milstein, C. (1977). *Cell* **12,** 663–673.
Williams, C. A. & Chase, M. W. (1967). *In* "Methods in Immunology and Immunochemistry", Vol. 1, pp. 307–335. Academic Press, New York and London.
Williams, C. A. & Schupf, N. (1977). *Science* **196,** 328–330.
Williamson, A. R. (1978). *In* "Handbook of Experimental Immunology", 3rd edn, Vol. 1 (D. M. Weir, Ed.). Blackwell Scientific Publications, Oxford.
Willingham, M. C. (1980). *Histochem. J.* **12,** 419–434.
Willingham, M. C. & Pastan, I. (1985). *In* "Endocytosis". (J. Pastan & M. C. Willingham, Eds), pp. 281–324. Plenum Press, New York and London.
Willingham, M. C., Yamada, S. S. & Pastan, I. (1978). *Proc. Natl Acad. Sci. USA* **75,** 4359–4363.
Wingerson, L. (1983). *Biotechnology* (March).
Winkler, H. (1976). *Neuroscience* **1,** 65–80.
Winkler, H. (1977). *Neuroscience* **2,** 657–683.
Winkler, H., Schneider, F. H., Rufener, C., Nakane, P. K. & Hortnagl, H. (1974). *In* "Advances in Cytopharmacology", Vol. 2 (B. Ceccarelli, F. Clements & J. Meldolesi, Eds), pp. 127–139. Raven Press, New York.
Wisdom, G. B. (1976). *Clin. Chem.* **22,** 1243–1255.
Witzemann, V. & Raftery, M. A. (1978). *Biochem. Biophys. Res. Commun.* **85,** 623–631.
Witzemann, V. & Walker, J. H. (1981). *In* "Cholinergic Mechanisms" (G. Pepeu & H. Ladinsky, Eds), p. 653. Plenum Press.
Wood, J. N., Hudson, L., Jessell, T. M. & Yamamoto, M. (1982). *Nature* **296,** 34–38.
Woodhead, G. S. (1892). *In* "Practical Pathology", p. 109, Young J. Pentland, Edinburgh & London.
Woodhead, J. S., Addison, G. M. & Hales, C. N. (1974). *Br. Med. Bull.* **30,** 44–49.
Wooten, G. F., Park, D. H., Joh, T. H. & Reis, D. J. (1978). *Nature* **275,** 324–325.
Wright, L. J., Feinstein, A., Heap, R. B., Saunders, J. C., Bennett, R. C. & Wang, M-Y. (1982). *Nature* **295,** 415–417.
Wulf, E., Deboben, A., Bautz, F. A., Faulstich, H. & Wieland, T. (1980). *Proc. Natl Acad. Sci. USA* **76,** 4498–4502.
Yalow, R. S. (1978). *Science* **200,** 1236–1245.
Yardi, L., Carnemollen, B., Siri, A., Santi, L. & Accolla, R. S. (1980). *Int. J. Cancer* **25,** 325–329.
Yelton, D. E., Diamond, B. A., Kwan, S-P. & Scharff, M. D. (1978). *Curr. Top. Microbiol. Immunol.* **81,** 1–7.
Young, R. A. & Davis, R. W. (1983a). *Science* **222,** 778–782.
Young, R. A. Davis, R. W. (1983b). *Proc. Natl Acad. Sci.* **80,** 1194–1198.

Young, W. W., MacDonald, E. M. S., Nowinski, R. C. & Hakomori, S-I. (1979). *J. Exp. Med.* **150,** 1008–1019.
Zacharius, R. M., Zell, T. E., Morrison, J. H. & Woodlock, J. J. (1969). *Anal. Biochem.* **30,** 148–152.
Zipser, B. & McKay, R. (1981). *Nature* **289,** 549–554.
Zomzely-Neurath, C. & Keller, A. (1977). *Neurochem. Res.* **2,** 353–377.
Zweig, M., Heilman, C. J., Rabin, H. & Hampar, B. (1980). *J. Virol.* **35,** 644–652.

# Index